D0909051

The Literature of Soil Science

Literature of the Agricultural Sciences

Wallace C. Olsen, Series editor

Agricultural Economics and Rural Sociology: The Contemporary Core Literature
By Wallace C. Olsen

The Literature of Agricultural Engineering
Edited by Carl W. Hall and Wallace C. Olsen

The Literature of Animal Science and Health
Edited by Wallace C. Olsen

The Literature of Soil Science
Edited by Peter McDonald

THE LITERATURE
OF SOIL SCIENCE

EDITED BY

Peter McDonald

Cornell University Press

ITHACA AND LONDON

This book was typeset from disks supplied by the staff of the Core Agricultural Literature Project, Albert R. Mann Library, Cornell University. Sharon Van De Mark prepared the machine-readable text, and corrections were made by Nicole Kasmer Kresock. The research was financially supported by Cornell Agricultural Experiment Station; National Agricultural Library, United States Department of Agriculture; and the Rockefeller Foundation.

First published 1994 by Cornell University Press.

International Standard Book Number 0-8014-2921-8
Library of Congress Catalog Card Number 93-27394
Printed in the United States of America
Librarians: Library of Congress cataloging information
appears on the last page of the book.

⊗ The paper in this book meets the minimum requirements of the American National Standard for Information Sciences—Permanence of Paper for Printed Library Materials, ANSI Z39.48–1984.

Contents

vi   Contents

The Literature of Soil Science

1. Trends and Developments in Soil Science

B. P. WARKENTIN

Oregon State University

Observations about soils must have predated recorded history, since soils have always been "under foot." Early peoples would have noticed the ability of soils to nourish different kinds of plants and the relative suitability of soils for ceramic uses or for building shelters. Recorded observations about soils exist from more than two thousand years ago in Europe and Asia. Information arising from millennia of practical experience has been augmented by scientific studies at an accelerating pace during the last two hundred years.

It is useful, in discussing trends and developments in soil science, to have a framework of different periods, even in full realization of the disadvantages of dividing any continuum. There is overlap between periods; the literature lags behind the ideas; not all individuals follow the general trends; and the periods chosen reflect a national or language bias. Nonetheless, the following periodization may be helpful in thinking about developments in soil science.

The first period in soil science may be said to have ended about 1850. This is a convenient date because a large number of changes occurred about this time. In 1847 A. Von Humboldt published the book *Kosmos*, in which he attempted a synthesis of all that was then known about the natural world.[1] The soils information included was fairly general, descriptive, and gained from experiments and intuitive thinking arising from practical soil management problems. The experimental development and use of the moldboard plow and planters, and experience with tillage, seed bed preparation, drainage, and crop growth had led to the accumulation of considerable information about soils. Before 1850, scientists from disciplines such as chemistry or geography contributed to our knowledge about soils, but most of the ideas were discussed by "gentleman practitioners" in meetings of agri-

1. A. Von Humboldt, *Kosmos*, trans. from German (New York: Harper Book Co., 1850).

1

cultural societies and published in their proceedings. These societies were formed in the 1700s in European countries and their colonies. They had a regional base, e.g., *Letters and Papers on Agriculture, Planting, &c. Addressed to the Bath and West of England Society for the Encouragement of Agriculture, Arts, Manufactures, and Commerce*, beginning in 1780. Some national journals were published; the *Journal d'Agriculture Pratique* started publication in 1838 in France.

The second period from about 1850 to 1910, was a time of much more intensive experimentation. Diffusion of the information in Justus Liebig's book *Organic Chemistry and Its Application to Agriculture and Physiology* during the 1840s can be taken as the watershed event.[2] It convinced people about the soil processes related to supply of nutrients to plants. Liebig's influence gave chemical theories the upper hand from about 1850 to 1870, while physical ideas predominated later and biological concerns were added still later.

Social and economic conditions also had a large influence on soil studies. In the eastern United States, as in other countries with recent European settlement, land was becoming exhausted, eroded, and abandoned. The phenomenon of worn-out soils was generally recognized, and an argument ensued over whether this loss of productivity was chemical or physical. Considerable movements of people, especially in Europe, the Americas, and Australia, had a large influence on soil use and the ideas that were developed in soil science.

Many other changes came quickly after 1850. The production of artificial phosphate fertilizers, the development of agricultural experiment stations with the concept of farm areas and laboratories, and the general development of experimental natural sciences were all important to soil science. By the end of this period, textbooks in the modern mode were available with chapter headings very similar to those of textbooks in use today. Their titles indicated an interest in use of soils. E. J. Russell's first edition of *Soil Conditions and Plant Growth*, published in 1912, is one example of many textbooks published in different countries in different languages.[3] Studies in all the subdisciplines of soil science except soil mineralogy were carried out during this period. The first volume of *Annales de la Science Agronomique* was published in 1884.

The third period, from approximately 1910 to 1945, featured the study of soil as an extensive resource. The classifications of soils before 1900 were

2. Justus Liebig, *Organic Chemistry in Its Applications to Agriculture and Physiology = Organische Chemie in ihrer Anwendung auf Agricultur und Physiologie*, ed. Lyon Playfair, 1st American ed. (Cambridge, Mass.: J. Owen, 1841); 1st German ed. (Brunswick: Vieweg, 1841).
3. E. J. Russell, *Soil Conditions and Plant Growth* (London: Longmans, 1912).

based on specific characteristics such as the crops they supported or grain-size distribution. Soil mapping in western Europe had become part of geological surveys, with soils seen as weathering products of rocks. Ideas on soil genesis and on soil classification based on landscape characteristics were introduced worldwide beginning in the 1900s based on studies done in Russia during the late 1800s. These were studies of soil variability on a macroscale. A. A. J. De'Sigmond dates the beginning of soil studies as a science to the early 1900s.[4]

The turn of the century brought national and international journals of soil science. *La Pédologie* began publication in St. Petersburg in 1899, printing papers in Russian with French summaries. From 1928 to 1947 the journal had dual titles, *Pochvovedenie* and *Pedology*; only the former continued after that. *Internationale Mitteilungen für Bodenkunde* started in 1911 and was succeeded in 1924 by the *Proceedings of the International Society of Soil Science* in the three official languages of the society. The first volume of *Mitteilungen* has the paper by Albert Atterberg, "Die Plastizitat der Tone," that became the basis for the extensive use in soil engineering of plasticity as an index property to predict strength of clay soils.

Soil scientists in the United States were planning a soil society in the early 1900s, but decided to join the American Society of Agronomy when it was formed in 1907 and began publishing the *Journal of the American Society of Agronomy*.[5] The ties of soil science to agronomy have always been stronger in North America than in Europe, where soil science had a more independent status or started as a part of agricultural chemistry. (The ties to agronomy were strengthened in Africa, Asia, and South America by the large role of United States funding and scientists in those areas after 1945.) International gatherings of soil scientists were held in Budapest in 1909, Stockholm in 1910, Prague in 1922, and Rome in 1924 when the International Society of Soil Science (ISSS) was organized. The first International Congress of the ISSS was held in Washington in 1927.

Journals reporting soil science research often began as journals of agricultural research. The primary Canadian agricultural journal began with soil science literature in 1921 as *Scientific Agriculture (Revue Agronomique Canadienne)*, then changed in 1953 to the *Canadian Journal of Agricultural Science*. The *Canadian Journal of Soil Science (Revue Canadienne de la Science du Sol)* was one of the products of the split into specialized journals after 1957.

4. A. A. J. De'Sigmond, "Development of Soil Science," *Soil Science* 40 (1935): 77–87.
5. T. L. Lyon, "History of the Organization of the American Society of Agronomy," *Journal of the American Society of Agronomy* 25 (1933): 1–9.

The period from 1945 to about 1980 was marked by a vastly increased number of experimental studies in soil science. The number of papers published grew dramatically, and the cost of journals rose to a level that prevented most libraries from maintaining all journals with soil science papers. Many new journals made their appearance at this time, most of them relevant to a particular aspect of soil science, for example, tillage or water. Regular series with review articles began to appear, for example *Advances in Agronomy* and *Advances in Soil Science*, published in the United States. National societies for soil science were formed in many countries after the 1950s (see Chapter 7).

Laboratory studies during this period were generally performed on samples of soils that were dried, crushed, and sieved, and field experiments were generally short-term, for one season of the growing crop. Statistical methods to eliminate the effect of soil variability became more sophisticated in the analysis of the results of experiments. Many of the ideas in soil science were worked out in greater detail during this time. Soil classification developed from systems that were based in soil genesis to systems that were based on observable or measurable characteristics of soils. The soil classification system of the Food and Agriculture Organization of the United Nations (FAO) and *Soil Taxonomy*[6] from the United States became systems for international communication of soil science information, even though most countries also retained systems that were developed to take into account local soil differences.

Toward the end of the 1970s application of general soils knowledge to the field and on a landscape basis began. It was found that many of the experimental results were not applicable because the processes occurring in the field had not been duplicated in the experiments. The concept of a uniform wetting front in soils had to give way to preferential flow of water and solutes. Short-term experiments on nutrition of crops had to evolve to studies of long-term changes in the biological, physical, and chemical characteristics of soils. Sustainable nutrient management in soils is now part of national and international research and development programs.

These changes led to the present era of soil science in which the questions are on a landscape basis, have an ecological nature, and ask about the sustainability of natural resources. Soil variability is seen as being intimately connected with soil processes such as components of the nitrogen cycle and effectiveness of leaching of soils, rather than a nuisance to be overcome in experiments. Geostatistical descriptions of variability are being developed.

6. Soil Survey Staff, *Soil Taxonomy* (Washington, D.C.: USDA Soil Conservation Service, U.S. Govt. Print. Off., 1975 [*USDA Agriculture Handbook* no. 436]).

Soil science studies are being reported in an ever wider range of journals and some are being carried out by scientists with a primary affiliation to other disciplines such as biology, forestry, or engineering. Textbooks that concentrate on processes, and especially on fluxes in soils, are replacing those that concentrate on static properties. The books by R. E. White,[7] Sheila Ross,[8] and Jorg Richter[9] are examples of the beginning textbook, advanced textbook, and research monograph levels.

A. The Scope of Soils

Forest Soils

The case has been made that forestry applications require an emphasis on aspects of soil science that is different from annual crop production. The early books on forest soils dealt extensively with the characteristics of humus layers. Obviously, concerns for tillage were not large because the management options were different. E. Henry wrote about the profound differences between agricultural and forested soils that legitimized the special study of forest soils.[10] G. Morosoff wrote the first paper in the Russian journal *La Pédologie* with the French title "La Pédologie et la Sylviculture."[11] Much of the early literature, however, dealt with the applications of soil science to both annual crop production and forestry.

C. Grebe[12] published a book on forest soil science in 1852, and C. G. Heyer[13] in 1860; both reference a book by J. C. Hundeshagen published in 1830.[14] The studies of forest soils by Peter E. Müller provided important ideas on the role of organic matter in soil development (see Chapter 2).[15] The books by Grebe and Heyer combined soil science and climatology, a fruitful association for soil science that would again be profitable after the 1950s, when soil physics had a major role in developing micro-climatology.

---

7. R. E. White, *Introduction to the Principles and Practice of Soil Science*, 2d ed. (Oxford: Blackwell Sci. Pub., 1987).

8. S. Ross, *Soil Processes* (London: Routledge, 1989).

9. J. Richter, *The Soil as Reactor* (Cremling, Germany: Catena Verlag, 1987).

10. E. Henry, *Les Sols Forestiers* (Paris: Berger-Lerrault, 1908).

11. G. Morosoff, "La Pédologie et la Sylviculture," *Pochvovedenie* 1 (1899).

12. C. Grebe, *Forstliche Gebirgskunde, Bodenkunde und Klimalehre* (Vienna: Eisenbach, 1852).

13. C. G. Heyer, *Lehrbuch der forstlichen Bodenkunde und Klimatologie* (Erlangen, Germany: Ferdinand Enke, 1860).

14. J. C. Hundeshagen, *Die Bodenkunde in Land-u. Forst Wirtschaftlicher Beziehuug* (Tübingen: Heinrich Laupp, 1830).

15. Peter E. Müller, *Studien uber die Naturlichen Humusformen und Deren Einwirkung auf Vegetation und Boden* (Berlin: Springer, 1887).

It can also be noted that the Bureau of Soils of the United States government was organized in 1894 as a part of the Weather Bureau, but was separated shortly thereafter.[16]

After the 1950s, textbooks on forest soils became common. H. J. Lutz and R. F. Chandler[17] and S. A. Wilde[18] were among the first to write such works in the United States. The basic discussions of soil physics and soil chemistry are similar to those in books relating to annual crop production, although the forest soils literature has always placed more emphasis on the biological characteristics of soils.

Because they were studying undisturbed soils, with minor net inputs of nutrients in the management systems, forest soil scientists led the work in the recycling of nutrients and the ecology of soil/root interactions. Soils under annual crop production in Europe were sustained by rotations and application of animal manures, but soils under forest were impoverished by removal of litter for application to annual crops. Preferential flow of water in large, continuous pores was accepted by forest soil researchers before it was common knowledge in agronomy.

Tropical Soils

Scientists from colonizing countries studied soils of tropical areas in Africa and Asia extensively in the first half of the twentieth century. They saw soil relationships that were different from their experiences at home. In Java, Paul Vageler described "peculiar clays . . . in wet tropical . . . strongly marked, jelly-like structure . . . [that] frequently exhibit the original crystal forms of the parent rocks. Any soil expert . . . (familiar only with temperate soils) . . . would reject crop growth . . . but drainage brings dehydration of soil colloids and good structure."[19] This is a good field description of wet andosols. F. L. D. Wooltorton described the particular swelling soil conditions that interfered with engineering structures in Burma and India.[20]

Some scientists considered that tropical soils had particular properties different from those of temperate regions. We now understand that these are differences only in degree, an idea Vageler stated in his early work: "To trace the characteristic features of tropical soils to the action of special

16. A. G. McCall, "The Development of Soil Science," *Agricultural History* 5 (1931): 43–56.

17. H. J. Lutz and R. F. Chandler, *Forest Soils* (New York: Wiley, 1946).

18. S. A. Wilde, *Forest Soils and Forest Growth* (Waltham, Mass.: Chronica Botanica, 1946).

19. Paul Vageler, *An Introduction to Tropical Soils*, trans. from German by H. Greene (London: Macmillan, 1933): 22.

20. F. L. D. Wooltorton, *The Scientific Basis for Road Design* (London: Arnold, 1954).

factors would, of course, be wrong. In the main general laws . . . hold in the tropics as in temperate climates: the same heat, light, water and air are at work in both places . . . Intensity of climatic factors is greater . . . in duration and degree."[21]

Most of this literature was in the language of the colonizing country. For example, early studies in Egypt were published in French and later in English. Now much of the tropical soils research literature is in the language of the country, but a large portion is still in English.

Monographs on land management in tropical climates had been published before the 1920s, but it was the late 1920s and early 1930s when books on tropical soils began to appear. Most colonizing countries established teaching and research centers for tropical agriculture. The Imperial College of Tropical Agriculture opened in Trinidad in 1922 to provide trained professionals and research results for several countries where Britain had interests. Since the 1960s there have been many good publications on tropical soils in most of the romance languages. Soil scientists have interpreted soils information from Central and South America.[22] Special mention should be made of the continued commitment of the Office de la Recherche Scientifique et Technique Outre-mer (ORSTOM), the Office of Scientific and Technical Research Overseas, in France to thorough and long-term studies of processes in tropical soils. Much of the work is reported in *Cahiers ORSTOM—Série Pédologie.*

The international agricultural research centers have, since the 1950s, contributed to soil science knowledge and to techniques for soil management in tropical areas. They have set examples for national research programs in many areas. The study of soils has always been more important in countries where fewer technological inputs are used in crop production. As production becomes more intensive, soil information is used less because soil behavior can be changed more readily. Horticultural crop production and rangelands illustrate two extremes of intensity of soil uses.

Soil Biology

Both macro- and microbiota are important in soils, but studies in soil biology came relatively late in soil science. Until the 1950s, most soil zool-

21. Vageler, *An Introduction to Tropical Soils*, p. 23.

22. (a) E. Bornemiza and A. Alvarado, eds., *Soil Management in Tropical America*, Proceedings of a Seminar, CIAT, Cali, Colombia, Feb. 1974 (Raleigh, N.C.: North Carolina State University, 1975). (b) G. H. Montenegro and C. D. Malagon, *Propiedades Fisicas de los Suelos* (Bogotá: Instituto Geográfico Agustin Codazzi, 1990). (c) Pedro A. Sanchez, *Suelos del Tropico: Características y Manejo*, trans. (into Spanish) by E. Camacho (San Jose, Costa Rica: IICA, 1981).

ogy was done outside of soil science. Soil microbiology has a hundred-year history in soil science. Soil biology has also been more susceptible than other areas to periods of inactivity. Theories on the importance of humus for plant growth were held before the 1850s, but were effectively discredited in the soil chemistry era of plant nutrition. The practical relationship of soil microbiology to plant growth was summarized by Felix Lohnis's work published in 1910.[23] More detailed studies followed, leading eventually to studies of soil biochemistry—for example, the work of S. A. Waksman a decade later.[24]

Charles Darwin discussed the role of earthworms in 1881, but the study of soil animals has been a minor component of soil science. The activities of soil animals are more obvious in tropical climates, and much of the information about them comes from studies of soils in tropical areas. As the appreciation of ecological relationships in soil science developed after the 1970s, studies on the role of soil animals in the decomposition process and in soil fertility have been more common. The recent interest in large pores in soils, resulting from a reevaluation of the role of tillage, has again focused more attention on earthworms and soil animal activity.

It is, however, soil microbiology that has been the main study in soil biology. Louis Pasteur's work became generally known after 1860. The role of soil microorganisms in decomposition of organic matter and in nitrogen fixation was generally appreciated one hundred years ago. Since the 1960s there has been renewed interest in microbiology and biochemistry of soils. Nitrogen derived from symbiotic or free-living nitrogen-fixing organisms regained importance. The addition of artificial chemicals to soils renewed the interest in decomposition of potentially toxic materials. This drew renewed attention to old ideas on production of toxins by microorganisms and the interference among different classes of microorganisms. Recent interest in using soil bacteria to control fungal diseases of plants points up a promising area of research and of development into practice.

Early soil microbiology understanding came from soil samples. Recent research, however, has concentrated on relating microbial habitat to the nature of soil surfaces and to the size, shape, and arrangement of pores in soils. From the 1940s to the 1970s soil biology studies were more visible in Europe and Asia than on other continents. This is directly related to the importance of purchased inputs such as fertilizer into crop production in North America. With the increasing interest in the function of soils in environmental processes and in global changes, the relationship of the physical

23. Felix Lohnis, *Handbuch der landwirtschaftlichen Bakteriologie* (Berlin: Gebrueder Borntraeger, 1910).
24. S. A. Waksman, *Soil Microbiology* (New York: Wiley, 1952).

and biological activities of soils has become increasingly important.[25] Specialized journals now exist for papers in soil biology.

Soils, Agronomy, and Crop Production

Crop production has always been one of the main uses of soils. The early literature, and even prehistoric experience in crop production, led to knowledge about soils. Literature is commonly quoted from Roman times[26] or from the early Chinese, but the same information would have been present on all the continents. The Roman philosopher Columella's studies were in agronomy; he was concerned about how the quality of soils influenced the plants that grew in them. Russell summarizes the early ideas on plant nutrients and how these developed before 1850.[27] The early soil classification systems were for crop production, often with the view to opening up new land areas for agricultural use, but also for soil conservation. Much of the information was based on which crops would grow in which soils.

By 1850, most of the present ideas on nutrition of crops had been set, and the applications needed to be worked out. Studies in agricultural chemistry led to the use of fertilizers and the maintenance of soil productivity. The approach was that of the bank balance, where plants were analyzed to measure the nutrients—such as phosphorus, potassium, and calcium—that were removed from soils, with the idea that these had to be added back to the soil in equal amounts to maintain crop production. The "limiting nutrient" concept was used at a later date.

The big increase in adding off-farm inputs into crop production came in the period 1940 through 1975, earlier in North America and somewhat later in Europe and the developing countries. The ideas were spread by international units such as the Food and Agriculture Organization. Emphasis was on assuring availability of nutrients for maximum growth and production, and timing for availability, rather than on the total amounts removed by crops. Tree growth and forest soil fertility experiments were also carried out. Soil tests were developed as index measurements of availability of nutrient elements in the soil and as calibration of the index of extractable nutrients with crop growth and performance. The use of tracers—first, radioactive P32 in the 1940s, and later, stable isotopes of nitrogen—allowed experiments that previously had not been possible in tracing nutrient ele-

25. C. S. Potter and R. E. Meyer, "The Role of Soil Biodiversity in Sustainable Dryland Farming Systems," *Advances in Soil Science* 13 (1990): 241–51.

26. L. Olson, "Columella and the Beginning of Soil Science," *Agricultural History* 17 (1943): 65–72.

27. Russell, *Soil Conditions and Plant Growth.*

ments from soils to crops, and in identifying specific nutrient fractions and sources in the soil.

From 1850 onward, there was a strong tie of crop production to soil chemistry and the chemical determination of a nutrient element in the soil. However, after the 1980s, the tie was closer to soil biology and to fluxes of nutrients in soils. Current emphasis is on the components of the cycle of nutrients in soils and on the transformations, e.g., sulfur, and on the flux and the kinetics of nutrients in soils. These changes can be followed in books from Liebig[28] to Russell[29] to C. A. Black[30] to Richter.[31]

Recent work has made extensive use of models that describe nutrient processes in the soil and processes occurring during plant uptake of nutrients. These models have been very useful in summarizing and integrating soils knowledge related to crop production, as well as pointing out those processes that are not well described or where further experimental work is required. International organizations such as the International Board for Soil Research and Management (IBSRAM) and the International Benchmark Sites Network for Agrotechnology Transfer (IBSNAT) have promoted the development of crop growth models and their use along with geographic information systems to transfer agronomic information for similar soils in different countries.

Soil Chemistry

Chemical studies of soils were the tradition after the 1850s. Analyses of soils, for example for their phosphorus content, were soon found to be imperfectly related to the availability of phosphorus to crops. This led to reevaluation and improvements in the study of chemical properties. Recently, the focus has been on the study of soil chemical processes.

Ion exchange properties have been a continuing topic for over one hundred years, with gradually improving knowledge of the process and its importance to uses of soils. The knowledge gained from the study of soil salinity has been the basis for reclamation of saline and sodic soils worldwide. The study of soil acidity, and especially that of aluminum chemistry, that began around the 1900s has explained many soil processes. Studies in the early 1900s with lysimeters measured losses of nutrients from soils.

Studies on the inorganic constituents of soils, especially of clay minerals, have become a separate area interfacing with geology and crystallography

28. Liebig, *Organic Chemistry in Its Applications to Agriculture and Physiology.*
29. Russell, *Soil Conditions and Plant Growth.*
30. C. A. Black, *Soil-Plant Relationships* (New York: Wiley, 1968).
31. Richter, *The Soil as Reactor.*

and with technology of various uses of clays. Soil chemists were an important group in the establishment of clay mineralogy, or clay science, as a separate field with its own national and international journals and societies, such as the Clay Minerals Society and the Association Internationale pour l'Etude des Argiles. The literature of crystallography, clay mineralogy, and ceramic technology contains useful information for soil science.

Inquiry into the nature of surfaces in relation to ion exchange started with the study of aluminum silicate surfaces, assumed to be amorphous, then shifted to the study of pure clay mineral surfaces, and then back to amorphous materials; from Sante Mattson[32] to C. E. Marshall[33] to K. Wada.[34] Recognition of the nature of water absorbed onto inorganic and organic surfaces has been important in defining soil chemical processes, especially with the application of recent measurement techniques used in the study of surfaces for electronics applications.

Study of the organic components of soil has been relatively neglected, in large part because of the difficulty of studying these fractions. M. Schnitzer was among those who used methods of organic chemistry to investigate the chemical structure of degradation products of soil organic matter and their interactions with metals and minerals.[35] More recently, the structure of humic substances has been investigated by spectroscopy tools such as nuclear magnetic resonance, infra red, and electron spin resonance.[36] The International Humic Substances Society was founded in 1983 with major assistance from soil scientists.

The trend now is away from static properties such as cation exchange capacity to dynamic biochemical processes from studies of simple absorption of organic compounds such as pesticides to the more complex sources and sinks interacting with fluxes in soils. Geochemical models have been useful in extending the information on phase equilibriums in soils to the chemistry of the soil solution.

The application of colloidal chemistry to soil science has always been important, from the early studies of "exchange complexes" to ion distributions at charged colloidal surfaces, and now to the importance of the movement of suspended colloidal inorganic and organic components and the

32. Sante Mattson, "The Laws of Soil Colloidal Behavior," *Soil Science* 28 (1929): 179–220.

33. C. E. Marshall, *The Colloid Chemistry of the Silicate Minerals* (New York: Academic Press, 1949).

34. K. Wada, "A Structure Schema for Soil Allophane," *The American Mineralogist* 52 (1967): 690–708.

35. M. Schnitzer, "Humic Substances: Chemistry and Reactions," in *Soil Organic Matter*, ed. M. Schnitzer and S. U. Khan (Amsterdam: Elsevier, 1978), pp. 1–64.

36. M. H. B. Hayes, P. MacCarthy, R. L. Malcolm, and R. S. Swift, eds., *Humic Substances II: In Search of Structure* (Chichester, England: Wiley, 1989).

partitioning of trace elements into different phases. Colloidal chemistry diverged from soil chemistry with the advent of applications to polymers, but merged again with studies of the detailed nature of surfaces. Trace element chemistry was originally important for crop production, but has recently become much more important in studying the movement of toxic materials and heavy metals for concerns about environmental quality.

Soil Physics

Physical characteristics and properties of soils are probably among the earliest observations and studies.[37] Color, consistency, and response to tillage and seed bed preparation are obvious characteristics. Physical properties often determine the limits of soil use. Measurements before 1900 were predominantly of static properties, with interest in fluxes coming later. Grain-size distribution was important because it was a good index property that allowed prediction of other properties more difficult to measure. Water retention and movement have always been large components of soil physics studies, from practical experience in irrigation and drainage to soil water potential concepts in the early 1900s and details of measurement after the 1940s. Robert Warington devoted half of his book to soil water.[38] Unsaturated flow has been a challenging study. The geometry of the pore space was relatively less studied, and the challenge now is to combine the thermodynamics of soil water with spatial structure and variability. Interest in soil water is shared by a number of disciplines including civil engineering and ecology. The current literature has many specialized journals in water research, where different disciplines share ideas.

During the first half of the twentieth century, soil physics developed in association with soil mechanics. Commission 1 of the International Society of Soil Science started as a soil mechanics and physics commission. Work in the 1920s and 1930s by Karl Terzaghi contributed to this study.[39] The two branches separated as soil mechanics became part of soil engineering, but they still overlap in the literature, with publications from soil sciences in the engineering literature and engineering publications in the soil science literature. The Highway Research Board (later the Transportation Research Board) in the United States fostered fruitful collaboration between soil science and soil mechanics. H. F. Winterkorn had a large influence because he

37. W. H. Gardner, "Early Soil Physics into the Mid 20th Century," *Advances in Soil Science* 4 (1986): 1–101.

38. Robert Warington, *Lectures on Some of the Physical Properties of Soil* (Oxford: Clarendon, 1900).

39. Karl Terzaghi, *Theoretical Soil Mechanics* (New York: Wiley; London: Chapman and Hall, 1943).

worked productively and easily in both fields.[40] His early work had been in the application of soil chemistry and soil physics to the problem of maintaining low-cost roads in the midwestern region of the United States. The Division of Soil Mechanics in the Commonwealth Scientific and Industrial Research Organization (CSIRO), Australia, and the Road Research Laboratory of the Department of Scientific and Industrial Research (DSIR), in Britain, are two other examples where important knowledge about soils was developed through this collaboration. Wooltorton's work in Asia has already been mentioned. The soil technology commission of the American Society of Agronomy began as an application of agricultural engineering to soil science.[41] This collaboration is still important in tillage, irrigation, drainage, and managing wastes on agricultural lands.

More recent studies in soil physics have concentrated on fluxes, for example of water in the liquid or vapor phase, gases such as oxygen and carbon dioxide or introduced gaseous pesticides, and solutes and colloidal-size solids both organic and inorganic. In these studies the physical, chemical, and biological studies of soils come together.

Sustainable Soil Use and Conservation

Different soil regions require different land use management to achieve sustainable production. Such systems have been worked out, for example, for paddy rice cultures in Asia, for shifting cultivation where population pressures were moderate in Africa and Asia, and for rotations in Europe. Each time, however, when people have moved into new areas, within a few generations they have had to face the problems of soil erosion and soil exhaustion, or worn out soils. Many soils in the tropics, as well as some in the southeastern United States, have low initial levels of nutrients and organic matter, which hastened soil exhaustion. Inherent properties of the soil, combined with crop production methods, determine how quickly these problems arise.

The pattern in the eastern United States is fairly typical and well documented in papers such as those of A. McDonald[42] and A. O. Craven.[43] Often, soil exhaustion is associated with single-crop systems such as the

40. H. F. Winterkorn, *The Science of Soil Stabilization* (Washington, D.C.: U.S. Highway Research Board, 1955 [*Bulletin* no. 108]): 1–24.
41. Lyon, "History of the Organization of the American Society of Agronomy."
42. A. McDonald, *Early American Soil Conservationists* (Washington, D.C.: U.S. Govt. Print. Off., 1941 [*USDA Misc. Pub.* no. 449]).
43. A. O. Craven, *Soil Exhaustion as a Factor in the Agricultural History of Virginia and Maryland, 1606–1860* (Urbana, Ill.: University of Illinois, 1925 [*University of Illinois Studies in the Social Sciences*, vol. 13, no. 1]).

tobacco culture in the southeastern United States. Economic conditions and world events, as well as tradition, influence agricultural practices. A. B. Hulbert discusses the importance of soils in determining migration patterns into central and western North America.[44] The soils lore and the information about which practices suit which soils was determined by the ethnic and regional background of the settlers coming to North America. Soils for meadows were of interest in a cattle economy, and later soils for grain were sought.

Soil exhaustion forced land use changes in the eastern United States by the 1850s. Contributing factors were single cropping, the frontier attitude that it was easier to abandon land and move to new areas than to worry about building soils, and international conditions that determined foreign markets for grain. Shifting cultivation was practiced. Land that became exhausted and eroded was abandoned; trees grew there and twenty to thirty years later the land would be cultivated again. Inputs of phosphate fertilizer and eventually guano were introduced about this time. They produced striking results on some soils, but were not widely used. The input of single factors also meant that the effect would not last many years, when another factor became troublesome. Limited use of crop rotation was attributed to economic necessity. These concerns were eventually overcome by the adoption of soil management methods suitable to the region.[45]

The concern for sustainability is increasing again due to some of the same factors: economics, human health, soil health, and changing social views on the use of resources. Decreased tillage and cultivation systems are being developed as the old statement "It is not the crops that exhaust the land, but the cultivation" is again being considered. Sustainable soil management is being explored in national and international programs. The close links between soil protection, economic conditions, and controlled markets is acknowledged in the concept of "sustainable agriculture and rural development."[46] "Land charters" and "world soil charters" are discussed at international meetings. International research centers and groups such as the FAO foster these programs. This will lead soil science in a different direction.

Soil quality concerns have a long history. Salinization of soils and degradation of the soil surface to alter water runoff and infiltration are old issues. Deterioration of soil structure and the accumulation of toxic metals and

44. A. B. Hulbert, *Soil, Its Influence on the History of the United States* (New Haven, Conn.: Yale University Press, 1930).

45. E. W. Hilgard, "Soil Waste and Soil Depletion as a National Danger," in *Encyclopedia of American Agriculture*, L. H. Bailey (London: Macmillan, 1909): 128–32.

46. C. Benbrook, "The Den Bosch Declaration: Grappling with the Challenges of Sustainability," *Journal of Soil and Water Conservation* 46 (1991): 349–52.

organic compounds are more recent. Soil quality is now seen as a concept of diversity of potential uses for a soil.[47]

Control of accelerated erosion is often viewed as the most prominent part of soil conservation. Along with salinization, accelerated erosion has made productive areas into sterile landscapes. Accelerated erosion usually occurs when new farming systems are introduced. Conversion of grassland to cereal production led to accelerated erosion by wind on several continents. Conversion of forested land to clean-tilled row crops led to excessive erosion from rainfall.

Soil conservation is a sufficiently important social concern that most countries have government agencies dealing with erosion control. All of the technological, social, and economic factors that enter into land use need to be considered in devising systems that control soil erosion. There has been a tendency in the past to consider soil conservation as a problem that would cease to exist if once solved. It is now realized that conservation has to be an integral part of any farming system, and must be changed each time the farming system is changed in response to technical, social, or economic changes. Developed and developing countries often cite economic factors as the reason for limited use of soil conservation measures, but attitudes probably play an equally large role.

There is a large literature on soil conservation in general publications and in refereed scientific publications. Soil conservation is an area where emotional and rational approaches to soil science meet. As an example, R. N. Sampson's *Farmland or Wasteland—A Time to Choose* considers different aspects of soil conservation, including soil erosion.[48] Conservation will continue to be a major concern in soil science.

B. Languages

Much of the literature is available in one of these languages: English, French, German, or Spanish. Translation among these languages is relatively easy, and even a cursory familiarity with a language allows the reader to obtain information from graphs and tables. The situation is very different for the Cyrillic and Arabic alphabets. Reading Japanese or Chinese is even more difficult for most non-Asian scientists. While the English language is

47. B. P. Warkentin and H. F. Fletcher, "Soil Quality for Intensive Agriculture," in *Proceedings of the International Seminar on Soil Environment and Fertility Management in Intensive Agriculture*, Tokyo, Japan, 1977 (Tokyo: Society of the Science of Soil and Manure, 1977): 594–98.

48. R. N. Sampson, *Farmland or Wasteland—A Time to Choose* (Emmaus, Pa.: Rodale Press, 1981).

used most frequently in soil science communication today, it is not certain that this trend will continue, since a local language is such an important part of a culture.

Various aids to using literature in other languages exist. Summaries or abstracts in more commonly used languages are helpful. Some of the Japanese journals include English legends for graphs and tables, and titles in English. English translations of many Russian journals became available after the 1950s. Textbooks and review articles written by scientists with a command of different languages are very useful. As one example, S. Iwata's familiarity with Russian, Japanese, and English brought together ideas on soil water retention and movement that were not generally available in English.[49] G. V. Jacks's *Multilingual Vocabulary of Soil Science* contains equivalents in English, French, Spanish, German, Portuguese, Italian, Dutch, and Swedish.[50] Russian was added for the second edition in 1960. *The Dictionary of Soils* by Georges Plaisance and A. Cailleux is a translation from the French.[51] Other language equivalents are also available.

The term "soil science" and its equivalents (e.g., ciencia del suelo) are used in most languages. The term "pedology" (pédologie) was introduced to include all aspects of soils and is used in that sense in some of the European literature. In North America, however, pedology is usually understood to include only soil genesis, morphology, classification, and mapping. The term "soil" is used in soil science for the unconsolidated material that has been altered by soil forming processes and supports plant growth at the upper layers of the earth's surface. In other disciplines, e.g., engineering and geology, it refers to any unconsolidated earth material.[52]

C. Peripheral Soil Science

Because of the importance of soil to life on earth, soil has been viewed subjectively as well as objectively, emotionally as well as rationally. Nearly 150 years ago in his book *Kosmos*, Von Humboldt wrote: "In order to comprehend nature in all of its vast sublimity, it would be necessary to present it under a two-fold aspect, first objectively as an actual phenome-

49. S. Iwata, T. Tabuchi, with B. P. Warkentin, *Soil-Water Interactions* (New York: Marcel Dekker, 1988).

50. G. V. Jacks, *Multilingual Vocabulary of Soil Science* (Rome: Food and Agriculture Organization, 1954).

51. G. Plaisance and A. Cailleux, *Dictionnaire des Sols* (Paris: La Maison Rustique, 1958). Trans. as *Dictionary of Soils* in 1981 for the U.S. Department of Agriculture (New Delhi, India: Amerind Pub. Co., 1981).

52. R. F. Leggett, *Geology and Engineering*, 2d ed. (New York: McGraw-Hill, 1962).

non, and next subjectively as it is reflected in the feelings of mankind."[53] There is a large literature on this subject, dating from very early times, of the four elements of earth, air, fire, and water.

Some of these feelings lie behind the literature that questions the soil management strategies dominant at any one time. A good example is E. H. Faulkner's *Plowman's Folly*.[54] In the introduction he states "The mold board plow which is in use on farms throughout the civilized world, is the least satisfactory implement for the preparation of land for the production of crops." These ideas find legitimacy in our present scientific concerns with no-till methods of crop production. Much of this literature is now being brought into the mainstream through studies in fields with such names as sustainable agriculture. This literature emphasizes the fragility of soils, management systems to support plant growth with fewer off-farm inputs, and conservation against excessive erosion. The general ideas so common in our heritage, of the unnaturalness of urban life, and the purity of rural life still have an appeal in most societies. In the United States, it is embodied in the concept of the independent yeoman farmer and the efforts to preserve the small farm. The current debate in the general literature on the importance of "sense of place" in the development of culture is another aspect.

The long tradition and debate about organic farming have also been peripheral to soil science. The justifications for organic farming generally include value statements that arise from feelings about soil and soil management. These ideas have always been an important component in human use of soil and are now coming closer to the mainstream in soil management in developed countries.

General literature and poetry are sources of information on the subjectively sensed value of soils. Several soil scientists, for example Hans Jenny, have drawn together the strings of soils and art.[55] A few lines from a poem entitled "Soil" by Sheila Weaver illustrate this feeling for soil and for soil management.[56]

> I am soil.
> From sea to sea
> on every continent and island
> with gossamer-thin mantle
> I cover mountain, plain, and valley

53. Humboldt, *Kosmos*, p. 22.

54. E. H. Faulkner, *Plowman's Folly* (Norman: University of Oklahoma Press, 1943), p. 7.

55. Hans Jenny, "My Friend, The Soil," *Journal of Soil and Water Conservation* 59 (1984): 158–61.

56. Sheila Weaver, "Soil," *The New Catalyst*, Gabriola, B.C., Fall 1987. Used by permission of author and publisher.

riverbank and crevice . . .
with black, red, brown, or yellow earth
soft and crumbly
porous and absorbent
bursting with life . . .
I am soil.
Look at me!
Smell me, touch me, feel me
walk on me with your bare feet
sing and dance on me
observe me closely . . .
as a living, changing, vital link
in a vast and ancient web
intricate and delicate
of which you too are part . . . I am soil.
Nurture me
plant in me
shelter me with trees
rescue me where
I am thin and worn
but above all
teach your children . . .
to know me
and to value me . . .

D. The Present

Now, at the end of the twentieth century, soil science is reaching out to a greater variety of applications and to syntheses with other disciplines. The concern with global change puts soil science at the center of ecological considerations. Soils have a major role in mediating such vital processes as the global carbon cycle. Modern technology, such as high-speed computing, makes it easier to assemble the different information about soils and to use integrating methods, such as models, to describe important soil processes. At the same time within the discipline there is a reaching toward a *soil science*, a core description of knowledge about soils which is more than the sum of all the information derived from diverse uses of soils.[57]

A quick glance at bookshelves shows that applications of soils knowledge are increasing. A new awareness and need for an agricultural ecology and geography, with soils as one of the major components is seen in, for exam-

57. W. R. Gardner, "Soil Science as a Basic Science," *Soil Science* 151 (1991): 2–6.

ple, the work of L. W. Canter.[58] After more than one hundred years of the dominance of plowing as a method of inverting the top layer of soil, there is now increasing emphasis on decreased tillage for crop production. This trend raises the importance of processes dominant in undisturbed soils, such as the increasing range of pore sizes and increasing continuity of pores. It brings forest and rangeland studies closer to agronomic studies. Soil quality continues to be an important theme, ranging from the practical effects of salinity or toxic pollutants in soils to more abstract concepts, such as the range of possible uses of soils. Current applications include use of brackish water in crop production, influence of agricultural chemicals like ammonia on soil processes, and enhancement of the natural processes of degradation that can occur in different soil environments. Soil conservation, while continuing to be largely a social problem, carries soil scientists to questions of hill slope farming and sustainable crop production systems. This concern is international, as shown by the new International Society for Soil Conservation and its activities in organizing forums for discussion of soil and water conservation under different regional conditions.

The rich and diverse literature of soil science during its first hundred years will undoubtedly lead to further interesting studies.

58. L. W. Canter, *Environmental Impacts of Agricultural Production Activities* (Chelsea, Mich.: Lewis Pub., 1986).

2. Early Soil Science and Trends in the Early Literature

JEAN BOULAINE

Institut National Agronomique de Paris

Translated from the French
by Linda Stewart

Historical studies of soil science are rare.[1] Since its New Delhi Congress in 1982, the International Society of Soil Science (ISSS)[2] has included a task force entitled "History, Epistemology, and Sociology of Soil Science," the members of which write detailed historical studies of aspects of the field. Some treatises, notably those by Louis Grandeau, Dimitri Prianichnikov, E. J. Russell, B. A. Keen, and J. S. Joffe involve historical introductions and bibliographies of authors' works from past centuries. There are also three major publications on the subject of the history of soil science: Fritz Giesecke[3] in German, I. A. Krupenikov[4] in Russian, and Jean Boulaine[5] in French. Publications of the Soil Science Society of America are also important.[6] This paper does not cite these sources each time they have been used. The majority of the information in this text comes from the works in the bibliography.

Only primary scholars have been cited. The names of those who have contributed most to the progress of soil science are followed by birth and death dates. Dates alone enclosed in parentheses refer to the publication of their major works. These are listed in the bibliography.

---

1. Modern French authors use the term *science des sols*, and the synonym *pédologie*, to designate what Anglo-Saxons call *soil science*, Germans *bodenkunde*, Russians *pochvovedenie*, and Spanish and Greeks *edafologia*.

2. In French, the Association Internationale de la Science du Sol (AISS).

3. Fritz Giesecke, chapter 2 vol. 1, in Edwin Blanck, *Handbuch der Bodenlehre* (Berlin: Springer, 1929–32).

4. I. A. Krupenikov, *Historija Pochvoviedenié*, transcription of recording by M. Peguet (Moscow: [INRA] Isdatielstvo Nauka, 1981).

5. Jean Boulaine, *Histoire des Pédologues et de la Science des Sols* (Paris: Masson, 1989).

6. *Soil Science Society of America Journal* 41 (1977): 221–65. Articles by W. H. Gardner, G. M. Browning, F. E. Clark, D. A. Russel, G. G. Williams, F. G. Viets, G. W. Thomas, and M. G. Cline.

A. Soil Science to 1870

Soil science has its beginnings in prehistory when humans began to accumulate knowledge about the soils, and Greek, Carthaginian, and Roman authors compiled this knowledge and added to it. The Byzantines of the ninth century and the Hispano-Arabs of the eleventh and twelfth centuries reworked and completed the texts of antiquity. The Middle Ages and the Renaissance added some details. H. Gregoire[7] and A. Dickson[8] outlined this ancient literature of soils and agriculture.

The eighteenth century saw a few interesting texts, but the first important modern publication on soils (1804) was written by the Swiss T. de Saussure (1767–1845) concerning plant nutrition. He borrowed the ideas of the scientists Antoine-Laurent Lavoisier, Ingen Housz, and Sennebier on the use of atmospheric carbon dioxide by plant leaves under the action of light and affirmed that plants all contained phosphorus and that this phosphorus came from the soil. Despite his demonstration, the great majority of savants of the period accepted the humus theory advanced by Johan Wallerius (1763), reworked by Jean Hassenfratz (1793), and popularized by Albrecht Thaer (1809), Humphry Davy (1813), and Baron Jöns Jakob Berzelius (1815).

The first treatise on soil science appeared in 1837 under the signature of Karl Sprengel (1787–1858). The term had been used a few years earlier by a German forester (Johann Hundeshagen, 1813). This book is the first in the literature dedicated to the study of soil.

Sprengel's work was superseded by that of the chemist Justus Liebig (1803–73). In 1837, the British Association for the Advancement of Science, meeting in Liverpool, solicited a work on the relationship between chemistry and agriculture from Liebig. Borrowing from a text published four months earlier as the introduction to his *Traité de Chimie Organique* (Paris, April 1840), Liebig wrote a book in German[9] which was immediately translated into English by Lyon Playfair in 1841, then into French by Gerhard in 1841. This was a genuine treatise on agriculture (origin of arable land, crop rotation, fallow land, plant decomposition, etc.). However, the major scientific advance was the refutation of the humus theory and the conclusion that plants are nourished from simple mineral substances in the soil.

---

7. H. Gregoire, *Essai Historique sur l'État de l'Agriculture en Europe au Seizième Siècle*, Introduction to the edition "Théatre d'Agriculture et Maison Rustique," Olivier de' Serres, ed. the Society of Agriculture of Paris.

8. A. Dickson, *From Agriculture of Antiquity—Edimbourg Traduit en Français en 1802: De l'Agriculture des Anciens* (Paris: Samson, 1788).

9. Justus Liebig, *Die Organische Chemie in ihrer Anwendung auf Agricultur und Physiologie* (Brunswick: Vieweg, 1841).

Liebig's personal contribution consisted of one theory and two laws. The theory dealt with the mineral nutrition of plants, which absorbed, through their roots and leaves, chemical elements in simple combinations (gases or salts of mineral acids). The organic components of the soil, specifically humus, were only reserves which provided mineral elements after their degradation and decomposition. Liebig then affirmed that particular mineral elements contained by the remains of plants after oxidation were indispensable and that all these elements entered plants as very simple compounds. His theories were confirmed by all subsequent experiments and are still the fundamental laws of agronomy.

Liebig's ideas provoked contemporary scholars to inquire about the necessity of elements found during analysis in plant remains. Were they impurities or indispensable constituents, despite their minuscule quantities? The University of Göttingen had set for its competition that same year, 1840, "the question of remains," which was resolved by A. F. Wiegmann and L. Polstorff (1842), thanks to cultures in a synthetic environment. Liebig, however, explained that plants received nutrition from nitrogen through absorption in the leaves, and from ammonia absorbed from water and the air. Later, struck by the enormous quantities of nitrogen in the humus of soils, he was persuaded by the French and English scholars who demonstrated in precise experiments that the nitrogen in plants came in the most part from soil.

In France, agronomic research was subsequently dominated by J. B. Boussingault (1802–87) and Adrien Gasparin with their studies related to phosphates and nitrogen. Gasparin (1783–1862) published in 1843 his *Cours d'Agriculture* in four volumes. The first volume, *Agrologie*, was devoted to soil science. This was the first treatise of this type in the French language. Gasparin later served as France's minister of commerce, agriculture, and public works for ten years.

Three years earlier, Jean B. Dumas, a chemist with the Academy of Sciences, in his *Statique des Êtres Vivants* (text from August 25, 1840, published in 1841) gave a remarkable synthesis of the formation of organic material in plants, but he limited the discussion to carbon, oxygen, hydrogen, and nitrogen. For him, the mineral elements in plant remains were simple impurities which entered through the roots and evaporated on the leaves. He accepted, with many reservations, Liebig's mineral theory. His collaborator and friend, Boussingault, was to devote a great part of his brilliant career in agronomy to the study of the nitrogen cycle, showing that nitrogen came from the soil and the degradation of humus, but that this phenomenon did not suffice to explain the global balance of this element. As the discipline of microbiology evolved between 1875 and 1895, this process was elucidated definitively.

On the other hand, Boussingault did not play a role in the phosphate debate that raged in the latter part of the nineteenth century, and Dumas, authoritative master in French chemistry, did not favor the use of phosphate fertilizers—in particular, superphosphates. French agriculture thus suffered a significant delay which was not overcome until the 1860s.

In England, Liebig's ideas were immediately adopted by the great agronomist and wealthy British landowner Sir J. B. Lawes (1814–1900). In 1837, he began experiments on fertilizers at his property in Rothamsted, outside London. With his chemist and friend J. H. Gilbert (1810–1901), he transformed the property into an exceptional research station, which he left to his country on his death. As the Scribner edition of the *Dictionary of Scientific Biography* states: "The work of Lawes and Gilbert was published entirely in the form of papers under both joint and separate authorship, mostly in the *Journal of the Agricultural Society*."[10] In the twentieth century, Sir A. D. Hall (1864–1942) carried on Gilbert's work. He was instrumental in synthesizing the station's work and in 1911 published the now famous *The Book of Rothamsted Experiments*. Important among the station's early pioneers were Sir John Russell (1872–1965) and his son E. W. Russell. *Soil Conditions and Plant Growth*, published in 1912, has had many editions, and the 11th (1988) is still widely used today. For 150 years, Rothamsted has been the capital of soil science in Great Britain.

The effect of farm manures on soils has long been recognized, and soils rich in humus are generally fertile. Historically, however, the exact role of the latter needed to be more clearly specified. Berzelius's student G. J. Mulder believed he could define a series of compounds: ulmine, humine, acides ulmiques, and humique crenique and apocrenique. Eilhard Mitscherelich (1794–1863), Louis Thenard, and Grandeau improved the analytic methods slightly. They pointed out certain functions of humus—in particular, that of supplying nitrogen—a role also discussed by Liebig and Boussingault. Around 1875, J. J. T. Schloesing (1824–1919) demonstrated the physical role of organic matter in soils and its effects on soil structure.

The Belgian De Bevnie had initiated, toward the end of the eighteenth century, work on particle-size distribution; the Swiss Albert Orth had already defined several physical parameters of soil. His compatriot Gustav Schubler (1787–1834) followed up this work; he studied temperature, humidity, structure, and other characteristics of soil (1831). Gasparin utilized Schubler's publications in his treatise *Cours d'Agriculture* (1843–48).

Many physicists sought in vain to display electrical phenomena in soil which they believed influenced plant growth. Research in this area, notably by M. E. Wollny, Grandeau, and Gasparin, encumbered scientific publica-

10. *Scribner's Dictionary of Scientific Biography 1970–1978*, vol. 8. (New York: Charles Scribner's Sons, 1978), p. 93.

tions of this period—even the correspondence between Karl Marx and Friedrich Engels shows an interest in the topic!

The chemists of this period did their best work in soil analysis, although they lost a lot of time specifying the conditions for extracting solutions for titration. Notable authors were Friedrich Mohr (1807–79) in Berlin, Jean C. A. Voelcker (1822–83) in London, Anselme Payen and F. Malaguti in France, I. P. Pavlov in Russia, and many others.

Two Italian agronomists pointed out the exchange capacity of soils, namely Giuseppi Gazzeri (1816) and Lambruschini (1830), as well as the German pharmacist Johann Bronner (1836). In 1848, a British farmer observed the transformation of ammonium sulfate into calcium sulfate in the soil and pointed this out to the chemist J. T. Way. This work was confirmed and extended by Rev. A. Huxtable, F. Brustlein, and then J. M. Van Bemmelen (1830–1911), a Dutchman whose work on humus also made him an authority: he showed that humus was composed not of defined chemical elements, but of very complex mixtures.

Practical use of the discoveries of Liebig and his contemporaries was long in gaining acceptance because serious opposition was raised, for example, by Berzelius (1779–1848), a celebrated Swedish chemist. In addition, the climatic years 1842–46 were severe in central Europe, and many experiments hastily performed to verify the new theories had limited results. In 1855, J. C. F. Hoefer claimed that many had found fault with Leibig's theory of plant life. It was only when, little by little, serious work was carried out, that the ideas of the Giessen scholar received resounding confirmation.

British agriculture had already seen the importance of using fertilizers. Imports of bone increased from £14,400 in 1823, to £254,000 in 1837. The first boat bringing guano from Peru arrived in Liverpool in 1835; in 1841, 1,700 tonnes were imported, and in 1847, 220,000. Lawes had obtained spectacular results in Rothamsted, and on May 23, 1842, he took out a patent for the manufacture of superphosphates (no. 9353) and created a factory in Deptford, a suburb of London, in 1843. Murray had experimented in 1818 with the use of bone treated with sulfuric acid as a fertilizer. (He asserted his rights to this process in 1857.) In 1836, he allegedly used the word "superphosphate" in a chemistry course. He also took out a patent for its manufacture (no. 9360) on May 23, 1842, the same day as Lawes. A legal quarrel took place because Lawes's patent had been written in ambiguous terms. From a strictly legal point of view, it is likely that Murray had the better claim.

German industry took longer to perceive the practical and commercial consequences of these discoveries for soil improvement. Not until 1855 were superphosphate factories recorded in Lehrte, and in 1857 at Heufelt. The Americans followed soon after their Civil War, and the French after the Franco-Prussian War in 1870 (Saint-Gobain: Chauny 1871 and Montlucon 1872).

Liebig knew that potassium salts were soluble, and he claimed that it was useless to add them to soil, for they would be eliminated by drying out. When Way, around 1850, demonstrated the absorbing power of soil and its exchange capacity, it was understood that potassium could be fixed on clay, and that this was the mechanism for potassium nutrition. The discovery of layers of potassium salts at Stassfurth put considerable quantities of these salts on the market at a low price. These potassium mines on the left bank of the Rhine were first discovered in the 1900s, and their exploitation began soon after World War I.

Liebig defended his ideas with his habitual energy; for example, in his 1846 brochure on chemical fertilizers he had harsh words for his French colleagues. Frederic Kuhlmann, who had demonstrated the importance of ammonia fertilizers, tried to reconcile the warring factions.

The principal source of fertilizer at this time remained guano from Peru, but by the 1860s the deposits had begun to be exhausted. The British and coastal Americans were the primary culprits in this depletion (an average of 300,000 tonnes per year for the 1860s), while France, Germany, the Netherlands, and Austria used about one tenth that amount. Nitrates from Chile replaced the Peruvian guano little by little, and, in turn, were superseded by the industrial manufacture of nitrogenous fertilizers after the First World War.

Liebig was one of the first to become aware of the paucity of fertilizers in Europe, which threatened to become more acute as his discoveries were applied. Starting in the 1850s, in Munich, he led a great campaign to economize fertilizer use and to recycle nutritive elements on European farms. He could not sense—although he imagined it to be a fortunate eventuality—the discovery of phosphate deposits at the end of the century in Florida and Tennessee, and later in Idaho, North Africa, Russia, and Oceania, and the potassium deposits in Germany, Alsace, the United States, and Russia. Processes for synthesizing nitrogenated products also changed the dimensions of the problem. Since the beginning of the twentieth century, many other sources of fertilizer have been discovered in the world.

Until his death, Liebig pursued vigorous actions in favor of an economical agriculture using fertilizing elements with a base of nitrogen-producing fodder and recycling all possible wastes from urban and industrial human activities. He was a precursor of today's ecologists with their most rational qualities.

B. Soil Becomes an Object of Study, 1870–1900

After the American Civil War (1861–65) and the Franco-Prussian War (1870–71), a new generation set to work: Vasilii V. Dokuchaev and P. A. Kostytchev in Russia; H. Hellriegel, M. E. Wollny, and Emil Ramann in Germany; Schloesing, C. A. Muntz, and Grandeau in France; and Eugene W. Hilgard in Berkeley, California, in 1875. Only the British lacked a soils leader of stature, which can be explained in part because free trade greatly reduced interest in agricultural issues, despite the activities of Lawes and Gilbert at Rothamsted and the value of the work of their young collaborator Robert Warington (1838–1907), whose works on chemistry (1881) and the physics of soil (1900) had a great influence on English-speaking countries.

In 1862, F. Albert Fallou (1794–1877) published *Pédologie oder allgemeine Bodenkunde*[11] This remarkable book was essentially theoretical and contained practically no examples to substantiate his claims, but it disclosed basic concepts which would be substantiated by his successors.

Toward 1875, the soil term, *chernozem* was commonly used in Russian economic circles.[12] It was a soil of the grain steppes which, from the Ukraine to southern Siberia, served as the foundation of the great Russian granary belt. English geologists who had studied these regions had recognized their riches; "The chernozem is to Russia what coal is to England" was a common refrain. Speculators bought up property, and swindlers sold other lands which had the sole advantage of being black in color but had none of the fertile qualities of chernozem. It became increasingly necessary to define chernozem scientifically. The work was entrusted to Dokuchaev (1846–1903), the conservator of the geology museum in St. Petersburg, who wrote a remarkable 500–page thesis entitled *Russkii Chernozem (The Russian Chernozem)*.[13] His friend and colleague Dimitri Mendeleyev was part of the jury who evaluated the work.

For Dokuchaev, soil was a body of nature like plants and animals. It was "independent and variable." It was the result, accumulated through the ages, of the action of climate and living beings on rocks, fine-tuned by relief and, in certain cases, underflow of water. The soil was "the mirror of nature": its characteristics, when one knew how to interpret them, provided information on the current dynamic of the surrounding environment and its

---

11. The word Pédologie is a German word created by Fallou from the Greek words *pedon* (soil) and *logos* (speech).

12. Chernozem was a word used by Ukrainian peasants to designate a soil which was black, deep, fertile, with a very developed structure, and rich in humus.

13. Vasilii V. Dokuchaev, *Russkii Chernozem. Otchet Imperatorskomu Bolnomu Ekonomicheskomu Obshchestvu* (St. Peterburg: Tipografiia Deklerona i Evodokimova, 1883).

past, and permitted a better prognosis on the possibility of transforming this environment, specifically with agriculture. Even more than ecology, soil science, as it was conceived of by the Russian school at the end of the nineteenth century, was the science best able to allow humankind some control of the environment.

The influence of these ideas in Eastern Europe was considerable. Besides his friends—Kostytchev (1845–95), F. Y. Levinsson-Lessig (1860–1939), and N. M. Sibirtzev (1860–99)—Dokuchaev had numerous students and disciples, among them such notables as: A. Ferkhmine, P. Ototsky, K. D. Glinka, Valérien Agafonoff, V. I. Vernadsky, and Serge Winogradsky (1856–1946), whom we will find again in the next period.

In the United States, as early as 1838, John Morton had published *On the Nature and Properties of Soils*, which supported many ideas that have modern counterparts. However, he was forgotten soon thereafter, and it was Edmund Ruffin (1794–1865) with his book *An Essay on Calcareous Manures* (1832), who was to influence soil science in the United States more deeply. But neither Morton nor Ruffin attained the stature of the great man of American soil science, Eugene W. Hilgard (1833–1916).

After studies in Europe and a brief career in the state of Mississippi in the United States, Hilgard arrived in Berkeley, California, in 1874 and spent the last forty years of his life there. His activity was prodigious. With his friend, the senior E. J. Wickson, and collaborators M. E. Jaffa and R. H. Loughridge, he enlivened Californian agriculture with his research and counsel. He taught at the University of California, Berkeley and had numerous students, including W. P. Kelley. Although his work was partly forgotten after his death in 1916, the extent of his scientific activity was brought back into the public eye in 1961, due to Hans Jenny's *E. W. Hilgard and the Birth of the Modern Soil Science*.

Hilgard shaped and popularized numerous analytical methods. In his work *The Cotton Census at 1880*, he characterized soil regions with subcategories and varieties. He also demonstrated the influence of climate on soil formation (1892) and became, at the end of his life, the specialist in saline soils. This latter was obscured by his controversy with Milton Whitney, who failed to understand the importance of chemical fertilizers and who aspired to restrict the study of soils to the study of their hydrous balance.[14]

---

14. (a) Hans Jenny, *E. W. Hilgard and the Birth of the Modern Soil Science* (Pisa, Italy: Agrochemica, 1961). (b) Milton Whitney believed that all soils contained enough nutrients for plant growth and that consequently soil chemistry was not important. He also believed that texture was the key to the moisture and temperature conditions of the soils and thus the key to soil productivity. This was stressed in his arguments with Hilgard, Hopkins, and others.

Hilgard was far from the only American scientist studying soil. Nathaniel S. Shaler wrote, in 1891, *The Origin and Nature of Soils* and was a teacher of C. F. Marbut. The soil physicist F. H. King (1849–1911) of the University of Wisconsin wrote several treatises on soil. G. P. Merrill (1854–1939), a geologist with the Smithsonian, initiated the study of the alteration of rocks. At Yale, T. B. Osborne (1859–1959) worked on granulometry before receiving the Nobel Prize for his work on vitamins. George N. Coffey, who had attended the First International Congress of Geology in St. Petersburg in 1897 and had come in contact with Dokuchaev's collaborators, was one of the first to see the importance of his ideas.

On the other hand, soil science in Germany and in bordering countries influenced by Germany was particularly influenced by geologists. The first soil specialists were C. Trommer, C. A. Girard, Biron Cotta, and E. E. Schmid. Cartographic work on soils then appeared, even before similar efforts by the Russians. Ferdinand Senft (1810–93) was the first to use the ideas of contour and horizon; he was followed by Orth (1835–1915) and later by Max Fesca, who directed, in 1881, the first cartographic survey of the soils of Japan, which was published as one of the earliest classifications schemes between 1885 and 1887.

One of the most remarkable scientists at the end of the century, whose work was ahead of its time, was the Dane D. E. Müller (1840–1926), who defined and interpreted the podzol group of raw humus soils starting in 1875. However, the most important soil scientist of this nascent era was Emil Ramann (1851–1926), who was first a pharmacist, then became a forester. He directed the agronomy research station of Bavaria and was professor at Munich from 1900 to 1925. Of the many contributions that he made, perhaps the most important was the notion of *braunerde* (brown soil). He was one of the first men in Europe to hold a chair in pedology at Eberswalde, second only to the scientist Sibirtzev of Novo-Alexandria in Poland, who can arguably claim to be the first scholar of pedology.

After the bacterial discoveries of Pasteur (1822–95), it was Schloesing and A. Muntz (1846–1917) who demonstrated the role of microorganisms in nitrification. Winogradsky in Russia caried out the two following stages of this process and isolated the responsible organisms. In 1886, Hellriegel (1831–95) and his collaborator H. Wilfarth revealed that fixation of nitrogen on the nodules of legumes was due to bacteria, which had been foreseen by H. W. L. Lachman (1858), M. S. Voronin (1866) and Wilbur O. Atwater (1881). The microbial agent of this fixation was discovered by the Dutchman Martinus W. Beijerinch in 1888. Finally, the direct fixation of nitrogen by certain soil bacteria was made evident in 1888 by Marcellin Berthelot and confirmed in other work, notably that of Beijerinck and the Russians V. L. Omeliansky and Winogradsky.

It was soon discovered that other microbial phenomena played an important role in soil: the dynamic of humus, the degradation of sulfur compounds, and the establishment of reducing conditions in waterlogged soils. These were summarized by Winogradsky, a refugee in France after the 1917 Russian Revolution, in his *Traité de Microbiologie* (1949), which also presented the global methodology for studying soil microorganisms that is still the basis for the majority of modern works on the subject. This book was published thanks to Sedman A. Waksman (1888–1973), a Nobelist and the author of several treatises on soil microbiology.

Since 1843 and the work of Wiegmann and Polstorff, it has been known that plant growth requires varied mineral substances. Little by little, the list and the order of importance of these elements were delineated by Wilhelm Salm-Horstmar (1856) and Julius Sachs (1860) for iron, and J. A. Knop, J. L. Raulin, and especially G. E. Bertrand (1867–1962), who showed, in 1903, the role of other elements in very minute quantities. These include arsenic, boron, copper, manganese, zinc, cobalt, and nickel.

The idea of conducting extensive, quantitatively controlled experiments came from Lavoisier (1743–94) at his Frechines property. He weighed crop yields and knew the exact surface area of his cultivated lands, the quantities of manure he applied, etc. Boussingault imitated him on his Pechelbronn property in 1834, as did Lawes at Rothamsted in 1843. From 1850 onward, state authorities took over these private properties and established in numerous regions agronomic research stations with laboratories and staff consisting primarily of chemists. The particular aim was quality control of fertilizers. Dozens of agronomic stations were thus founded in each large European country: in Germany at Mochern and Tharandt; in France at Nantes in 1852 and Vincennes in 1860; in Italy at Modena in 1870; as well as in Russia, Austria-Hungary and North America (Illinois in 1879, Pennsylvania in 1881). The directors of these tiny organizations often were able to rise to the ranks of true scholars. Of these P. A. Bobierre and Hellriegel are excellent examples.

This was the era when men of the laboratory began to consult with men of the land and with researchers from the agronomic stations, laying the foundation for the true beginnings of soil science which would come early in the twentieth century. Their principal preoccupations were with methods of physical measurement and quantification of chemical elements in soils, plants, and fertilizers. Important in this synthesis of methodologies was Voelker, who was the director of the laboratory of the Royal Agricultural Society of England. In France, Grandeau (1834–1911), inspector general of agronomic stations, and his student Edmond Henry integrated these principles in studying forest soils. Also in France, the professors Georges Ville and P. P. Deherain (1830–1902), as well as Schloesing and Muntz in Germany, were promoters of chemical fertilizers in farm management, and they were instrumental in studying physicochemical mechanisms for soils in the field. In Denmark, I. G. Kjeldahl focused on nitrogen as a fertilizer. In Estonia, Iohan Lemberg organized soil analyses for Dokuchaev. A. Cossa in Italy, Ramon de Luna in Spain, and A. J. Petermann in Gembloux (Belgium) were examples, among many other laboratory scientists, who contributed to the early progress of soil science.

By 1900, there were numerous agronomic research stations worldwide. Table 2.1 lists them by country.

In particular, soil physics at the end of the nineteenth century was dominated by Wollny (1846–1901) of the University of Munich, who created and enlivened one of the first specialized periodicals in agronomy, namely, *Forschungen auf dem Gebiet der Agricultur-physik*, which published articles by the greatest world specialists, notably Kostytchev and Hilgard.

C. The Era of Specialists, 1900–1930

Around 1900, the conditions for agronomic research changed in almost all countries, particularly in soil science. New mineral resources had been discovered at the end of the century, and it was necessary to analyze both the fertilizers suggested to cultivators and the soils in which they were introduced. Analytical chemistry resolved these problems. In countries like the United States and Russia, exploration of soils and their cartography was essential in order to control their development. This would also be necessary, later on, in tropical zones, especially to direct irrigation in arid regions. The development of the mechanics and hydrous control of the soil necessitated research on soil physics and microbiology, which had explained certain mechanisms but would in the future extend its field of expertise.

At the Universal Exposition in Paris in 1900, Dokuchaev's team presented the first national soils map: that of European Russia. In 1909 at Budapest, they made a tour of Eastern Europe to present the map of its soils. At the same time in many large countries, cartographic services were

Table 2.1. Agronomic research stations

France	82	Russia	42
Germany	81	Sweden	29
Austria	50	Italy	26
United States	46	Great Britain	21

established. Whitney (1860–1927) can be credited with starting the U.S. Soil Survey, which took different administrative forms from 1898 to 1901. A first attempt at a soil map was made in Montana in 1898, which convinced the authorities to extend the Survey considerably.[15] Several notable scientists worked on this with Whitney, including Coffey, Briggs, and Cameron (see Chapter 11).

The practical and theoretical work conducted by the U.S. Soil Survey was already considerable. The idea of a series of county maps was introduced in 1909. Marbut (1863–1935) succeeded Whitney in 1915 and was able to make the U.S. Soil Survey a powerful resource for research and soil inventories. An admirer of Glinka's books, Marbut had them translated into English, and by 1927, an information exchange between the Russian schools and American universities was facilitated by an act of the U.S. Congress.

From 1906, the efforts of soil scientists in Russia were concentrated on the development of Siberian soils maps under the direction of Glinka. Numerous important soil specialists conducted their first work here. L. I. Prassolov (1876–1951) later became a great organizer and directed the Dokuchaev Institute from 1937 to 1949. S. S. Neustruev (1874–1928) became one of the first creators of soil geography and developed the idea of combining soils. Also notable were A. I. Bezsonov; A. A. Yarilov (1868–1947); B. B. Polynov (1887–1952), explorer of Soviet Asia and specialist in salty soils, who oriented the study of soils to the study of surfaces (geomorphology); N. A. Dimo (1873–1959), who studied the soils of arid zones, and many other soil scientists who contributed to this great scientific adventure. Tens of millions of hectares were thus mapped.

In the more densely populated countries, soil inventories took place later on, some beginning as late as the 1950s. The following important figures in European soil research and surveying should be mentioned: E. Huguet del Villar (1871–1951) in Spain, Paolo Principi in Italy, A. J. Oosting, D. J. Hissink and C. H. Edelman in Holland, Gregoire and Rene Tavernier in

15. Trial soil surveys were made in two counties in the state of Maryland during the summer of 1898 by Thomas H. Means and Clarence W. Dorsey. Each area consisted of about 250 square miles. One was near Hagerstown, in Washington County, and the other northeast of Baltimore, in Cecil County. Both of these surveys showed soil types, unlike the survey near Billings, Montana, which showed only salinity classes. No information except that the field work was done and that maps were hand-colored remains from the area near Hagerstown, but additional mapping and some re-mapping were done during the summer of 1899 in Cecil County, enough so that the survey was published in 1901. (See Chapter 11.)

The first men to map soils for the Bureau of Soils were Means and Dorsey plus Frank D. Gardner and J. A. Bonsteel. All of these men eventually left the Bureau of Soils for other work. M. H. Lapham began work with the Bureau of Soils in 1900; George N. Coffey, two years later. Coffey was a junior man in the soil survey of Tama County, Iowa, in 1903.

Lyman Briggs was a physicist and F. K. Cameron a physical chemist. Both men worked in the laboratories of the Bureau of Soils in Washington, D.C., and neither took part in the soil surveys.

Belgium, G. M. Murgoci (1872–1925) in Romania, Nicolas Pouchkarov in Bulgaria, P. Treitz (1866–1934) in Hungary, C. Miklczewsky (1874–1949) in Poland, and Benjamin Frosterus in Finland. Russian émigrés arrived after 1919 to make Dokuchaev's concepts well known in the West: A. I. Stebut (1876–1952) worked in Yugoslavia, Agafonoff taught in France, and Joffe published his famous work *Pedology* while at Rutgers University in the United States. In Germany, Ramann and Edwin Blanck (1877–1953) orchestrated research on the genesis of soils. In 1929, Blanck published the *Handbuch der Bodenkunde*, one of the most important early works on the emerging issues in soil science.

After the great discoveries of the end of the nineteenth century, microbiology became a part of university curricula by the early years of the twentieth century, and Felix Lohnis (1874–1931) wrote in German the first treatise, *Handbuch der Landwirtschaftlichen Bakteriologie* (1910). He came to the United States to teach this science from 1913 to 1923, and he had as students E. B. Fred, N. R. Smith, Charles Thom, and the Canadian G. A. Lochhead. In particular, an inventory of microbial species in the soil and the study of their roles were undertaken. Jacob G. Lipman (1874–1939) broadened the field of study to include all fauna in the soil. A professor at Rutgers University in New Jersey, he founded the journal *Soil Science* (1915) and recruited Waksman, who conducted, along with others, research on humus and who received the Nobel Prize in 1952 for having isolated a microorganism in soil which furnished the antibiotic used in the treatment of tuberculosis.

Charles Darwin (1809–82) devoted part of his life to the study of earthworms and their influence on soil formation. He declared that the term vegetal land was incorrect and that it would be more appropriate to say animal land. The tradition of studying soil fauna was then well established in Great Britain. At Rothamsted, following a laboratory accident, Sir John Russell and his collaborator H. B. Hutchinson demonstrated the effects of soil sterilization and developed, after 1918, an entire series of projects on earth-dwelling fauna and microfauna. These had important applications in agriculture, particularly in horticulture.

Soil physicists were effective during this period in detailing the relationships of soil to water and accompanying structural dynamics. Keen and his student R. K. Schofield (1901–60) had made Rothamsted illustrious in this area. However, the Americans in particular made progress in their conception of soil water. L. J. Briggs (1874–1963) defined the "wilting coefficient and moisture equivalent," and his colleague E. Buckingham (1867–1940) defined capillary potential.

G. Boyoucos and especially Leonard D. Baver (1901–80), author of the treatise *Soil Physics*, should be cited among a group of remarkable researchers, while W. H. Gardner (1883–1964) renewed conceptions of water dynamics in soil. He directed the agricultural research station in Utah, where he focused on the sphygmomanometer and trained numerous students, such as L. A. Richards and L. B. Lindford. Gardner was perhaps the most notable of a remarkable team of soil physicists at the University of Utah who made great progress in their discipline of soil-water management. Devoted to helping their compatriots survive in an arid climate, they were the pioneers of irrigation, the use of salty soils and water, and dry cultivation. J. A. Widstoe (1872–1952) published his book on dry farming (1911), which reworked older Mediterranean techniques reintroduced after 1840 by French colonists in Algeria.

A new awareness of the importance of the process for extracting elements for titration caused successive techniques to be developed; for example, for soil phosphates, the techniques of Schloesing and A. A. J. De'Sigmond and then of Emil Truog (1884–1969) were subsequently taken over by the carbonate method. Truog had been a lecturer on soil chemistry at the University of Wisconsin; he had worked in almost all areas of soil research but especially on soil acidity and phosphorus dynamics. An excellent professor, he had 175 students, among whom W. H. Pierre also became an excellent educator. The American chemists thus took over from Europeans diverted from their work by World War I. Among them were W. H. Macintire (1885–1962), Cyril G. Hopkins at Urbana (Illinois), F. P. Vietch, and many others.

In Russia, Kostychev's successor was P. S. Kossovitch (1862–1915), who oriented soil studies in a dynamic direction. His student Sergei Sakharov (1878–1949) sustained his thesis in 1906 on soil solutions; he thus led the way to studies on the physicochemical behavior of soil as a function of humidity. Another of Kossovitch's students, K. K. Gedroits (1872–1933), profoundly influenced soil chemistry: he published important works on methods of analysis (1909 and 1933), but he is especially known for his work on the absorbent complex and on the consequences of ion exchange in soil, in the areas of fertility and alkalinity. He presided at the Second International Congress in Leningrad in 1931 and was, with Glinka, the lifeforce of soil science in the USSR during the 1920s.

In Budapest, De'Sigmond (1873-1939) also contributed to the work on exchange phenomena, along with the Swede Sante Mattson (1886–1980). The German Georg Wiegner (1883–1936), who worked in Switzerland, and the American W. P. Kelley (1878–1965) were also active in this area. These four scientists revolutionized our concepts of soil physicochemistry. In the more classical domains, the effectiveness and precision of analyses were greatly improved by men such as the American F. E. Bear (1884–1968) for fertilizing elements and the Swedes Olof Tamm for humus and A. M. Atterberg (1846–1916) for particle-size distribution, which was also detailed in the United States by Boyoucos.

In his memoirs, Waksman recounted his European trips of this period. His diagnosis explained the considerable havoc caused in the area of science by the 1914–18 war. The urgent need for solutions to technical problems had reduced the interest of the public powers in theoretical studies and geared the efforts of scientists toward practical issues.

From 1930 to 1950, soil science specialists, despite political and economic crises, were preparing for the flowering which soil science was to experience during the years 1950–90.

D. Internationalization of Soil Science, 1930–1950

At the end of the nineteenth century, all scientists felt the need for regular meetings, for choices, and for group decisions, as well as for international networks.

The First International Congress of Agronomic Station Directors was organized in Paris by Grandeau in June 1881. This congress gathered about 170 registrants and seventy participants from numerous countries. Two other meetings took place in Paris, in 1889 and 1900, at the same time as the Universal Expositions where several countries displayed their research projects. Dokuchaev, in particular, showed his work, which he also sent to Chicago in 1893. The scientist V. R. Williams, whose father was an American engineer working in Russia and married to a Russian, was a presenter at this exposition; Whitney also attended.

In these meetings of agronomists, however, the subjects largely extended beyond that of soil studies into all aspects of agronomy. The 1881 congress proposed the creation of an international publication, which was established by Grandeau in 1883 with the title *Annales Agronomiques Francaises et Étrangeres* and which disseminated the numerous projects of researchers, among them Wollny, Hilgard and Kostytche.

The first meeting devoted exclusively to soil issues was organized in Budapest in 1909 by the Romanian Murgoci and by Timko and Treitz, two Hungarians. This First Conference on Agrogeology was very successful, attracting one hundred participants. Austria, Belgium, Germany, Hungary, Italy, Norway, Romania, and Russia were officially represented. Some scientists (Hilgard, for example) wrote to the organizers to express their personal approval.

The following year, another meeting took place in Stockholm, where it was decided to create an international journal committed to German, *Forschungen in Bodenkunde*. The frequency of the conferences was set at four years, and the next was scheduled for Moscow in 1914. Toward the end of 1913, the Russians decided not to hold it, and not until 1922 in Prague did the soil specialists meet again. Fifteen countries were represented, and nine others sent their agreements in principle. In Prague, the delegates decided to meet again in June 1924 (Fourth International Meeting on Pedology), where it was voted during the June 19 morning meeting to create the International Society of Soil Science. Its first meeting, hosted by Lipman of Rutgers University, took place in Washington, D.C., in 1927.

On this occasion, an important Soviet delegation directed by Glinka came to reestablish contact and to "reveal" pedology to the conferees. Four years later, in Leningrad and Moscow, meetings were particularly fruitful for the delegations which had not attended the first ISSS congress. In Oxford (1935), mostly Europeans were present. International relations in the field of soil science were thus multiplied between 1927 and 1935. Unfortunately, after 1935, the combined effects of the economic crisis and the preparations for the Second World War caused a reduction in the membership of the ISSS, and the 1939 congress planned for Berlin, to be presided over by F. Schucht, did not take place. After World War II, it was the initiative of Albert Demolon (1881–1949) that brought about international meetings on Mediterranean soils at Montpellier and in Algiers. This gave several celebrities (Charles Kellogg, Georges Aubert, Edelman, Tavernier) the chance to meet and restart the society, thanks to its new secretary general, Hans Van Baren, who organized the fourth congress in Amsterdam in 1950. Since then, the congress has met regularly every four years. The growth of the society has been spectacular, although it was affected by the oil crisis of 1975.

The president of the ISSS serves from one congress to the next, and is, therefore, chosen by the national association hosting it. The secretary general, on the other hand, remains in office for a much longer time (there have been four in sixty-six years). He thus ensures the continuity and maintenance of the society's administration. A trimestrial bulletin informs the members on organization life, the activities of the national associations, and publications and career events of its members. Between congresses, meetings of the principal commissions, subcommissions, and task forces take place on current topics. At each congress, copious minutes are published, and each host country presents in detail the state of the art of its national science. Members of Honor are named and the list of these, as well as that of the presidents, includes most of the great names in soil science. The

ISSS in 1990 was composed of seven commissions, four subcommissions, sixteen task forces, and five permanent committees.

At the end of the 1940s, a debate began among soil specialists as to the true nature of soil.[16] This conflict of ideas can be summarized by the following doctrines on the nature of soil:

(1) Soil is considered to be a milieu definable by independent variables. They can be studied in isolation; any reference to genetic determinism is useless and of no interest. This was the position of members of the Soil Conservation Service in the United States, but also, implicitly, of many researchers with laboratory specialties working all over the world (Wollny, etc.).

(2) Fallou, Hilgard, Müller, and the Soviet pedology school presented an essentially genetic notion, shaded by the progressive discovery of oblique movements in soil. From this point of view, the soil results from a natural determinism which asserts itself as a function of the increase in the duration of evolution and the aggressiveness (heat and humidity) of the climate. The result is the existence of units of soil, simple or complex (the notion of combination was detailed by Neustruev at the end of the 1920s), which can be mapped and in which the variables that define the soil are either linked together or are mutually exclusive. Around 1950, Jenny represented this philosophy in the United States, and this point of view was shared by G. W. Robinson, Demolon, W. L. Kubiena, Edelman, J. K. Taylor, and many others.

(3) Under the influence of Kellogg, the U.S. Soil Survey and most of the soil inventory services took an intermediate position, determined by the characteristics of agricultural applications. One could not provide overly complicated information to cultivators who, in any case, would use it to cultivate parcels of land in a homogeneous way. The solution, easy to apply in plains regions, was to divide land into map units which were made as homogenous as possible by combining relatively simple analytical data (texture, structure, pH, humus, chalk, etc.) with as clear an awareness as possible of the factors influencing soil formation, especially when these determined unambiguous limitations (cliffs, interruptions of gradient, edges of bodies of water, etc.). The variations which diverged too much from the definition of the unit were considered impurities (less than 15% of the surface area in the soil map series).

(4) Toward the 1960s (although the pioneers probably understood this well beforehand), in the hilly zones, mountains, and regions of recent accumulation of the original soil materials, gradients, hazardous distributions, and special forms due to climatic agents determined a morphogenesis whose variations covered all scales and intensities. V. M. Fridland, at the Dokuchaev Institute, explored techniques for representing soils having these characteristics.

16. (a) M. G. Cline, "The Changing Model of Soil," *Soil Science of America Proceedings* 25 (1961): 442–46. (b) Roy W. Simonson, "What Soils Are," *Yearbook of Agriculture* (1957): 17–31. (c) Simonson, "Concept of Soil," *Advances in Agronomy* 20 (1968): 1–47. (d) Jean Boulaine, *Pédologie Appliqué* (Paris: Masson, 1980).

In 1950, awareness of this problem was still limited to a few researchers, notably heads of large services like Kellog or Demolon, or academics like Jenny, Cline, Robinson and Fridland. After 1950, the debate was masked by the rapid growth of institutions and by the arrival in laboratories and on the land of young men interested in resolving simple and immediate technical problems. It remained to be seen whether the interest, for the future of research, was to use as perfected a conception as possible, or whether the demands of technical applications would lead us to limit the content of the concept of soil.

The intensity of international activity demonstrated that, despite the Depression of 1929 and its aftermath, the Stalinist drift in the USSR, and World War II, men in most countries maintained and even developed soil research.

In the USSR, the nefarious influence of V. R. Williams (1863–1939), linked with T. D. Lysenko, was fortunately counterbalanced by the whole group at the Dokuchaev Institute, which rallied around Prassolov, I. V. Tiruin (who died in 1962), and A. A. Rode (1896–1979). V. N. Sukachev (1880–1966) was permitted to pursue his studies of marshy soils, while D. N. Prianichnikov (born in 1865) continued his projects on soil fertility. Others who should be mentioned are Polynov, Y. N. Afanasiev, A. N. Sokolovsky, E. N. Ivanov, C. P. Kravkov, V. V. Akimtsev, and A. A. Zavalishin, and more recently I. P. Gerasimov, V. A. Kovda, and Fridland. This rapid enumeration gives only a slight idea of the immense amount of research and inventory work accomplished on soils by the Soviet Union: approximately 5,000 specialists could be counted there in 1960.

There were even more researchers and specialists in the United States (about 7,000). In that country, one of the greatest openings in soil science was the awareness of the gravity of soil erosion (made evident by the dust bowl) and the choice by President Franklin Roosevelt, as one remedy for the Depression, to use the protection and conservation of soils as a way of temporarily reducing cultivated areas in the West and lowering the level of agricultural production.

Consultants and men of action such as Marbut, Kellogg, and H. H. Bennett (1881–1960) were at the president's side to create what would gradually become the Soil Conservation Service. The projects on erosion begun in the 1890s by Dokuchaev in Russia, by A. C. Surrell in France around 1850, and by innumerable farmers in Asia, Mediterranean Europe, and the American South became from then on the object of rational studies, as illustrated by the work of W. C. Lowdermilk, W. H. Wischmeier, and H. J. Harper.

France had been ravaged by the First World War; its economy and population had both suffered; one fourth of the agronomists had been killed. The

Treaty of Versailles, however, had made the country the greatest European power in the manufacture of chemical fertilizers. The creation in 1922 of the Centre des Recherches Agronomiques de Versailles (which became the INRA in 1946) would provide a scientific basis for the promotion of fertilizing elements and for an exemplary agricultural recovery after 1950. However, first it was necessary to train researchers, and this was done under the direction of Demolon and his collaborators G. Barbier, Henri Lagatu, C. V. Garola, and others. He hired S. Henin, Aubert, and G. Drouineau at the same time that A. Oudin recruited Philippe Duchaufour at the forestry school in Nancy. These were the great craftsmen of the soil science recovery in France after 1950.

The British, less affected by the torment of the Great War, had realized how precious agriculture was for a country vulnerable to isolation from sources of supply; they developed their own research as a way of supporting the second shock of the 1939–45 war. Sir John Russell and his son E. W. Russell, as well as K. D. Hall (1864–1942), were the artisans of an activity which was extended to the entire British community overseas. Schofield and Keen in soil physics, G. W. Robinson (1888–1950) and Alexander Muir in pedology and Frederick Hardy (1889–1977), G. V. Jacks, C. E. Green, G. Milne, and A. F. Joseph in tropical zones deserve mention.

In Australia, J. A. Prescott, J. K. Taylor (1898–1982), C. G. Stephens, and K. H. Northcote began a soil inventory, while, in New Zealand, H. Taylor enlivened this research.

In South Africa, C. F. Van der Merve performed analogous work. In Canada, geographer Henri Prat was able to link his knowledge of soils to that of human geography. Adolph Reiffenberg began soil studies in Palestine. Adolpho Matthei, from Chile, explored South American soils, while, Leonardo Vettori founded soil chemistry in Brazil. That same role was played by Luis Bramao in Portugal; by first N. R. Dhar and then S. P. Raychaudhuri in India; and by T. Seki and Yoshiaki Ishizuka, eminent specialists in soil acidity, in Japan.

The Belgians owe much to George Waegemans (1913–76), G. Manil and Tavernier, who trained the future explorers of the soils of the Congo and many other tropical countries. The inventory of Dutch soils was begun by Oosting, with Edelman and Hissink, while Mohr and Van Baren devoted themselves to the Malay Archipelago. In Italy, A. Comel, Ugo Pratolongo, Principi, Azzi, and Alberto Oliva maintained the tradition of scholarly agronomy. Del Villar and Albareda Herrera established the basis for a soil science in Spain which would be decimated by the Civil War, but their young students would take up the work again after the war.

Thus, between 1930 and 1950, at different intensities and speeds, men, institutions, and publications devoted to soil science were put into place or were reinforced in all the countries mentioned. This intellectual and human patrimony passed somehow through the events of 1939–49, but when scientific life was definitively resumed in 1950 with the International Congress of Soil Science in Amsterdam, the infrastructure for vigorous growth was in place.

The progress of physicochemical science between 1930 and 1950 had important repercussions on subsequent projects in soil science. The determination of clays and even the notion of clays are remarkable examples. To the size definition of clays was added, in the 1930s, the mineralogic definition rapidly perfected by x-ray techniques and thermic analysis. S. B. Hendricks (1903–81) and numerous students of the mineralogists M. Jackson and E. Grim in the United States and G. Millot in France perfected analytical techniques.

The notion of pH, which appeared during the 1920s, gave better definition to the physicochemical milieu; the development of measurement techniques made possible the determination of acidity in the field as well as in laboratory experiments (Ishizuka, Truog, Mitscherlich, etc.). Ion theory clarified exchange phenomena and enabled the process of ion exchange to be treated in a more rational fashion. The definition and treatment of saline and alkaline soils owe much of their progress to these developments.

Aerial photography, which began in the 1920s, made enormous progress after the 1930s due in large part to military operations. It placed at the disposal of cartographers some exceptional documents which permitted great precision in the plotting of contours and facilitated the identification of land units for soil surveying.

The techniques for measuring chemical elements experienced a gradual improvement: miniaturization and automation of quantities and the progressive introduction of more rapid and precise physical methods ensued. The transformation of laboratories, however, was particularly an accomplishment of the 1960s, with the wide introduction of the electron microscope, among other advanced equipment. In 1950, the techniques in use practically everywhere required a long apprenticeship; laboratory men were still primarily chemists.

By the end of the war, the use of all-terrain vehicles transformed the daily lives of cartographers; the exploration of tropical and desert zones was facilitated. New types of soils were discovered, and the importance of geomorphology would be revealed to the explorers of these relatively unknown environments. Great development projects, agrarian reforms, the expansion of irrigation, the conquest of newly cultivated land, attempts at rehabilitating salty soils, the drainage of marshes, and the fight against erosion required technical inventories, periodic surveillance, and predictions concerning the eventual evolution of the land. These needs had repercussions on research programs and on the ways of defining pedological variables.

All these evolutions were in full development in 1950. A remarkable flowering of soil science marked the forty years to follow.

Bibliography

Atwater, Wilbur O., and C. F. Langworthy. *A Digest of Metabolism Experiments.* Washington, D.C.: U.S. Govt. Print Off., 1898.

Baver, Leonard D. *Soil Physics.* New York: Wiley; London: Chapman and Hall, Ltd., 1940.

Beijerinch, Martinus W. *Beobachtungen und Betrachtungen uber Wurzelknospen und Nebenwurzeln.* Amsterdam: J. Muller, 1886.

Berthelot, Marcellin P. E. *Chimie Végétale et Agricole.* Paris: Masson, 1899.

Blanck, Edwin. *Handbuch der Bodenkunde.* Berlin: Springer, 1929– 32.

Boulaine, Jean. "Deux Siècles de Fertilisation Minérale," in *Deux Siècles de Progrés pour l'Agriculture et l'Alimentation.* Paris: Lavoisier, 1990 (Bicentenaire de l'Académie d'Agriculture de France no. 14).

Boulaine, Jean. "La Bataille des Phosphates au XIXème Siècle," *Bull. Intérieur de l'INRA* (sous presse).

Bronner, Johann P. *Der Weinbau im Konigreich Wurtemberg.* Heidelberg: C. F. Winter, 1837.

Davy, Humphry. *Elements of Agricultural Chemistry, in a Course of Lectures for the Board of Agriculture.* Philadelphia: John Conrad and Co., 1815.

Dumas, Jean B. *Essai de Statique Chimique Desetres Organisés.* 2d ed. Paris: Fortin, Masson, 1842.

Fallou, F. Albert. *Pédologie: Oder, Allgemeine und Besondere Bodenkunde.* Dresden: G. Schonfeld [D. A. Werner], 1862.

Fesca, Max. *Beitrage zur Kenntniss der Japanischen landwirthschaft.* Berlin: P. Parey, 1890–93.

Gasparin, Adrien. *Cours d'Agriculture.* 3d. ed. Paris: Bureau de La Maison Rustique, 1843–48.

Gazzeri, Giuseppe. *Compendio d'un Trattato Elementare di Chimica.* Florence: Nella Stamperia Piatti, 1828.

Gedroits, Konstantin K. *Chemische Bodenanalyse: Methoden und Aleitung zur Untersuchung von Boden im Laboratorium.* Berlin: Gebrunder Borntraeger, 1926.

Hassenfratz, Jean H. *Traité Theorique et Pratique de l'Art Calciner la Pierre Calcaire, et de Fabriquer Toutes Sortes de Mortiers, Cimens, Betons, etc., Soit à Bras d'Hommes, Soit à l'Aide de Machines.* Paris: Carilain-Goeury, Mme. Huzard, 1825.

Hilgard, Eugene W., special agent in charge. *Report on Cotton Production in the United States: Also Embracing Agricultural and Physico-Geographical Descrip-*

tion of the Several Cotton States and of California. Washington, D.C.: U.S. Govt. Print Off., 1884. (Tenth Census of the United States [1880], v. 4 & 5.)

Hundeshagen, Johann C. *Encyclopadie der Forstwissenschaft, Systematisch Abgefasst.* 4th ed. Tübingen: H. Laupp, 1842–59.

Jamieson, Thomas. "La Science Agricole dans la Grande-Bretagne," *Ann. de la Sc. Agr. T.* 2 (1901): 320–92.

Kellogg, C. E. "Conflicting Doctrines about Soils," *Science Monthly* 66 (1948): 475–87.

King, F. H. *A Text Book of the Physics of Agriculture.* Madison, Wisc.: von Alexander Classen, 1910.

Kubiena, W. L. *Claves Sistematicas de Suelos.* Madrid: CSIC, 1953.

Mohr, Friedrich. *Lehrbuch der Chemisch-Analytischen Titrirmethode* . . . Brunswick: Vieweg, 1862.

Pallmann, Hans. *Über die Geschichliche Entwicklung der Bodenkunde.* Zurich: Berichthaus, 1933.

Poggendorf, J. C. *Biographisch-Literarisches Handwörterbuch der Exacten Naturwissenschaften.* Berlin: Akademie Verlag, 1976.

Ramann, Emil. *Bodenkunde.* 3d ed. Berlin: Springer, 1911.

Rohna, A. Rothamsted. "Un Demi Siècle d'Expériences Agronomiques," *Ann. Sc. Agr. Fr. et Etr.,* T1 (1900): 30–490.

Ruffin, Edmund. *An Essay on Calcareous Manures.* Petersburg, Va.: J. W. Campbell, 1832.

Russell, E. J. *An History of Agricultural Science in Great Britain: 1620–1954.* London: G. Allen and Unwin, 1966.

Russell, E. J. "The Rebirth of Soil Science in Great Britain," *Soil Science* 94 (1962): 204–14.

Salm-Horstmar, Wilhelm F. K. A. F. *Versuche und Resultate Uber die Nahrung der Pflanzen* . . .. Brunswick: Vieweg, 1856.

Sastriquez, Garcia. *A Proposito del Centenario* de la Obra "el Chernoziom Russo" de V. V. Dokuchaev. Cuba: Sciencias de la Agricultura, 1983.

Saussure, Nicholas Théodore de. *Recherches Chimiques sur la Végétation.* Paris: Chez la Ve Nyon, 1804. (Translated into German 1890: *Chemische Untersuchungen Über die Vegetation.*)

Schubler, Gustav. *Grundsatze der Agricultur-Chemie in Naherer Beziehung auf Land- und Forstwirthschaftliche Gewerbe.* Leipzig: Baumgartners Buchhandlung, 1831.

Shaler, Nathaniel S. *The Origin and Nature of Soils.* Washington, D.C.: U.S. Geological Survey, 12th Annual Report, 1890–91.

Simonson, R. W. *Historical Highlights of Soil Surveys and Soil Classification With Emphasis on the United States (1899–1970).* Wageningen, Netherlands: ISRIC, 1989.

Sprengel, Karl. *Die Bodenkunde, Oder, Die Lehre vom Boden* . . . Leipzig: I. Muller, 1837.

Stefanovits, P. *Brown Forest Soils of Hungary.* Budapest: Académiai Kiado, 1971.

Thaer, Albrecht D. *The Principles of Agriculture,* trans. by William Shaw and Cuthbert W. Johnson. New York: Greeley and McElrath, 1858.

Voelcker, Jean C. A. *Chimie Appliquée à l'Agriculture*. Paris: Berger-Levrault et cie, 1886–88.

Waksman, S. A. *Ma Vie avec les Microbes*. Paris: A. Michel, 1964.

Wallerius, Johan G. *Mineralogie, Oder Mineralreich, von ihm Eingeteilt und Beschrieben*. Berlin: Bey Friedrich Nicolai, 1763.

Warington, Robert. *Lectures on Some of the Physical Properties of Soil*. Oxford: Clarendon, 1900.

Warington, Robert. *Note on the Appearance of Nitrous Acid During the Evaporation of Water: A Report of Experiments Made in the Rothamsted Laboratory*. London: Harrison and Sons, 1881.

Wiegmann A. F., and L. Polstorff. *Ueber die Anorganischen Bestandtheile der Pflanzen, Oder Beantwortung der Frage*. Brunswick: Vieweg, 1842.

Winogradsky, Serge. *Microbiologie du Sol: Problèmes et Methodes* (Paris: Masson, 1949).

3. Characteristics of Soil Science Literature

PETER MCDONALD

Cornell University

Soil science does not stand alone. Historically, the discipline has been integrated with all aspects of small farm management. The responsibility of maintaining a good crop yield over a period of years was laid upon the soil. Research into soil fertility reflected this production-oriented emphasis during most of the nineteenth century. To fulfill the promise of production, farmers and soil experts had to encompass all facets of agriculture, from engineering to farm management to basic concepts of crop nutrition in the field, but the focus of their efforts remained, and to a large extent still remains, to benefit overall harvests.

The early literature of soil science reflects this interconnectedness. Between 1865 and the second decade of this century, when the constituents of soil were first studied, the written record of soil science was inextricably bound to that of agronomy. The term "agronomy" is less used today, but well into the 1950s agronomy was a term in wide circulation, with a precise meaning and with practical applications. It was defined in Webster's second edition (1934) as, "that branch of agriculture which deals with the theory and practice of field crop production and soil management." The word is derived from the Greek *agros* meaning "field," and *nemein* or *nomos*, which meant "management" or "overseer." The difficulty of ascertaining where crop science ended and soil science began is problematic and is dealt with more fully in Chapter 14.

The diversity and interconnectedness of early soil science is reflected in the soil types themselves, which differ in their character and in their ability to sustain a seasonal crop. While all soils will have some properties in common, soils of different regions require different care, with treatments as diverse as their geographical genesis (see Chapter 5). It is well to remember that even in a discipline as broad as agriculture, the study of soil has been widely seen as the basis upon which all farm management must build. Consequently, soil research at the close of the last century held practical applications foremost. The predominant concern was with soil nutrients. Much of the literature included prescriptions on how to increase fertility through wise management of the "food factory" out of which these crops grew, namely, the soil of the local farm. As the analysis in Chapter 14 shows, the early literature dealt heavily with problems of soil fertility.

The history of soil science is well covered in other chapters in this volume (see Chapters 1, 2, 11, and 14). Early treatises on farm management tended to describe soil as half of a grand equation with crops as the equal partner, and the idea that soil research might also have applications beyond the practical scope of agriculture was slow to gain scientific respectability. The early years of agricultural chemistry, which began with the work of the German chemist Justus Liebig, whose *Organic Chemistry In Its Application to Agriculture and Physiology = Organische Chemie in ihrer Anwendung auf Agriculture und Physiologie*, one of the first modern treatises on soil science, produced vigorous debate within conflicting camps of agriculturalists.[1] As Margaret Rossiter points out, "spokesmen for the field [of agricultural chemistry] could respond to the changing moods . . . of its own practitioners by shrewdly stressing its practical applications at one moment and then its contributions to pure science the next."[2] Despite scathing criticisms by laboratory scientists, farmers in the field were slow to adopt these "latest ideas" and early agricultural chemistry had many false starts. The debate did lead to wide recognition that agricultural problems needed long-term experimentation and rigorous scientific investigation. In the United States, this realization spurred the formation of the agricultural experiment stations under the Hatch Act of 1887, which, in turn, laid the foundation for the modern science of experimental agriculture. The work begun at these stations instigated the great age of American soil research. This might reasonably be called its nascent period, which culminated with the founding of the Soil Science Society of America in 1936, a year before the U.S. Department of Agriculture, Division of Soil Survey, published its first edition of the *Soil Survey Manual.*

Soil science as a whole came into its own in the 1920s and 1930s, beginning with soil chemistry. From a meager handful of serial publications in 1920, by 1938 Europe and the United States had many journals that could be classified as soil related. The 1930s also saw the first comprehensive serial bibliography on soils. In England, the Imperial Bureau of Soil Sci-

1. Liebig's work is considered by many to be the first scientific study of soils. Justus Liebig, *Organic Chemistry in Its Applications to Agriculture and Physiology = Organische Chemie in ihrer Anwendung auf Agricultur und Physiologie*, ed. Lyon Playfair, 1st American ed. (Cambridge, Mass.: J. Owen, 1841); 1st German ed. (Brunswick: Vieweg, 1841).
2. Margaret Rossiter, *The Emergence of Agricultural Science* (New Haven, Conn.: Yale University Press, 1975), p. xi.

ence (later named Commonwealth Bureau of Soil Science) began publishing its *Bibliography of Soil Science, Fertilizers and General Agronomy* in 1931. It was published triennially until 1962 and served in its heyday as the preeminent bibliographic resource for the discipline. It had four sections: the first was arranged in numerical order according to a decimal classification system, the second consisted of a complete listing of the abstracted journals (about seven hundred titles in its last iteration), and the third and fourth were author and subject indexes. This bibliography was followed in 1938 by *Soils and Fertilizers*, one of many abstract publications issued by the Commonwealth Agricultural Bureaux (now CAB International) and to date the most comprehensive soils abstracting journal. Between 1931 and 1950, the number of citations in each volume of the *Bibliography* remained essentially static, at about seven thousand.[3] It was only after the Second World War that a marked increase in the literature occurred. One interesting statistic mentioned by G. V. Jacks is that the next decade saw a doubling of the number of authors indexed, from 5,400 in 1950, to over 12,000 in 1962.[4] However, Jacks does not say whether this statistic matches a similar increase in the number of citations or whether multiple authorship became more prevalent.

A. Early Modern Trends

The year 1950 is, therefore, a logical starting point for the analysis of modern trends in soils literature. In 1951, the U.S. Bureau of Plant Industry, Soils, and Agricultural Engineering published a revised *Soil Survey Manual*, one of the first attempts in the United States at a truly comprehensive guide "intended for use by soil scientists engaged in soil classification and mapping."[5] It was in 1949 that the first *Advances in Agronomy* appeared, under the auspices of the American Society of Agronomy, with many articles concerning soils. Five years earlier, in 1944, one of the first full-length treatises on tropical soil was published by Edward Mohr,[6] opening up whole new avenues of study for students of agronomy. In the same year, the French founded their Office de la Recherche Scientifique d'Outre-Mer (now the Office de la Recherche Scientifique et Technique Outre-Mer,

3. G. V. Jacks, "The Literature of Soil Science," *Soils and Fertilizers* 29 (3) (1966): 227–30.
4. Ibid., p. 227.
5. Soil Survey Staff, *Soil Survey Manual* (Washington, D.C.: U.S. Govt. Print. Off., 1951 [USDA Agricultural Handbook no. 18]), p. 2.
6. Edward C. Mohr, *The Soils of Equatorial Regions with Special Reference to the Netherlands East Indies* (Ann Arbor, Mich.: J. W. Edwards, 1944).

ORSTOM), which began its long history of pedological research both in Europe and in Africa. In 1945, the Food and Agriculture Organization (FAO) was founded. Indeed, in the years following the Second World War, with the expansion of the United States highway system, the ease of international travel by air, and because of increased mechanization on large farms with their demand for petroleum fertilizers, specialization in soil science increased and was soon reflected in a growing body of literature that shows no sign of slowing.

Soils respect no political boundaries. As a consequence, modern soil science has become increasingly a global enterprise. Table 3.1, taken from an early bibliometric study conducted by Dan Yaalon, shows the geographical breakdown of soil research in 1962.[7]

Soil chemistry, which included both soil fertility and fertilizers at this time, accounted for almost one third of the citation counts in *Soils and Fertilizers* in the early 1960s, as given in Table 3.2.

Table 3.1. Geographic distribution of soil research citations, 1961–62

Region	1961	1962	Percent
U.S. and Canada	381	523	21.9
Western Europe	420	495	22.2
USSR	470	491	23.4
Eastern Europe	232	221	11.0
Mediterranean	73	83	3.8
India and Pakistan	63	78	3.3
Japan	69	77	3.5
Australia and New Zealand	89	97	7.2
Third World[a]	91	173	6.4

Source: Based on data from *Soils and Fertilizers*, various issues.
[a]All developing countries excluding India and Pakistan.

Table 3.2. Percentage of citations by subject in *Soils and Fertilizers*, 1961–62

Soil Chemistry	34.4%
Soil Biology	21.8%
Soil Physics	19.6%
Soil Genesis	17.6%
Soil Analysis	16.1%
Soil Technology	6.6%

Source: Based on data from Dan H. Yaalon, "Has Soils Research National Characteristics?" *Soils and Fertilizers* 26 (1962).

7. Daniel Yaalon, "Has Soils Research National Characteristics?" *Soils and Fertilizers* 26 (1962): 89–93.

In the last forty years, the number of citations in *Soils and Fertilizers* has risen dramatically, from 2,667 in 1950, to the current level of over 15,600, a sixfold increase. A similar rise in soil citation counts from the major agricultural databases over the last twenty years reveals a trend of similar dimensions. In AGRICOLA, the database of the National Agricultural Library (NAL) in Beltsville outside Washington, D.C. with coverage of worldwide journal literature and monographs in agriculture and related subjects, the numbers reflect a threefold increase, from 2,868 soil citations in 1975, to 8,962 in 1985.

Figures such as these suggest that research in soils is changing at a dizzying pace. Aided by satellite technologies and computer-driven systems for the dissemination of information, it will not be long before tremendous quantities of soil data are collected and accessed via telecommunication networks with the use of alphanumeric and imaging systems, and less so in books or journals. Local soils database systems are already widely used for manipulating data with geographic information systems technology. This phenomenon is examined more fully by D. L. Anderson and J. Dumanski in Chapter 8.

The discipline has become multifaceted and technical, with wide-ranging applications for accompanying economic and international research goals. In the chapter on "Soil Science" in *The Uses of Earth Science Literature*,[8] D. A. Jenkins and R. Tully list the following disciplines to show the range of expertise in the study of soil: mineralogy; physical, inorganic, and organic chemistry; biochemistry and microbiology; agronomy; ecology; geomorphology; and microclimatology and its broader discipline of meteorology. In what was to become a rallying theme of twentieth-century science, the editorial of the first issue of *Soil Science* (1916) proclaimed, "Specialization must follow expansion in every field of knowledge . . . classified, divided and sub-divided."[9] Such had not always been the case, certainly not in nineteenth century agriculture. Early in the fourteenth century, William of Ockham wielded his philosophical razor with the admonition that one should not multiply entities needlessly.[10] Modern science has not paid Ockham much heed. This constant subdivision of "entities" in science has an impact on bibliometric studies such as these, complicating analysis and making quantification increasingly difficult, while limiting the efficacy of forecasting long-term trends.

8. D. A. Jenkins and R. Tully, "Soil Science," in *The Uses of Earth Science Literature*, D. N. Wood, ed. (New York: Archon Books, 1973).

9. *Soil Science* 1 (1) (1916): 1.

10. William of Ockham, *Philosophical Writings* (Edinburgh, Scotland: Nelson and Sons, 1957).

In a commentary published in *Current Contents* (1990), John Hodge, of the USDA Cereal Science and Foods Laboratory in Peoria, Illinois, discussed the seminal impact of his 1953 *Journal of Agricultural and Food Chemistry* paper on the chemistry of browning reactions, one of the most highly cited agricultural papers of all time. He said: "Today's tightly organized research seems to leave little room for serendipity. [In the 1950s], research was planned as it progressed; dollar values were not preassessed. Now . . . chance discoveries suffocate."[11]

Today, scientists have become specialists in narrow fields of expertise, which, with budgets tightening, might well describe the evolution of soil science in this century. Accompanying this specialization, the sanguine precepts evinced by the editors of *Soil Science* in 1916 have made the study of soil literature, in particular, a difficult task to quantify, even with sophisticated bibliometric tools such as the *Science Citation Index* (*SCI*) compiled and published by the Institute for Scientific Information (ISI) since 1961. As Yvon Chatelin and Rigas Arvanitis point out, the parameters of soils research are fluid at best and research specifically oriented to the understanding of agricultural sciences is rare.[12] They observe that soil science literature has followed a pattern of division, becoming increasingly segmented, with much of the research published in nonagricultural journals. Any review of the literature of soil science will necessarily reflect this fragmentation and scattering.[13]

As noted earlier, soil science does not stand alone. It retains close ties to applied chemistry and agricultural engineering, with its emphasis on irrigation and field machinery. Where soil research receives scant attention is in the parent disciplines of geology and the earth sciences. This is exemplified by the independent paths followed by civil engineering (soil mechanics) and geology, which are only of marginal interest to agronomists. Earth scientists tend to ignore the first meter of the surface of the globe, which is the soil layer of interest to farmers. Jenkins and Tully point out, "the main incentive in understanding the nature and properties of soil remains agricultural."[14] In a bibliographic study such as this, it is important to keep in mind the diverse relationship between soil sience and the parent discipline. More than half the cited journals in the discipline have no direct bearing on

11. Eugene Garfield, "Journal Citation Studies 53: Agricultural Sciences: Most Fruitful Journals and High Yield Research Fields," *Current Contents* 33 (51) (1990): 11; reproduced in his *Journalogy, KeyWords Plus, and Other Essays* 13 (1990): 455–467.
12. Yvon Chatelin and Rigas Arvanitis, "National Scientific Strategies in Tropical Soil Sciences," *Social Studies of Science* 18 (1) (1988): 113–45.
13. Chatelin and Arvanitis, *Stratégies Scientifiques et Dévelopment—Sols et Agriculture des Régions Chaudes* (Paris: ORSTOM, 1988).
14. Jenkins, "Soil Science."

soil research per se (see Chapter 10), clearly showing the wide net used by soil scientists in pursuing their work. This is especially true in the area of agroclimatology and the broader discipline of meteorology, as well as with the earth sciences and geology.

B. The Current Trend

In the spring of 1989, the Albert R. Mann Library of Cornell University began a project to identify the current core literature of the agricultural sciences. The Core Agricultural Literature Project covers seven major subjects in agriculture. Two major objectives were accomplished by this project. The first was a multivolume series on the agricultural sciences, of which this volume on soils is the fourth. The seven subject areas and volumes are

(1) Agricultural Economics and Rural Sociology
(2) Agricultural Engineering
(3) Animal Science and Health
(4) Soil Science
(5) Food Science and Human Nutrition
(6) Crop Improvement and Protection
(7) Forestry and Agroforestry

The second objective is to transfer the texts of the core literature of these seven disciplines onto compact disk (CD-ROM), primarily for distribution to academic institutions and research stations in the Third World. This full-text library on disk will include nearly nine thousand monographs and about four hundred journals covering five years. This project is unique in that it is the first time that such a systematic and comprehensive analysis of the literature of a large discipline has been undertaken in the sciences. How these core lists of monographs and journals were derived is explained in Chapters 9 and 10, respectively.

In preparing the *Core List of Soil Monographs* (see Chapter 9), the discipline of soil science was divided into seven major categories. These include:

(1) Soil Genesis and Classification; Pedology
(2) Soil Physics
(3) Soil and Water Management; Conservation and Erosion
(4) Soil Chemistry
(5) Soil Fertility and Fertilizers

(6) Soil Biology
(7) General Soil Science, Sustainable Agriculture, Agronomy, Agricultural Engineering.

This subject taxonomy was derived from a combination of sources, primarily *Soils and Fertilizers* and the *Soil Science Society of America Journal*. For the purposes of this investigation, the discipline was further divided into two major categories, the literature of soil science pertaining to developed regions and that pertaining to developing countries, the latter most commonly concerned with the tropics.

The examination of soil literature by subject categories divided into two broad geographical strata has helped to facilitate analysis statistically. Most agricultural databases, such as AGRICOLA and CAB Abstracts, divide the discipline by subject codes, enabling researchers to gather data specific to narrowly defined subsets. Second, the needs of researchers in developed countries, compared to those in the Third World, differ both in scope and application, though the gap is closing. By analyzing these corpora separately, it is easier to compare and contrast historical antecedents of development and thereby make correlations to the evolutionary trends in the literature.

AGRICOLA and CAB Abstracts are only partly successful in outlining the totality of the literature. The universe of published material in soils is immense and defies a precise count. Data from CAB Abstracts and AGRICOLA, while indispensable for this sort of analysis, reflect citations from journals which are primarily research-oriented. Third World material literature from Asia and the Eastern Bloc, and works in non-Roman alphabets are indexed very selectively. As many studies have shown, particularly those of Y. N. Rabkin and H. Inhaber, the tools bibliometric researchers use "are clearly biased in favor of central nations," i.e., the United States and Europe.[15] The ISI does not analyze monographs per se (although recently they have included analysis of the proceedings of selected conferences), thus analysis of this sort remains an area of study where more hard data are needed. Furthermore, established subjects which are no longer hot topics and which are covered in publications like textbooks, are generally underrepresented in the citation counts of current research. These and similar caveats represent typical skews which must be accounted for at the outset. Finally, the subject categories of the various databases do not align in soil science, especially when comparing AGRICOLA and CAB Abstracts, as can be seen in Table 3.3 below.

15. Y. M. Rabkin and H. Inhaber, "Science on the Periphery: A Citation Study of Three Less Developed Countries," *Scientometrics* 1 (3) (1979): 261–74.

Table 3.3. Comparison of subject code categories in AGRICOLA and *CAB Abstracts*

CAB Abstracts		AGRICOLA	
Os11	Soil Chemistry	J200	Soil Chemistry and Physics
Os16	Soil Biology	J100	Soil Biology
Os14	Soil Classification	J300	Soil Classification
Os15	Soil Fertility	J500	Soil Fertility
Os17	Soil Surveying	J400	Soil Surveying
Os12	Technique and Analysis	J600	Soil Resources
Os13	Soil Physics	J700	Soil Cropping Systems
Os18	Soil Morphology	J800	Soil Conservation
Os22	Fertilizers and Soil Amendments		

The analysis of the soil science literature using data collected from on-line or other bibliographic sources is often imprecise or incomplete for the reasons outlined. This problem is partially overcome by the sheer quantity of data available on agricultural bibliographic data bases. CAB Abstracts, for example, has over three million citations to date. When these numbers are considered in aggregate they stand up to rigorous examination, revealing clear patterns in the literature over time. These data are most useful when put in scale ranked by percentages.

Table 3.4 shows a numerical analysis by database of soil science citations, over a ten-year period, divided into subject categories. While there is some overlap of subject definition, notably in the first four categories, the remaining subject areas become problematic, especially when trying to correlate CAB Abstract terms with those used in the AGRICOLA database. The AGRIS database, also in Table 3.4, is a cooperatively gathered, computerized index, which corresponds, in part, to the print *Agrindex*, published monthly by the AGRIS Coordinating Centre of the FAO.[16] Its coverage is primarily of worldwide agricultural literature that reflects research results, food production and rural development.

Table 3.4 demonstrates that in all three databases, soil chemistry, the oldest soil discipline, ranked second only to soil fertility. Because crop production relies directly on the latter, it is easy to see why fertility of soils comes out on top, especially when the influence of fertilizer companies toward fertilizer research and promotion are included.[17] The problem of comparing subjects in each database is clearly seen when the total number

16. *Agrindex* (Rome: AGRIS, Food and Agricultre Organization, 1973–).

17. (a) "Fertilizer Data Summary," National Fertilizer Development Center, Tennessee Valley Authority, 1988. (b) Judith Soule and Jon Piper, *Farming in Nature's Image: An Ecological Approach to Agriculture* (Washington, D.C.: Island Press, 1992), pp. 62–83.

Table 3.4. Database citation counts in soil science by subject and year

	1980	1981	1982	1983	1984	1985	1986	1987	1988	%	Total 1980–88
AGRICOLA	159,329[a]				125,700[a]				79,601[a]		
Soil Chemistry and Physics	2,719	2,686	2,614	2,477	2,416	2,040	1,946	1,854	1,458	23%	20,210
Soil Biology	1,504	1,617	1,762	1,685	1,910	1,513	1,420	1,214	1,140	16%	13,765
Soil Classification	585	452	551	579	502	444	325	346	281	5%	4,065
Soil Fertility	4,579	4,333	3,829	3,562	3,189	2,512	2,466	2,040	1,773	33%	28,283
Soil Resources	240	226	213	160	198	183	119	74	51	−2%	1,464
Soil Erosion	1,093	967	1,030	859	885	1,017	766	913	629	9%	8,154
Soil Surveying	331	284	260	183	209	226	216	198	148	+2%	2,055
Soil Cultivation	948	968	1,215	1,060	1,114	986	820	808	612	10%	8,531
Soil Science, General	136	139	79	59	64	41	27	23	32	−1%	600
Soil Totals	12,135 (7.6% of 1980 total)[b]				10,423 (8.3% of 1984 total)[b]				6,124 (7.6%)[b]		87,127
CABI Abstracts	148,251[a]				132,720[a]				144,233[a]		*1980–88*
Soil Chemistry	1,918				1,921				1,391	16%	20,378
Soil Biology	1,788				2,186				1,732	16%	20,621
Soil Classification	1,750				2,016				1,888	15%	19,665
Soil Fertility	3,017				3,254				3,151	26%	33,777
Soil Physical Properties	1,280				1,408				814	12%	14,716
Soil Analysis	878				923				765	8%	9,713
Soil Formation	597				755				697	7%	8,359

Soil Totals	11,228 (7.5% of 1980 total)[b]	12,463 (9.3% of 1984 total)[b]	10,438 (7.2%)[b]		127,229
AGRIS	144,360[a]	100,624[a]	108,082[a]		*1980–88*
Soil Chemistry and Physics	2,617	2,165	2,177	33.9%	20,871
Soil Biology	1,524	1,478	1,332	23.1%	12,996
Soil Classification	552	371	406	5.8%	3,978
Soil Fertility	4,538	1,024	899	16.3%	8,654
Soil Resources	252	N/A	N/A		N/A
Soil Erosion	752	970	981	15.1%	8,019
Soil Surveying	385	182	257	2.8%	2,466
Soil Science	102	193	176	3%	1,413
Soil Totals	10,722 (7.4% of 1980 total)[b]	6,383 (6.3% of 1984 total)[b]	6,228 (5.7%)[b]		58,397

[a]Total citations for given year.
[b]Soil citations as percent of total database for given year.

of citations for Soil Classification in AGRICOLA, 4,065, is compared with that in CAB Abstracts, 19,665. This is a fourfold jump in numbers. It is hard to make sense of this discrepancy between CAB Abstracts and AGRI-COLA without first studying the indexing methodologies outlined in their online search manuals. Both have unique subject emphasis; both use unique terminology; and subject headings, though similar, may mean different things in each database. There is no foolproof way to circumvent these shortcomings in data analysis. These and other peculiarities in the data should alert the reader that correlations such as these can be only rough guides at best.

C. Trends in Soil Literature in Developing Countries

Analysis of the data shows that soils citations from developing countries are a considerably higher percentage of all indexed agricultural citations than those in either the United States or Europe. For the period 1980–90, in AGRIS, the percent of soils citations published in and about India to indexed agricultural literature is 15.6%, twice the average of the major agricultural databases. For Egypt, it is 14.7%; Brazil, 13.1%; Indonesia, 8.7%; and the Philippines, 8.2%. In all these countries, the research emphasis on Soils Fertility (9.4% average), and Soils Cultivation and Cropping (7.7% average) carried almost twice as many citations as in other disciplinary subsets (e.g., Soils Chemistry (5.1%), Soils Biology (3.7%), and Soils Classification (3.5%)).

This trend toward higher citation counts is reflected in tropical soil research in general. The counts listed in Table 3.5 are from AGRICOLA and AGRIS and reflect this trend.

While whole numbers such as these in Table 3.5 are suspect because the indexing methodologies and the literature coverage of the databases involved are so uneven, what the citation counts do show is that materials published on the subject of tropical soils and indexed online has risen over the last fifteen years. The sixtyfold increase derived from the AGRIS counts may have more to do with better indexing over time than with an increase of this magnitude in the literature. More sobering is the analysis by Yaalon: "The share of all the Third World countries in soil research increased from 9% to 11% in twenty-one years (1963–1984). This indicates only a small relative increase above the annual increase of about 5%, so the overall picture [in the Third World] is really not so rosy."[18] It is a certainty that the

18. Daniel Yaalon, "Book Reviews," *Scientometrics* 15 (3–4) (1989): 315.

Table 3.5. Tropical soil citations

Selected tropical soil citations in AGRICOLA 1970–88	
1970	35 "tropical soil" citations
1972	43 "tropical soil" citations
1974	63 "tropical soil" citations
1976	47 "tropical soil" citations
1978	25 "tropical soil" citations
1980	89 "tropical soil" citations
1982	117 "tropical soil" citations
1984	118 "tropical soil" citations
1986[a]	123 "tropical soil" citations
1988	103 "tropical soil" citations
Selected tropical soil citations in AGRIS 1975–88	
1976	0 "tropical soil" citations
1978	2 "tropical soil" citations
1980	5 "tropical soil" citations
1982	63 "tropical soil" citations
1984	86 "tropical soil" citations
1986	127 "tropical soil" citations
1988	130 "tropical soil" citations

Note: Some Third World soil citations will *not* appear in these counts because the descriptor may name the country or location and not use the term "tropic?." The citations include journal, pamphlet, and monographic titles from all sources such as commercial, university, and government publishers. Only those titles with "soil?" *and* "tropic?" in either the title or the descriptor were included. The ? is the symbol for truncation, thus "tropic?" would search for "tropic," "tropics," and "tropical."

[a]Overall AGRICOLA indexing declines at this date.

data show a very basic *trend* toward greater soil research worldwide as local agronomic institutes in the tropics become more sophisticated and capable of sustained self-sufficiency. As a corollary, in a study of scientific publishing trends in sub-Saharan Africa, C. H. Davis makes the following comparisons.[19] Authors in universities accounted for 65% of all published output 1975–80; those belonging to national or international organizations, 19%; government, 15.2%; and researchers in the private sector in sub-Saharan Africa accounted for only 1.8%.

Major players in Third World soil literature are identified in Table 3.6, based on counts of monographic works (twenty-five pages or greater) published within the designated country and indexed online. It should be kept in mind, as Eugene Garfield clearly states, that the United States still pub-

19. C. H. Davis, "Institutional Sectors of Mainstream Science Production in Sub-Saharan Africa," *Scientometrics* 5 (3) (1983): 163–75.

Table 3.6. Soil monographs published in selected Third World countries

South America	Brazil	332
	Argentina	106
	Mexico	28
Asia	India	427
	Philippines	149
	Malaysia	92
Africa	Egypt	218
	Nigeria	18
	Kenya	5

Source: AGRIS 1980.
Note: J. D. Frame lists the eight leading developing nations in science research as India, Argentina, Egypt, Brazil, Mexico, Chile, Nigeria, and Venezuela, six of which appear on this list: Frame et al., "The Distribution of World Science," *Social Studies of Science* 7 (1977): 501–16.

lishes the greatest share of literature on Third World agriculture, of which soil is an important component.[20]

In his work on sub-Saharan Africa, Davis notes that the percentage of studies in agronomy in relation to total counts of scientific publications in these countries reaches 22.3% for the years 1970–79[21] The fact that agriculture should play such a large role in the overall scientific output of developing nations does not seem surprising, given their high ratios of people involved with farming to the total population. The pressing need of food production in tropical countries to sustain burgeoning populations dictate, as a consequence, the high profile agriculture will receive in research priority. Furthermore, Garfield points out that, of the twenty-five scientific journals of developed countries that give the most citations to articles from the Third World, none covers the subject of agronomy, including soil science.[22] This lag in subject analysis for soils in the *SCI* is reflected in the fact that, of the top ten soils journals evaluated for the Core Agricultural Literature Project, only five correspond to titles in the *SCI* ranked by impact factor (1988). This discrepancy is discussed in detail in Chapter 10. Agronomic analysis of Third World research in the *SCI* database (and elsewhere) is heavily skewed toward research that is published outside the tropics. This caveat is reiterated by Chatelin and Arvanitis, that there is no way for "all the literature concerning the soils and agriculture of developing countries [to] be obtained by computer (i.e., database) inquiry."[23] Elsewhere, these

20. Garfield, "Mapping Science in the Third World," *Science and Public Policy* 10 (3) (1973): 112–127.
21. Davis, "Institutional Sectors of Mainstream Science Production."
22. Garfield, "Mapping Science."
23. Chatelin and Arvanitis, *Stratégies Scientifiques et Développement*, p. 25.

authors observe: "To deal with the Third World, one has to solve conceptual issues, rather than statistical."[24]

One conceptual issue relating to scientific output in the Third World is the difference regional factors such as political and economic stability have on the publishing aggregate of a given region, particularly with regard to language barriers and cooperative trade agreements.[25] This can be seen in South America, which is relatively homogenous and has a common cultural heritage and a language (Spanish), which predominates throughout the continent; the exceptions are Brazil (Portuguese), Guyana (French), and Suriname (English). Several governing bodies add to this regional cohesion, for example, the Organization of American States and the Centro Latinoamericano de Fisca.

Africa and much of Asia, by contrast, represent continents with greater language barriers and less cultural cohesion. The scientific output of these regions has suffered accordingly. There are also fewer cooperative mechanisms to overcome many of their regional and transnational ecological problems, particularly in the areas of water resources and soil management. The number of soil citations from South America, in any case, is far higher in whole numbers and per number of inhabitants than in either Africa or Asia.

Another way to look at this is by the number of research stations or university agronomy departments in countries with similar populations. Brazil has a population of 153,000,000 (1990), with eighty-one research facilities or centers of higher learning where agricultural sciences are taught.[26] The most famous of its research enterprises is the Empressa Brasileira de Pesquisa Agropecuaria (EMBRAPA, the Brazilian Agricultural Research Enterprise) where much soil research is currently undertaken. EMBRAPA has a strong publishing record in soils symposia and workshops, which are indexed by the major databases. A glance at Table 3.6 above shows that Brazil is second only to India in soils monograph publications. Its membership in the International Society of Soil Science (ISSS) stands at 317 (1986), second highest in the Third World.

By contrast, Indonesia, with a population of 191,000,000, has only thirty-eight agricultural research centers or universities, with a low citation count in both CAB Abstracts and AGRIS and only forty-five members in the ISSS.[27] Only one Asian country (outside of Japan and China) stands out

24. Chatelin and Arvanitis, "National Scientific Strategies," p. 115.
25. J. Blickenstaff and M. Moravcsik, "Scientific Output in the Third World," *Scientometrics* 4 (2) (1982): 135–69.
26. (a) *Agricultural Research Centers: A World Directory of Organizations and Programs*, 8th ed. (London: Longman, 1986). (b) *Agricultural Information Resource Centers: A World Directory 1990* (Urbana, Ill.: IAALD, 1990).
27. Ibid.

with a strong record in soil literature, namely, the Philippines. Several of the top-ranked monographs on the *Core List of Soils Monographs* (Chapter 9) are works published by the International Rice Research Institute at Los Banos.

Yaalon posed the question in 1962: "Has Soil Research National Characteristics?" and answered this inquiry largely in the affirmative.[28] Twenty years later, in reviewing the work of Chatelin and Arvanitis,[29] Yaalon reiterates his early claims "that conclusions reached by me some twenty years ago in studies from the worldwide CABI database are supported by Chatelin and Arvanitis."[30]

Chatelin and Arvanitis concluded that of the 9,398 soils citations which they examined in the French indexing service PASCAL (an online bibliographic database for the sciences), 2,040, or 21.7%, were relevant to the tropics (see Chapter 4).[31] Of this percentage, just over half of these citations (51%) originated in the developing countries themselves. This finding suggests that the contribution by Third World countries to soil-related studies in the tropics may be higher than for other sciences. There are probably several influencing factors: soil management and crop production carry greater weight to the overall GNP of the respective countries and soil research is decidedly local in application, for soil constituents are often unique to a region. Both in the analysis performed by Chatelin and Arvanitis and in numbers taken from AGRIS, the three top-producing soil citation countries are the same—India, Brazil, and Egypt—all of which have higher publication counts in soils than many countries belonging to the European community, including Portugal, Switzerland, Italy, Ireland, Belgium, Denmark, and Greece.[32]

Another method of ranking soil research output by country is by "fixation power," which is examined more fully in the next chapter. Chatelin and Arvanitis describe this methodology as "the proportion of studies published in a country that are carried out by scientists of that country."[33] The computations done by Chatelin and Arvanitis in 1983 show that, in the Third World, Brazil, Egypt, and Argentina come out on top. These figures may show possible skews which are hard to define precisely because correlating data from indexing services in the Third World are virtually nonexistent. The authors examine some of these skews more fully in Chapter 4.

28. Yaalon, "Has Soils Research National Characteristics?"
29. Chatelin, "National Scientific Strategies."
30. Yaalon, "Book Reviews," p. 4.
31. Chatelin, "National Scientific Strategies."
32. From counts in CAB Abstracts, AGRIS, and AGRICOLA.
33. Chatelin, *Stratagies Scientifiques et Développement*, p. 50.

Another indicator of journal publication strength by country is the NAL's computerized *List of Journals Received*.[34] What the NAL list shows is that Israel has only one active soil journal and Nigeria none—at least insofar as journals acquired by the United States government are concerned. It is acutely difficult for scientists to publish soil research in these countries; this does much to explain their low standing. The Netherlands, on the other hand, has strong publishing possibilities and is ranked high.

The corollary of fixation power is "attraction power," which posits the question: of the studies published in a country, how many are written by foreigners? Here, the numbers are somewhat reversed as Table 3.7 below indicates. The percentages refer to the number of papers in the area of tropical soils published nation-wide by nonresidents of that country. The skews mentioned in the section on fixation power above can be said to apply in this ranking.

Obviously, the publishing possibilities in the Netherlands make it a haven for scientists from the Third World seeking recognition of their work. This seems equally true of the United States and Great Britain, which have many major journals publishing tropical soil research.

In the studies undertaken by Chatelin and Arvanitis, one finding deserves mention: the high citation counts in tropical soil research undertaken by the countries in what they call the "periphery" (i.e., Australia, New Zealand, South Africa, and Israel), which they describe as countries "culturally similar to the central countries (i.e., the United States, Europe and the former USSR); peripheral countries have highly developed scientific communities and a mainly tropical or sub-tropical ecological environment."[35] Almost 27% of all publications in tropical soils is done in these countries, most of it

Table 3.7. Publication attraction power in tropical soil research

Netherlands	91.9%	West Germany	66.2%
Great Britain	89.5%	France	35.6%
United States	73.1%		

Source: Yvon Chatelin and Rigas Arvanitis, *Strategies Scientifiques et Développement— Sols et Agriculture des Régions Chaudes* (Paris: ORSTOM, 1988).

34. This is a computerized list kept by the U.S. National Agricultural Library in Beltsville, Md. which contains the complete listing of their serial holdings. Entries on the list are subject coded, and selected subsets of journals in particular fields can be extracted. For the Core Project, NAL supplied the Soil Science subset of 847 titles from its full "List of Journals Received." Of this subset, only 419 titles matched the project's definition of a journal.

35. Chatelin, *Stratégies Scientifiques et Développement,* p. 58.

within their own borders. In peripheral countries, the ratios of scientific articles to numbers of inhabitants are some of the highest in the world, with Israel surpassing the United States by a good margin.[36]

D. Language Concentrations

English remains the predominant language of published soil literature worldwide. The percentage of soil citations in English fluctuated between 78% and 82% in CABI Abstracts for the years 1984–90, depending on subjects. Chatelin and Arvanitis found a similar correlation for English as the *lingua franca* of soil research using the PASCAL database 75%. However, both the CABI figures and Chatelin's fall far below the estimates postulated by Garfield, that English predominates in the Third World at 92%.[37] Soil research, because of the distinct characteristics of local soil types, may have a strong local bias, and this may account for the high count of locally published, non-English publications. Spanish and Portuguese literature are especially prevalent in South America.

Another approach in understanding language distribution is to look at a comprehensive collection of worldwide journals. The NAL's *List of Journals Received* shows that, of its 419 titles classified as soils related, 197 (47%) are published in English. Russian publications are second at thirty-five, or 8.3%. Publications in French are next at thirty-two, or about 7.6%, followed by Spanish at twenty-five, or 5.9%, and German, twenty-one, or 4.9%. These numbers suggest that although 53% of the soil journals received by NAL are currently non-English, the remaining 47% account for fewer than one fourth the number of citations indexed by AGRICOLA, 1980–90. This seems especially true of Russian and Eastern Bloc publications, which are selectively indexed outside their respective countries. Of even greater concern is the poor coverage given to Japanese and Chinese publications in all major databases, a point clearly delineated by J. Frame and Narin in their study on the growth of Chinese scientific research.[38] The vast majority of non-English research in soils goes unindexed in AGRICOLA and CAB Abstracts. For this reason, and because of the language difficulty inherent in using such material, it is rarely cited.

The numbers in Table 3.8 are from CABI, given as percentages of citations by selected language and by subject.

36. Yaalon, "Book Reviews."
37. Garfield, "Mapping Science," p. 116.
38. Frame et al., "The Growth of Chinese Scientific Research: 1973–1984," *Scientometrics* 12 (1–2) (1987): 143–147.

Table 3.8. Soil citations by language in CAB Abstracts

Soil Subject	Language by %							
	Fr.	Gr.	Sp.	Ru.	Eng.	Ch.	Pr.	Jp.
Chemistry	2.2	6.2	3.0	2.3	77.9	1.9	2.2	0.3
Analysis	2.6	5.8	2.7	0.9	82.9	0.3	1.7	0.5
Physics	2.4	6.3	1.5	1.0	81.8	0.7	1.2	1.0
Classification	3.5	7.1	7.9	1.3	71.3	2.2	1.6	1.1
Fertility	3.5	4.7	4.3	4.8	73.4	1.4	1.7	.03
Biology	3.2	4.0	1.7	1.3	85.8	0.6	0.8	0.3
Survey	2.5	5.9	1.7	.01	85.0	.03	1.3	.01
Morphology	3.9	5.1	3.1	0.8	82.0	0.7	1.3	1.0
Fertilizers	2.1	5.4	1.1	1.7	77.0	.05	0.9	.02

Source: CABI, 1984–91.

Fr. = French; Gr. = Germany; Sp. = Spanish; Ru. = Russian; Eg. = English; Ch. = Chinese; Pr. = Portuguese; Jp. = Japanese.

Dutch counts, which are not shown, were so low as to be negligible (.02 average). However, the low count in the Dutch language does not indicate a lack of soil publications from the Netherlands. Rather, that country publishes most of its soil literature in English. Indeed, like many of the major soil journals of the world, Dutch publications such as *Geoderma* and *Plant and Soil*, although published in a non-English-speaking country, almost exclusively publish in English. A majority of the publications of both the Centre for Agricultural Publishing Documentation (PUDOC), and the International Soil Reference and Information Centre (ISRIC), both in Wageningen, are in English. In the case of PUDOC, all their soils and tropical agriculture monographs for 1990 and 1991 were published in English. Finally, the Dutch have several distinguished commercial publishing houses, notably Elsevier and Nijhoff, which have strong records in soil publication. Publishing houses such as these, which seek the widest distribution, publish predominantly in English for a worldwide audience.

The high count of Spanish citations in the area of classification reflects the current efforts of Latin American countries to classify their unique soils systematically.[39] The problem of industrialization in Europe and its impact on agricultural lands is clearly seen in the high citation count for Germany under "Survey." Almost 78% of these citations are concerned with soils that have either been disturbed or polluted by urban industrial sprawl, especially in the former East Germany and the Eastern Bloc.[40] The German language

39. J. R. Landon, ed., *Booker Tropical Soils Manual* (London: Longman, 1984).
40. From CABI database analysis.

predominates in Table 3.9 below, a comparison of soil science with other disciplines.

English, again, is the overwhelming language of publication in the 941 distinct titles of the *Core List of Soils Monographs* (Chapter 9) compiled by the Core Agricultural Literature Project. For monographs, titles in English comprise roughly 90%, which is close to Garfield's estimates.[41] Of the foreign core titles, French predominates at forty-five; followed by Russian, thirty-seven; German, thirty-two; Spanish, eleven; and Portuguese with five. Several languages have one title apiece. The 161 titles from the Netherlands were all in English. No title in either Chinese or Japanese made the list, unless translated, confirming the difficulty that works in non-Roman alphabets face in gaining wide recognition. Chapter 9 includes a more detailed analysis of the core monographs.

Data such as that in Table 3.9 show that English is the language of most scientific publications in soil science. The high count of Russian language materials does not necessarily reflect a correspondingly wide readership outside the former communist bloc. While influential in earlier years, Russia's low citation counts dealing with current tropical soils is an indication that the country's current influence in soil science outside its borders is slight, particularly in the Third World.[42] While the German language had higher overall counts in CABI than French, the percentage of French citations dealing with tropical soils to those in German was higher by a margin of two to one, indicating that French still has an impact in Africa.

Table 3.9. Percentages of English and non-English literature in four subject areas

Subject	English	Second most common language
Agricultural Economics and Rural Sociology	87.6%[a]	French 2.5%[a]
	64.1[b]	German 8.0[b]
Agricultural Engineering	88.4[a]	Russian 5.0[a]
	70.9[b]	German 8.9[b]
Animal Science and Health	73.0[a]	German 4.9[a]
	67.0[b]	German 7.1[b]
Soil Science	86.8[a]	German 3.3[a]
	79.6[b]	German 5.6[b]

[a]AGRICOLA 1984–90.
[b]CAB Abstracts 1984–90.

41. Garfield, "Mapping Science."
42. Garfield, "Journal Citation Studies 53."

E. Format of Publication

In developed countries, it is difficult to make clear distinctions between scientists's professional affiliations and publishing trends. Authors not uncommonly belong to several organizations and, even while employed by a university, may do grant-funded research for the government. Also, in the developed world there are more opportunities to publish commercially which in no way suggest an affiliation with the private sector. Therefore, the measure most likely to yield pertinent data about scientists from developed countries is rarely their work affiliation, but, rather, the format in which they publish.

As Wallace Olsen has pointed out, journal articles account for almost 80% of all agricultural literature cited in the major databases, with the remainder divided equally between government documents and monographs (about 10% apiece).[43] In soils, these neat categories seem more difficult to quantify, in part, because of the strong federal government publications in the areas of surveying and classification.

Table 3.10, with data from AGRICOLA (1979–91), reveals fluctuations of format according to subject areas. The percentages of USDA publications in each subject are listed in the column on the right. AGRICOLA codes all United States government publications, including those of the USDA, but does not code for other publishers, such as commercial, organi-

Table 3.10. Publication type by subject code from AGRICOLA

Soil subject	Format %		% of total USDA pubs.
	Article	Monograph	
General	46.4	42.4	9.9
Biology	95.3	4.4	0.3
Chemistry/Physics	88.5	10.6	2.3
Classification	85.1	12.2	4.6
Surveying	36.4	62.5	51.3
Fertility	86.0	12.7	12.0
Resources	35.0	63.7	47.3
Cultivation	84.9	13.9	2.5
Conservation	78.8	20.1	10.5
Average	70.7	26.9	15.6

Note: Publications missing which make the percentages in the first two columns short of 100% are maps, some census material, pamphlets, and ephemera.

43. Wallace Olsen, *Agricultural Economics and Rural Sociology: The Current Core Literature* (Ithaca, N.Y.: Cornell University Press, 1991).

zational, or university. Where there is a high percentage of monographs, there is a correspondingly high percentage published by the USDA. This should come as no surprise, since the United States government is the primary publisher of monographic material on surveying by the Soil Conservation Service (SCS) and soil resources (U.S. Forest Service, Bureau of Land Management). The relatively high count of USDA publications in the area of soil fertility is due, in part, to the work of the National Fertilizer Development Center. Soil biology and chemistry, more "academic" pursuits, show low government counts, with the majority of research being carried out at the land-grant universities and published in journals. The high percentage of monographs in the category of general soil science is due primarily to texts on general agronomy, gardening, and broad overviews of soil science.

A more thorough analysis of publication formats is pursued in Chapters 9 and 10. Data on publishers of soil-related serials are available by analysis of the NAL *List of Journals Received* and the 100 top-cited journals identified by the Core Agricultural Literature Project analysis in Chapter 10. These comparisons are given in Table 3.11.

Project data indicate that top-cited soil journals are predominantly published by independent organizations. By contrast, the NAL's *List of Journals Received* is heavily weighted with foreign and domestic government serials. The record of library journal holdings from the NAL in Table 3.11, given by type of publisher, may reflect the general pattern of library collections in the United States, but does not reflect journal use or journal citation patterns. Only careful analysis of the literature can assist in defining and ranking core journals. The percentages on the right reflect actual use and citation patterns of the top core journals of the discipline and better represent the types of material which are, in fact, used in the scholarly publishing process. Chapter 10 explains the methodologies involved in arriving at this core.

Two examples of commercial journals of standing are *Plant and Soil* (Nijhoff Publishers) and *Soil Biology and Biochemistry* (Pergamon Press).

Table 3.11. Journal publisher

Type of publisher	NAL journals received	Top 100 core journals
Commercial	19%	24%
Government	30%	15%
Organization	42%	51%
University	9%	11%

A typical example published by a government body is the *Cahiers OR-STOM—Série Pédologie* from France. The most numerous and the highest ranked journals in soils remain those published by societies, primarily the American Society of Agronomy and the Soil Science Society of America. The journals of both consistently rank in the top five by impact factor among agricultural journals in the *SCI* (see Chapter 10).[44] While universities are strong in monographic publishing, their journal output and quality are low. The most prestigious and oldest current United States university journal in agronomy, with a strong soils coverage, is *Hilgardia*, published since 1925 by the University of California, Berkeley. The highly regarded *Journal of Agricultural Science*, from Cambridge University, is a British example. While it covers all of agriculture, it, too, has a strong record in soils.

Soil Surveys and Maps

The National Cooperative Soil Survey of the Soil Conservation Service (SCS) is a primary publisher of soils literature whose format is unique. McCracken et al. discuss U.S. soil surveys in depth in Chapter 11. The survey work of the SCS continues, and they have been publishing soil surveys since the early 1900s. In 1989 alone, almost 37,000,000 acres were surveyed in the United States, and a further 8,000,000 acres updated, with the work carried out primarily in Montana (5,440,000 acres), Missouri (2,218,000 acres), North Dakota (2,937,000 acres), and Illinois (1,943,000 acres). Recent maps are based on aerial photographs and are usually on a scale of 1:20,000 or 1:24,000.[45] Yearly, the SCS updates its *List of Published Soil Surveys*, which serves as the single most up-to-date listing of United States soil surveys.[46]

In the United Kingdom, as late as 1979, 75% of England and Wales lacked soil mapping at a scale other than 1:1,000,000.[47] Before that date, much of the work had been done from the Rothamsted Experimental Station in Harpenden. Under the supervision of the Soil Survey of England and Wales (SSEW), established in 1939, a number of soil maps were prepared

44. Garfield, "The Effectiveness of American Society of Agronomy Journals: A Citationist's Perspective," in *Research Ethics, Manuscript Review, and Journal Quality*, eds. E. F. Mayland and R. E. Sojka, Proceedings of a Symposium on the Peer Review-Editing Process (Madison, Wis.: ASA, SSSA, CSSA, 1992).

45. J. R. Blanchard and Lois Farrel, *Guide to Sources for Agricultural and Biological Research* (Berkeley, Calif.: University of California Press, 1981).

46. U.S. Dept. of Agriculture, Soil Conservation Service, *List of Published Soil Surveys* (1990).

47. P. Bullock, "Soil Mapping in England and Wales," in *Soil Survey—A Basis for European Soil Protection. Proceedings of the Meeting for the European Heads of Soil Survey*, December 1989, Silsoe, U.K., ed. J. M. Hodgson (Brussels: Commission of the European Communities, 1991), pp. 27–38.

at various scales, using a variety of methodologies. Not until the *Soil Survey Record* began publication in 1970 did a concise, published record of survey work exist in the United Kingdom. In 1990, the *Record* began its 115th installment. On a parallel track, the SSEW *Technical Monographs* (1969–), published jointly by the Macaulay Institute for Soil Research in cooperation with Rothamsted, have covered many subjects dealing with surveys and soil classification over the last twenty years. Today, the SSEW has transferred out of the government domain to become a public-sector organization under the supervision of the Cranfield Institute of Technology and is known as the Soil Survey and Land Research Centre (SSLRC).

In Canada, work began in Ontario in 1914, but efforts were suspended because of the war. Several provinces followed suit in the 1920s in part to find land suitable for settlement by returning World War I veterans. National efforts began in the 1950s at the Soil Research Institute (SRI) in Ottawa. Then in 1965, the ongoing Canada Land Inventory (CLI) began publishing data on the groundwork laid by the SRI. Many maps and surveys have been published in series on Canadian land resources by the CLI, with soils an important component.

Both India and Brazil have extensive survey efforts. The work in India falls under the jurisdiction of the National Bureau of Soil Survey and Land Use Planning, and much of their current work is published in the government's *NBSS Publications*, a set of pamphlets in series generally running to about seventy pages, much like the American surveys with an explanatory text of the maps. The ongoing work in Brazil currently is being performed by RADAMBRASIL, using satellite technology. The original mapping was done in the early seventies by the Ministry of Agriculture. But perhaps the best map series of Brazil was the work undertaken by EMBRAPA in the early 1980s generally at scales 1:444,444–1:500,000 (see Chapter 12).

A German classification system was devised and published by Eduard Muckenhausen in his investigation *Entstehung, Eigenschaften und Systematik der Boden der Bundesrepublik Deutschland*.[48] Although Muckenhausen is given credit for popularizing the need for a national soil survey, his work was essentially a revision of that done by W. L. Kubiena in 1953.[49] However, Muchenhausen's work set in motion the surveying by several groups now being done in Germany, including the Geologisches Landesamt (the soil surveys of the individual federal states) and the work being undertaken by prominent professional organizations. A soil map of the republic has been completed by the national geological survey (1986) at a scale of

48. Eduard Muckenhausen, *Entstehung, Eigenschaften und Systematik der Boden der Bundesrepublik Deutschland* (Frankfurt: DLG Verlag, 1962).
49. W. L. Kubiena, *The Soils of Europe* (London: Thomas Murby and Co., 1953).

1:1,000,000, with comprehensive explanatory notes. Many states are now currently being mapped at scales from 1:5,000 to 1:25,000.[50]

In France, soil surveying began with the work of the French pedologist Albert Demolon in 1935, with his *La Carte des Sols de France*.[51] Today, following on the work of Demolon in the 1930s and the efforts of P. Duchaufour and Jamagne in the 1960s, the ongoing survey work is being completed under the auspices of the Soil Survey Staff of the Institut National de la Recherche Agronomique (INRA). The first sheets of the series, *The Soil Map of France* at scale 1:100,000, were published in 1976. Work continues today, notably in the mountainous regions of the Alps.

At almost the same time that INRA was formed in 1946, the Commonwealth Scientific and Industrial Research Organization (CSIRO) of Australia began soil survey work under its Division of Soils. Since 1949, CSIRO has published its *Soils and Land Use Series*, and 1989 marked its sixty-first issue.

China got an early start and, in 1930 in Peking, began publishing its survey work in the soil bulletin *Tu Jang Chuan Pao*, with abstracted translations in German and English. It has since been superseded by various government publications. Earlier still were Russia's monumental efforts under Vasilii Dokuchaev, who published his first Russian soils maps in 1879 (see Chapter 11). Much of the work done in Europe today is based upon Dokuchaev's genetic concepts of soils.

Soil maps are dealt with more fully in Chapter 11, "Soil Surveys and Maps," as well as in Chapter 12, "Major Soil Maps of the World."

Miscellaneous Publications

Between 1900 and 1950, two major publication sources of soil research were the Bulletin series of the USDA and of the state agricultural experiment station system. These were essentially monographic series of pamphlet length, with an in-depth look at a single topic per issue. Citation analysis completed by the Core Agricultural Literature Project of several important soil titles from the 1930s and 1940s reveals that, on average, 50% of the monographs cited were of university experiment station publications, and, in the separate case of Selman A. Waksman's *Humus: Origin, Chemical Composition, and Importance in Nature*, experimental station publications

50. K. H. Oelkers, "Soil Mapping in the Federal Republic of Germany," in *Soil Survey*, Hodgson, pp. 57–60.

51. Jean Boulaine, *Histoire des Pédologies et de la Sciences des Sols* (Paris: INRA, 1989).

account for over 70% of the monographs.[52] Commercial houses were cited less than 5% of the time.

Since 1950, the decline in Bulletin citations, as well as other types of manuals, workbooks, and pamphlets, has been precipitous. One reason is that these publications have limited application in the academic tenure process. Fewer career-oriented scientists are, therefore, willing to expend time publishing them. Second, refereed journal articles and soil monographs have grown in number so dramatically in the last forty years that they have eclipsed the older, more traditional agricultural experiment station pamphlets. One can suggest that the decline of the USDA Cooperative Extension Services as a strong outlet for farm information dissemination has had its effect, too, and that how-to books and manuals are less likely to find wide access through this central distribution source than they were before the Second World War. Finally, as Olsen points out, "Today's bibliographic databases do not index . . . the traditional trade or application literature thoroughly."[53] This implies that much of the bulletin work simply becomes invisible, rarely cataloged at the item level in online university catalogs. Only the major databases, such as AGRICOLA, seem to index it.

Pedro Sanchez's *A Review of Soils Research in Tropical Latin America* in North Carolina's Agricultural Experiment Station *Bulletin* 219[54] is a good example of an important, well-cited bulletin as is California's *Bulletin* 766, *Analytical Methods for Use in Plant Analysis*,[55] both of which have been highly cited in the literature of the last thirty years. The latter is not yet out of date. Auburn University, Pennsylvania State University, Texas A & M, University of Nebraska, Washington State University, and Iowa State are traditionally also strong in agriculture experiment station soil-related publications.

Some government publications are heavily cited. A noteworthy example is the USDA Agricultural Handbook *Soil Taxonomy*.[56] This remains one of the most highly cited publications in the Core List of Monographs (Chapter 9). Similarly, the FAO *Soils Bulletins* are also highly cited and are pub-

52. Selman A. Waksman, *Humus: Origin, Chemical Composition, and Importance in Nature*, 2d ed. rev. (Baltimore: Williams and Wilkins, 1938).

53. Olsen, *Agricultural Economics and Rural Sociology*, p. 21.

54. Pedro Sanchez, *A Review of Soils Research in Tropical Latin America* (1975 [North Carolina Agricultural Experiment Station Bulletin no. 219]).

55. Clarence Johnson and Albert Ulrich, *Analytical Methods for Use in Plant Analysis* (1959 [California Agricultural Experiment Station Bulletin no. 766]).

56. United States Soil Conservation Service, Soil Survey Staff, *Soil Taxonomy: A Basic System of Soil Classification* . . . (Washington, D.C.: Soil Conservation Service, USA, 1975. [USDA Agricultural Handbook no. 436]).

lished in handbook format primarily for use in developing countries; the FAO's *Shifting Cultivation and Soil Classification in Africa* is the most commonly cited.[57]

Developing countries also have extensive publishing mechanisms outside the mainstream press. Citation analysis of a work such as *Soils and Rice*, published by the International Rice Research Institute (IRRI) in the Philippines, reveals that over 28% of the 500 monographic citations are manuals and bulletins distributed by government sources and agriculture research centers in such places as Indonesia, Thailand, Malaysia, and, of course, from IRRI itself.[58] This suggests that locally published manuals and how-to books in the Third World are some of the most important literature resources to local farmers and academicians. Trends in Third World publication are covered more extensively in the next chapter.

Graduate Level Theses

Dissertations have little influence on the literature of soil science, though they have some in tropical soil research. Analysis of the literature for this project revealed that, for the developed world, dissertations accounted for 4.1% of all monographic citations. For the Third World, the number stood slightly higher, at 6.1% (see Chapter 9).

Much has been published recently about the demise of small-scale agriculture throughout North America, best seen in the decrease in the number of family farms. The statistics do not lie. When the first soil surveys began in the United States at the turn of the century, almost 85% of the population was in some way involved with production agriculture. Today, that figure has dropped to about 5%, and the decline of small family farms is well documented. This suggests that the demographics of the traditional pool of candidates from which agricultural graduates were chosen has changed drastically.

Tallies of advanced degrees awarded by major land-grant institutions in the United States and universities in other countries tell a story of expanding interest in the agricultural sciences, despite the demographic drop in the numbers of farmers. It is an increase that can be measured, in part, by the numbers of university graduates in agriculture. The population of the United States has not yet doubled since the census count of roughly 150,000,000 citizens in 1950, but the number of agronomy dissertations awarded by

57. Food and Agriculture Organization, *Shifting Cultivation and Soil Conservation in Africa* (Rome: FAO, 1974 [FAO Soils Bulletin no. 24]).
58. International Rice Research Institute, *Soils and Rice* (Los Banos, Philippines: IRRI, 1978).

United States land-grant universities has doubled since 1945. Despite this trend, the USDA Higher Education Programs (HEP) in their report *Employment Opportunities for College Graduates in the Food and Agricultural Sciences 1990–1995*, paint a less than rosy prognosis for agriculture as a whole.[59] The increase in agricultural graduates is not keeping pace with employment demands. It appears that the sciences, both pure and applied, are well covered. The disparity is found in the area of "marketing, merchandising, sales, and financial management positions."[60] How this will effect the growth of graduate level enrollment in soil science is not clear. One in-depth report which analyzed the future of land-grant colleges saw a need for expanded recruitment to attract the brightest students, for promotion of multidisciplinary research, and for broadening of curricula to include a strong environmental component in agricultural policy and application instruction.[61]

Growth in Third World agricultural dissertations is not uncommon either. Mexico, which has one of the highest rates of population growth in the world and faces corresponding pressures on food production, has seen the number of advanced degrees (Master and Ph.D.) awarded in soil science jump from sixty-one in the period 1968–70, to 154 in 1978–80—almost a threefold increase.[62] In India, with similar population pressures, the number of Ph.D.s in soils has risen from fifty-three in the academic year 1975–76, to 111 ten years later in 1984–85.[63] These are exponential jumps which are mirrored in dissertation counts in agriculture in many regions of the world, notably Brazil, Egypt, and the Philippines, each with their own population problems.

Table 3.12, with limited data, charts this trend among selected United States land-grant universities.

Each of these institutions has a strong soil research program and all show increases in advanced degrees in soils at a pace greater than the increase in the general population. For the United States as a whole, 1982–91, it is not surprising that most degrees were awarded in the area of General Soil Science, whereas Soil Management and Fertility were second. According to

59. Higher Education Programs, *Employment Opportunities for College Graduates in the Food and Agricultural Sciences 1990–1995* (Washington, D.C.: USDA, Higher Education Programs, 1990).

60. Ibid., p. 13.

61. James H. Meyer, *Rethinking the Outlook of Colleges Whose Roots Have Been in Agriculture* (Davis, Calif.: University of California, Davis, 1992).

62. *Lista de Testis Presentadas en las Escuelas de Agricultura, Ganader ia y Medicina Veterinaria de la Republica Mexicana, 1976–1980* (Chapingo, Mexico: La Biblioteca, 1983).

63. *Bibliography of Doctoral Dissertations Accepted by Indian Universities* (Delhi, India: Association of Indian Universities, 1972–).

Table 3.12. Advanced soil science degree counts from five United States land-grant institutions

Date	Institution				
	Cornell	Auburn	North Carolina	Texas A & M	Washington State
1950	9	n/d	5	n/d	n/d
1960	14	1	6	2 ('65)	n/d
1970	10	2	13 ('75)	10	2
1980	14	3	16	8	5
1990	10	8	13	15	5

Source: Dissertation Abstracts and Core Agricultural Literature Project survey.
All columns include both masters and Ph.D.s awarded.

the Food and Agriculture Education Information System, total doctoral degrees in soils rose from fifty-six in 1982, to ninety-six in 1990, with over half awarded in the general soils category and about 19% in Soil Chemistry, the latter posting the greatest increase over the last decade. Lowest ranked was doctoral work in Soil Conservation, whereas at the master's level it was the highest overall, after General Soil Science.[64]

F. Soils Citations from the *Science Citation Index*

Citation analysis completed by the Institute for Scientific Information, covering the past thirty years has identified the nine most cited soil science articles in the literature. These are given in Table 3.13.[65]

Several of these authors were the subject of "Citation Classic" commentaries in *Current Contents*. Garfield says: "In several of these commentaries, the authors account for the high impact of their classic papers by noting the multidisciplinary nature of agronomy and soil science. For example, F. N. Ponnamperuma, [of IRRI], stated that his . . . 1972 *Advances in Agronomy* review article on the chemistry of submerged soils involved identifying and reviewing research in: biochemistry; electrochemistry; physical, inorganic and organic chemistry; thermodynamics and bacteriology."[66]

It is this "multidisciplinary nature of soil science" which best describes the field today. As Chapter 9 on soils journals points out, almost 90% of

64. Food and Agriculture Education Information System, College Station, Texas, personal communication.

65. Garfield, "Journal Citation Studies 53."

66. Ibid., p. 13.

Table 3.13. The nine most-cited soils articles, 1945–88

(A = 1945–88 citations; B = 1989 citations only)

R. H. Bray and L. T. Kurtz, "Determination of Total Organic and Available Forms of Phosphorus in Soils," *Soil Science* 59 (1945): 39–45. (A = 534; B = 47)

S. C. Chang and M. L. Jackson, "Fractionation of Soil Phosphorus," *Soil Science* 84 (1957): 133–144. (A = 356; B = 12)

R. W. F. Hardy, R. C. Burns, and R. D. Holsten, "Applications of Acetylene-Ethylene Assay for Measurement of Nitrogen Fixation," *Soil Biology and Biochemistry* 5 (1973): 47–81. (A = 461; B = 26)

V. J. Kilmer and L. T. Alexander, "Methods of Making Mechanical Analyses of Soils," *Soil Science* 68 (1949): 15–24. (A = 409; B = 12)

W. L. Lindsay and W. A. Norvell, "Development of a DTPA Soil Test for Zinc, Iron, Manganese and Copper," *Soil Science Society of America Journal* 42 (1978): 421–28. (A = 427; B = 34)

J. P. Martin, "Use of Acid, Rose Bengal and Streptomycin in the Plate Method for Estimating Soil Fungi," *Soil Science* 69 (1950): 215–32. (A = 430; B = 13)

F. N. Ponnamperuma, "The Chemistry of Submerged Soils," *Advances in Agronomy* 24 (1972): 29–96. (A = 344; B = 38)

A. Walkley and I. A. Black, "An Examination of the Degtjareff Method for Determining Soil Organic Matter and a Proposed Modification of the Chromic Acid Titration Method," *Soil Science* 37 (1934): 29–38. (A = 752; B = 52)

F. S. Watanabe and R. S. Olsen, "Test of an Ascorbic Acid Method for Determining Phosphorus in Water and NaHCO3 Extracts from Soil," *Soil Science Society of America Proceedings* 29 (1965): 677–78. (A = 356; B = 30)

---

the journals cited by soil scientists are not soils specific and range far afield into disparate subjects, including those mentioned by Ponnamperuma. If this multidisciplinary approach is the future of soil science, studies such as this on its literature increasingly will encompass a broader category of subjects, and the prediction of Charles Kellogg in 1940 may come true: "Our soil scientist has become a general reader!"[67]

67. Charles Kellog, "Reading for Soil Scientists," *Journal of the American Society of Agronomy* 32 (11) (1940).

4. Bibliometrics of Tropical Soil Sciences: Some Reflections and Orientations

RIGAS ARVANITIS AND YVON CHATELIN

Institut Français de la Recherche Scientifique pour le Développement en Coopération (IFRSDC), France

This chapter presents some methodological problems in realizing a bibliometric study in soil and agricultural sciences in a tropical environment. Soil science research in the tropics is a complex field and, therefore, needs to be analyzed cautiously. This is especially true where new scientific communities from the Third World are showing an important and active participation in science. We believe that the factors operating in this arena are far from unique. They remain largely overlooked, and the influence of tropical researchers in the world of science is often underestimated when the quality of science or the impact of scientific literature is evaluated. Moreover, we feel that scientists would benefit from being more aware of the forces which orient the scientific production in their spheres of knowledge.

A. Focusing on Tropical Soil Sciences

Bibliometrics, the statistical analysis of bibliographic material, is of recent development.[1] This is especially so in soil science.[2] The intent of bibliometric analysis is to grasp an overview of a science, including the work of laboratories, institutions, and scientists working in a determined field. This analysis can be useful to scientists who sometimes lack a comprehensive

This chapter is a précis of the authors' *Stratégies Scientifiques et Développement: Sols et Agriculture des Régions Chaudes* (Paris: ORSTOM Editions, 1988). IFRSDC is a division of the Office de la Recherche Scientifique et Technique Outre-mer (ORSTOM).

1. For a global introduction to bibliometrics, see J. P. Courtial, *Introduction à la Scientométrie* (Paris: Anthropos, 1990).

2. The first bibliometric analysis of soil sciences was Daniel Yaalon, "Publication as a Measure of a Nation's Research Effort," *Geotimes* 11 (3) (1966): 20–21. To our knowledge, our 1988 study is the only more recent one. Lea M. Velho has also worked on the scientific production of Brazilian university soil scientists (Ph.D. diss., 1985).

perspective of their discipline. It can also be used in a more pedagogical way to explain the patterns of organization of knowledge in a field of learning. The scientific community can profit from systematic bibliometric analysis with a global perspective. It should be noted, however, that the uses to which these scientific indicators are put can often be in conflict with prescribed beliefs, and the mere possibility of drawing such a large picture is itself problematic.

The authors of this essay are, respectively, a researcher in the sociology of science and a tropical soil scientist. In this investigation, we wanted to know, first, how the study of soils in tropical areas is organized. Some questions regarding this organization are obvious: what share of the scientific literature is produced in the tropics and how much in northern latitudes? What countries contribute most to this body of scientific knowledge? How does the contribution of northern influence that of southern countries? Evident as they may appear, these questions have never been extensively investigated. Few have researched the geopolitics of research, and rarely have the inherent biases of such an investigation been analyzed. In many instances, the research tools for such undertakings are unavailable. Since we are going to examine facets of tropical soils research in some detail, we must mention briefly some of the political problems in a bibliometric analysis of this sort.

Science is universal in so far as the corpus of knowledge it creates is held to be scientifically valid in all countries. But it is not universal in the way the science itself is carried out, nor in its application, and still less in the motivations of the individual scientists doing the work. This disparity is true of all science, but perhaps more so in the natural sciences where the medium of investigation is in precise and diverse environments. Soil, or a landscape, is just such an environment, particular to a specific locale. One can decide to study a soil because of its properties or because of its uses in crop production. The results obtained may serve other scientists. Allocations toward that research, the timing of the research, and the ways by which results are disseminated and accessed in the literature, all depend on a variety of contributing factors, not least the initial input and affiliation of the scientists or scientific body doing the work.

Science is also a dynamic process which remains fragile in many developing countries, despite recent improvements. Today, scientists from the Third World have been recognized not only in relation to the developed world, but also for their work as original contributors. This is particularly true with soil science. Despite the fact that its concepts and science were developed in the north, the emphasis of much current research depends on choices made by local scientists. Pressures exerted on scientists of developing countries are often much stronger than those felt by their counterparts from industrialized countries. Developing country scientists are required to do applied research in areas of national interest, especially research which contributes to the country's self-sufficiency; they are requested to help the growth of their educational systems; often they are asked to participate actively in politics.[3] These pressures translate into stresses and conflicts which may cause pressures within the local scientific community, stresses which must also be dealt with in conditions often less amenable than those faced by their northern counterparts.[4] These scientists are also more dependent on outside expertise and lack many information resources.[5] It is inevitable that any investigation into the world distribution of scientific knowledge will enter immediately into an arena of opposing interests and motivations. This is especially true in the arena of scientific funding and production, which is increasingly organized by a multiplicity of national and international players.[6]

These observations are not abstract but arise every day in dealing with funding and scientific policy in the international arena.[7] Why does a laboratory receive funding for research that will benefit some users whose activities may be in conflict with that of the funding body? What type of knowledge should international bodies fund? Is there a need to redirect funding in order to benefit more rapidly the welfare of nations and people, such as research in agricultural and industrial activities? In soil science, these sorts of questions are generally absent from the scientific literature, yet seem to be integral to the scientific process of soils research: funding, designing, executing, and disseminating scientific research is often an exercise in geopolitics. This is not known to all scientists because most of them work on a limited scope. Tropical soil science is no different, for it is a discipline

3. (a) H. M. C. Vessuri, "O Inventamos, O Erramos: The Power of Science in Latin America," *World Development* 18 (11) (1990): 1543–53. (b) S. Schwartzman, *A Space for Science; The Development of the Scientific Community in Brazil* (University Park, P.A.: Pennsylvania State Univ. Press, 1991). (c) J. Fortes and L. A. Lomnitz, *La Formación del Científico en México; Adquiriendo una Nueva Identidad* (Mexico, D.F.: Siglo Veintiuno, 1991). (d) R. Arvanitis, "De la Recherche au Développement. Les Politiques et Pratiques Professionnelles de la Recherche Appliquée au Vénézuéla," Ph.D. diss., Paris VII, published in Spanish (Caracas: Fondo Editorial Acta Científica Venezolana, 1992). (e) G. Argenti, C. Filgueira, and J. Sutz, "From Standardization to Relevance and Back Again: Science and Technology in Small Peripheral Countries," *World Development* 18 (11) (1990): 1555–67.

4. (a) M. Schoijet, "The Condition of Mexican Science," *Minerva* 22 (3) (1983): 381–413. (b) J. Gaillard and R. Waast, "La Recherche Scientifique en Afrique," *Afrique Contemporaine* 148 (1988): 3–30.

5. M. Roche and Y. Freites, "Producción y Flujo de Información Científica en un País Periférico Americano (Venezuela)," *Interciencia* 7 (5) (1982): 279–90.

6. A. J. J. Botelho, "Struggling to Survive: The Brazilian Society for the Progress of Science (SBPC) and the Authoritarian Regime (1964–1980)," *Historia Scientiarum* 38 (1991): 45–63.

7. J. Gaillard, *Scientists in the Third World* (Lexington, Ky.: University Press of Kentucky, 1991).

which depends both on the scientific progress in the field and on local scientific communities in developing countries. Moreover, soil science has been the locus of many scientific controversies which have developed along national lines, the most famous being that of soils classification, where the United States, Russia, France, and the Food and Agriculture Organization (FAO) have all competed to supply the world's classification scheme.

These questions may help explain why the Institut Français de Recherche Scientifique pour le Développement en Coopération, a division of ORSTOM (Office de la Recherche Scientifique et Technique Outre-Mer), in France, created a team dedicated to the understanding of the scientific process in developing countries. Bibliometrics is one of our principal tools of analysis.

B. Using Bibliographic Databases

Bibliographic databases are a unique analytical resource which help us to access the literature of scientific research more efficiently. However, to some extent, all bibliographic databases are problematic. Each database offers a unique coverage of the literature: none can claim to be comprehensive of all the literature of a subject. A classic misunderstanding among researchers is believing that what is indexed in a database is mainstream science.[8] "Mainstream science" would be a valid concept in this regard if every database retained the same set of journals as a core, but this is not ususally the case. What is indexed in databases is often skewed by commercial imperatives, including document acquisition cost, literature coverage policies of the database, and choices by the indexers at the citation level.[9] These biases are apparent in the coverage of locally produced articles from developing countries, which traditionally have had less representation in the major bibliographic databases. The factors which govern these selections have not been extensively examined in print.[10]

Soil science involves both agriculture and geology; it is important in international development; it is a physical science unto itself; and as soils are both versatile and dynamic (more a process than an object), they are, there-

8. The notion has acquired quite a different sense from that first proposed by J. D. Frame, "Mainstream Research in Latin America and the Caribbean," *Interciencia* 2 (3) (1977): 143–47.

9. G. Whitney, "Access to Third World Science in International Scientific and Technical Bibliographic Databases," in *Proceedings of the International Conference on Science Indicators in Developing Countries*, eds. Rigas Arvanitis and J. Gaillard (Paris: ORSTOM Editions, 1992 [A selection of papers from this conference has also appeared in a special issue of *Scientometrics* 23 (1) (1992)]), pp. 391–409.

10. Recently, many authors tried to investigate some of these biases. See *Proceedings of the International Conference on Science Indicators in Developing Countries*, ed. Arvanitis.

fore, in a constant state of flux. This may explain the growing claim in soil science circles that it is becoming increasingly multidisciplinary.[11]

Bibliometric researchers are advised to state at the outset of their work that the bibliographic databases used for analysis are problematic and prone to skews. Some databases are biased toward English and index only specific sets of sources. Others only analyze certain types of published material and exclude others. Skews such as these should alert the researcher to be careful when studying Third World literary output. This is particularly true of the *Science Citation Index* (SCI). For instance, Cuba has a citation rate sixteen times higher in BIOSIS, the biological database of the BioScience Information Service, and *Chemical Abstracts* than in the *SCI*.[12] With Brazil, *Chemical Abstracts* lists 201 chemical reviews, whereas only six are listed in the *SCI*.[13] Again, of the seventy-six veterinary reviews listed by the *SCI*, only four come from developing countries.[14] The point is not to dismiss the *SCI* as a research tool, but to reveal its shortcomings. The same is true of many other databases, but only the *SCI* provides a yardstick for measuring the impact of specific research by citation analysis.

Given these caveats, one approach for bibliometric studies is to work with one database, either specialized or general, and try to understand the trends that appear in that single source.[15] We made a choice to use a French database, PASCAL, which indexes some 400,000 references per year and currently contains almost 8,000,000 records. PASCAL is a multidisciplinary database covering the years 1973 to the present, which indexes literature from some 9,000 international journals, with coverage of all the sciences, including physics, applied technologies, psychology, medicine, the life sciences, biotechnology, earth sciences, astronomy, civil and mechanical engineering, computer science, transportation, energy, and agriculture. It also abstracts monographs, master's and doctoral theses, proceedings of conferences, as well as reports and many patents. The two specialized subfiles on agriculture and earth sciences cover all aspects of these disciplines.

11. (a) Yvon Chatelin, *Une Epistémologie des Sciences du Sol* (Paris: ORSTOM Editions, 1979). (b) Chatelin and G. Riou, eds., *Milieux et Paysages; Essai sur Diverses Modalités de Connaissance* (Paris: Masson, 1986).

12. R. Sancho, "Misjudgments and Shortcomings in the Measurement of Scientific Activities in Less Developed Countries," in *Proceedings of the International Conference on Science Indicators*, ed. Arvanitus, pp. 411–23.

13. M. A. Cagnin, "Patterns of Research in Chemistry in Brazil," *Interciencia* 10 (1985): 64–77.

14. J. M. Russel and C. S. Galina, "Research and Publishing Trends in Cattle Reproduction in the Tropics: Part 2. A World Prerogative," *Animal Breeding Abstracts* 55 (11) (1987): 819–28.

15. This does not satisfy the need for a comparative analysis of the content of databases. That type of work should be done by librarians or documentation specialists, not by policy analysts or scientists in search of a good description of the science they work in. For recent efforts, see *Proceedings of the International Conference of Science Indicators*, ed. Arvanitis.

Extraction of pertinent data from these files occurred in two phases for this analysis. First, we extracted all references that were related to soil science and to agriculture as a whole. Our core concept was "soil," but we also looked at those terms which can be applied to soil science and the work of soil scientists in general. We tried to place ourselves in the position of a soil scientist who would want to know what is produced in one year in the discipline which has a direct impact on his field (see Appendix for the disciplinary coverage of the file). Second, we limited this set with key-words on tropical topics. For the year 1983, PASCAL contained 9,398 references in soil science. This can be considered as a sample of world literary output. However, to date, no one has determined what is the exact statistical value of such a sample, which is true for any set of articles used as a proxy of published world literature.

C. Selecting the Useful Science for Tropical Environments

Our sample needed to reflect the reality of research for tropical countries. Bibliometric research of a specific field in tropical areas, meaning tropical and subtropical zones, is further complicated because it involves not only scientists from tropical countries, but also scientists in developed countries who have worked extensively in the tropics. Most tropicalists, (scientists from developed countries working on tropical environments) come from the United States, France, the United Kingdom, and Germany. It is noteworthy that some industrial countries are practically absent from the tropical sciences, notably countries of Eastern Europe, although their importance for the discipline as a whole may be extensive. How then does one select in a database what could be construed as useful literature for developing counties? We made our selection of citations carefully, by including only those documents which have a tropical object in the title, the abstract, or the keywords. Thus, we gathered 2,040 tropical soil citations out of 9,398 references, which amounts to 21.7% of the entire "Agriculture and Soil Science" file. These are given in Table 4.1. With this selection, we obtained a corpus of literature that has direct application to soils research in developing countries, as well as for the scientists from the north who work on tropical soils.

We further distinguish three world areas, which correspond to geographical and cultural areas. First, publications authored by scientists in northern countries (hereinafter the north) include North America, the former USSR, and Europe. These represent 21% of the tropical soil and agriculture production in 1983. The principal producers are France (6.7%) and the United

Table 4.1. Subject categories with citations in the total file and the tropical file

	Category	Total	Tropical
1	General, Conference Reports, and Bibliographies	467	96
2	Geomorphology	1,686	255
3	Superficial Soil Formations	1,857	332
4	Satellite Imagery and Remote Sensing	48	9
5	Soils and Agriculture, General	171	28
6	Cartography and Soil Classification	199	39
7	Physicochemistry of Soils	382	93
8	Organic Matter	245	32
9	Physical Properties	215	36
10	Water Dynamics	284	70
11	Microbiology of Soils	1,150	363
12	Fertilization (Mineral and Organic)	1,821	529
13	Uses of Wastes	161	22
14	Hydroponic Culture	77	5
15	Soil-Plant Relations	239	64
16	Soil Conservation	152	30
17	Soil and Irrigation Management	120	28
18	Soil Pollution	124	9

Source: PASCAL, 1983.

States (5.8%). The strong presence of French literature doubtless reflects a common skew in databases such as PASCAL, which will index the literature of their respective country more thoroughly. By contrast, AGRICOLA reflects a United States skew and CAB Abstracts, an emphasis on the United Kingdom. The percentage in PASCAL is still remarkable because the United States is a more important producer in the overall Agriculture and Soils file (tropical and nontropical). This strong showing may be partly explained by the colonial past of France, and its heavy tropical scientific interests forged toward the end of the nineteenth century and the beginning of this century.[16]

Second were publications authored by scientists in large peripheral countries (the periphery). These include Australia, Israel, New Zealand, and South Africa, which are countries culturally similar to the North, which have a highly developed scientific community and a mainly tropical or subtropical ecological environment. These countries also share another common characteristic: their production of publications surpasses what one

16. C. Bonneuil, *Des Savants pour l'Empire. La Structuration des Recherches Scientifiques Coloniales au Temps de "la Mise en Valeur des Colonies Françaises," 1917–1945* (Paris: OR-STOM Editions, 1991).

would expect using an economic criterion such as the gross national product (GNP) as measurement.[17] These four countries represent 26% of the tropical file, largely dominated by Australia (14%).

The third and largest group was publications authored by scientists in southern countries (the south), including the remainder of the world belonging to tropical areas. They represent 51% of tropical soils literature production. The two biggest producers in this group are India (15.2%) and Brazil (9.2%), followed by Egypt (4.9%).[18]

We emphasize that, for the time being, there exists no automated search strategy that would satisfactorily retrieve data of use only for the tropical world. Future advances in searching tools may give better answers. Our two stage approach seems realistic.

D. Analyzing Tropical Scientific Literature

It is important in planning a bibliometric study to make clear the distinctions of country, discipline, type of research work, and, when possible, the type of research institution doing the work.[19]

Bibliometric studies are done as if all scientists had equal access to publishers. Another implicit assumption is that all scientists agree on what constitutes their best publication strategy. This assumption states that all scientists wish to publish in big-name scientific journals, in order to gain a foothold in the international scientific community. Therefore, "good science" is included in the pages of journals like *Soil Science*, *Geoderma*, *Soil Science Society of America Journal*, and science that is less good is often relegated to secondary journals, mainly locally edited reviews.[20]

This *ceteris paribus* approach does not reflect the actual situation. First,

---

17. J. D. Frame, "National Economic Resources and the Production of Research in Lesser Developed Countries," *Social Studies of Science* 9 (2) (1979): 233–46.

18. Our analysis is limited in scope and time. Because of their poor coverage by PASCAL we did not retain Japan and China in further analysis. It is also probable that other Asian countries are poorly covered. Most of our comparative work retains only the fourteen biggest producers, covering more than 30% of the 1983 publications. A wider time span would have allowed us to analyze changes that occur in the strategies of each country and within disciplines.

19. This last aspect is a difficult task, since one needs to have information on the scientific institutions of all countries represented.

20. This is a sketchy and oversimplified view. The Institute for Scientific Information has proposed a more sophisticated tool for the evaluation of quality of science: citation analysis. After much debate, it appears that citations reflect not the quality of science, but rather the impact of journals or articles. Recently, J. L. MacLean and M. J. M. Vega indicated that a better test of the quality of a journal is not the number of times it is cited, but the number and type of citations it emits. MacLean and Vega, "Citation Behaviour of Philippine Biological Scientists," in *Proceedings of the International Conference on Science Indicators*, ed. Arvanitis.

in many developing countries, there exist a set of journals that are considered by local scientists to be as valuable as foreign titles. In fact, in the natural sciences, at least half of the scientists publish locally;[21] in some cases the proportion can be as much as 92%, as is the case of soil science in Brazil.[22] It is important to bear in mind that a national publishing capacity is not free of costs, nor free of political, sociological, or other factors that affect editorial selection. Second, even if local journals are a second choice, they serve a different purpose than mainstream international journals. They are not read by the same public, and they do not contain the same type of information.[23] It appears that a large proportion of scientists tend to publish simultaneously in both local and international journals.[24]

Since we had access to few local publications, we worked hard to determine the proportion of totally autocentered publications. These are publications that simultaneously satisfy three criteria: (a) the research is carried out by a national laboratory; (b) it treats a local agricultural problem; and (c) it is published in a local journal or book. We computed the autocentered research in nine countries that do not belong to the north, since northern countries cannot satisfy to the criteria of the definition. These included India, Brazil, Egypt, Nigeria, and Argentina for the south, and all the periphery countries (Australia, Israel, New Zealand, South Africa). Autocentered soils publications for the world excepting northern countries represent 20% of all publications; this figure rises to 30% for the nine top publishing countries. A percentage of totally autocentered research that fluctuates between one-fifth and one-third of tropical agricultural citations indicates a high degree of fluidity; that is, in the vast majority of cases, it is customary for a study to be done in one place, concern research in another, and for the results to be published in a third. This fluidity indicates a wide integration in international networks.

The target of interested readers is largely determined by the publication

21. Gaillard, *Scientists in the Third World*.

22. (a) Lea M. Velho, "Science on the Periphery: A Study of the Agricultural Scientific Community in Brazilian Universities," Ph.D. diss., SPRU, University of Sussex, 1985. (b) Velho, "The Meaning of Citation in the Context of a Scientifically Peripheral Country," *Scientometrics* 9 (1–2) (1986): 71–89.

23. (a) MacLean and Vega, "Citation Behaviour of Philippine Biological Scientists," pp. 557–68. (b) T. O. Eisemon and C. H. Davis, "Publication Strategies of Scientists in Four Peripheral Asian Scientific Communities: Some Issues in the Measurement and Interpretation of Non-Mainstream Science," in *Scientific Development and Higher Education*, ed. P. G. Altbach (New York: Praeger, 1990).

24. For ecological research in Argentina, apart from the above references in note 23 see J. E. Rabinovitch, "Publications of Scientists in Developing Countries: National and International Production of Argentinian Ecologists," pp. 467–77, or R. Meneghini, "Brazilian Production in Biochemistry: International vs. Domestic Publication," pp. 457–65, in *Proceedings of the International Conference on Science Indicators*, Arvanitis.

capacity of the researcher's country and field. In tropical soil science, few truly tropically oriented research journals exist. Therefore, soils articles will be dispersed in a variety of journals, covering not only soil sciences but other agricultural sciences as well, which may be local agricultural periodicals. They also may appear in general science journals such as *Science* or *Nature*. This explains, in part, the different publishing patterns at national or regional levels. Some countries/regions have a high local writing capacity; these researchers will publish a higher percentage locally. For example, Asian and Latin American scientists publish more nationally than their African colleagues.

The presence or absence of a local publishing capacity, can be measured by what we term the fixation power of a country, that is, the proportion of studies published in a country that are carried out by scientists of that country. It is also an indicator of scientific autonomy. A fixation power ratio of 100% would indicate autarky. The best publishing situation is one that maintains a balance between inter national and national or local publications. Countries with a high fixation rate are those having a ratio of over 70%.

The fixation power ratios of the Netherlands (76%), Israel (26%), and Nigeria (6%) illustrate local publishing opportunities. When these ratios are low, scientists are publishing abroad (Israel and Nigeria). The 76% fixation power ratio attributed to the Netherlands, however, does not represent the contrary situation, because it is unlikely that this high percentage reflects the work solely of Dutch authors, but rather the fact that the country has numerous publishing possibilities for scientists from around the world; this also explains why the Netherlands has an extremely high attraction power, which is the proportion of articles produced by foreign authors in the total of journal articles published within a country. There may well be other factors at work. However, we believe that the respective position of each country is a deliberate choice to be part of the international publishing world or, on the contrary, to stimulate mainly national output, bearing in mind the two limitations, language and publication possibilities.

It is in these terms that the different positions of Brazil (77%), Egypt (71%), and India (35%) can be understood. Brazil and Egypt show a policy deliberately oriented toward national publication (note that Egypt publishes mostly in English); in contrast, India is probably more internationally oriented.

A second group of countries with a low fixation power include South Africa, Great Britain, New Zealand, Australia, India, Israel, and Nigeria. The case of Great Britain (55.8%) is of interest. While there exists a large number of English journals dedicated to soil science and tropical agriculture, British tropical soil scientists seem to prefer publishing abroad. The United States is positioned in the middle, reflecting both its very powerful publishing capacity and the immense research that United States scientists perform. They are also important organizers of international meetings. These and the other reasons explain why 30% of United States production was published in non-United States journals.

Further distinctions can be made in looking at the type of work published. Soil microbiology is more internationally oriented than fertilization studies. In the agricultural sciences, we know that animal production seems more locally oriented than crop science.[25] Other studies seem to indicate that locally published research is not necessarily more applied.[26] The differing patterns of publication seem difficult to reduce into a dichotomy between applied and fundamental research.

The way soil science is studied and presented is a more workable distinction. A study titled "Some Aspects of the Action of Termites on Clays" has a more general character than a study on "Soil-slope Relation in the Lowlands of Selangor and Negri Sembilan, West Malaysia." Clearly, mainstream publications emphasize general work, rather than that which is specific to a locale. The question of when does a local topic become more universal is not easily answered. Most local descriptions of soils rapidly become the general inheritance of the discipline, as is the case, for instance, of andosols first discovered and described in Japan, then in New Zealand, and then in the rest of the world.

It is beyond the scope of this bibliometric study to analyze this dynamic.[27] Distinguishing general types of articles from those with more specialized topics seems more realistic than distinguishing papers of fundamental or applied research. We found the proportion of local to general papers to be very high for the Netherlands, France, and Great Britain, and less so for the United States and Germany. Two southern countries, Brazil and Nigeria, also have a high ratio of local to general studies, whereas New Zealand and Egypt are midratio and big producers, like India, Australia, and Israel have a higher proportion of general studies. These countries share a mainstream orientation.

25. Gaillard, *Scientists in the Third World*.

26. (a) Gaillard, *Scientists in the Third World*. (b) MacLean, "Citation Behaviour of Philippines Biological Scientists." (c) Eisemon, "Publication Strategies of Scientists in Four Peripheral Asian Scientific Communities." (d) J. E. Rabinovitch, "Publications of Scientists in Developing Countries," pp. 467–77. (e) R. Meneghini, "Brazilian Production in Biochemistry," pp. 457–65.

27. Some authors try to discover if there exists a general pattern of growth of the disciplines in bibliometric terms, for instance, Courtial and B. Michelet, "A Mathematical Model of Development of a Research Field," *Scientometrics* 24 (2) (1990): 123–38.

Analysis at the subject level in the discipline is instructive and corroborates these results. We studied the distribution of the world literary output by subject, and the many factors which influence this output. Some subjects in soils can be influenced by local ecological conditions (for example, research on dunes in Saudi Arabia). Portability of research is also important; many geographical studies on southern countries are done in European laboratories with the help of aerial photographs or satellite images and samples taken from the field.

Four of eighteen soils subjects in PASCAL have very high rates. These are fertilization (mineral and organic), microbiology of soils, superficial soil formation, and geomorphology. These tropical subjects each makeup between 15% and 20% of the whole and are work done by developing countries. Others, like cartography and soil classification, soil conservation, satellite imagery and remote sensing, physical properties, and geomorphology, are done by developed countries. Tropical soil subjects with less research are soil pollution, uses of wastes, and hydroponic cultures. Interestingly, these are also subjects with little research production at the world level. The lack of local interest in the study of organic matter, which represents 13% of the world production, is more striking. It is a subject where there are many Russian studies which do not concern tropical areas. In 1964, Daniel Yaalon noted this Russian presence and the strong position of the United States on physical properties and mineral elements of the soil. The evidence seems to indicate that there is a lack of interest in this area in tropical agriculture. We must stress that our co-word analysis of the same literature showed that the two biggest poles of research, in terms of content, were nitrogen fixation and mycorrhiza, themes that clearly involve organic matter.[28] This is an analysis of the strength of relations between words used for indexing in several documents.

Third World countries choose subject categories that bear directly on their current agricultural needs, which confirms our observation that the pressure from government exerted on southern scientists is often strong and is reflected in the area of their work. A comparison of the tropical literature in microbiology of soils and fertilization illustrates these choices and the contrasting pattern of production of southern countries. Microbiology research is a subject with a high percentage of general studies in many foreign publications—this is a typical mainstream domain—but each country chooses to be more (India) or less (Brazil) a part of the mainstream. Much fertiliza-

28. Chatelin and Arvanitis. *Stratégies Scientifiques et Développement; Sols et Agricultures des Régions Chaudes* (Paris: ORSTOM Editions, 1988).

tion research is site-specific to local environments. Brazil ranks high in local studies published nationally.[29]

Another way of understanding the scope of research in a country is to look at the keywords used to locate a document by its content. In Table 4.2 are reproduced the keywords that characterize 30% of the production for India, Brazil, and French-speaking countries (Belgium and parts of Africa and Canada, but excluding France). Each keyword is preceded by the number of documents in which it appears.

Indian production was indexed with 765 keywords and a total of 2,365 occurrences. Of these keywords, 2.5% allow access to a third of the sample and 8.2% of the words correspond to half the Indian documents. Proportionally, more keywords were used to index the Brazilian literature: 444 keywords and 1,154 occurrences. One can call up 30% of the Brazilian documents with 4.7% of the keywords; 13.5% of the keywords characterized half the sample. There is an even wider span of themes in French literature. Some 668 keywords occur 1,218 times. A third of the sample can be characterized with 6.7% of the keywords, and half of it, with 15%.

Different strategies can be observed. For example, India has many studies on relatively few scientific subjects, Brazil has a slightly greater range, while French-speaking countries have the widest range. Indian keywords often indicated general interests; very few words point to regional or site-specific subjects. There were a few cartographic studies, some of natural formations. The main keywords indicated the names of plants and nitrogen fixation themes. Brazil also dealt with these subjects, but in a wider range of plants. There was also more emphasis on soils, and this tended to correspond to specific research besides nitrogen fixation or fertilization. In the French-speaking group, more emphasis was given to cartography, regional studies, satellite imagery, and local studies of natural environments.

However, using subject categories or keywords reveals little about the dynamics of research. We wished to examine in detail how the content of one publication is organized in relationship to the contents of a set of publications, which is done by co-word analysis. Fourteen thematical clusters were identified as the general orientation of research in 1983 in tropical soils and agriculture.[30] The list in Table 4.3 reproduces the names of the

29. The figures agree well with Velho's results on the production of four university research centers in Brazil (Velho, "The Meaning of Citation in the Context of a Scientifically Peripheral Country"), and with Y. Texera's results on Venezuela (Texera, "Publicación Científica: Análisis del Caso del al Agricultura Vegetal en Venezuela," *Interciencia* 7 (5) (1982): 273–78.

30. No automatic clustering method was used, rather we used the direct cartography of the network of keywords. The links are calculated by a simple equivalency indicator that calculates the ratio of the co-occurrence of a pair of words relative to the occurrence of each word of the pair. These indicators vary from 0 to 1 (or 0 to 100). The clusters chosen were the more frequent

Table 4.2. Top-ranked keywords constituting 30% of soil science literature within each geographic group

INDIA

87	sol	37	oryza sativa	26	azote
77	plante céréalière	33	oligoélément	23	gramineae
45	Inde	32	nutrition	22	rhizobium
43	rendement	30	zinc	21	absorption
38	bactérie	29	phosphor	21	microflore
38	fixation azote	27	triticum aestivum	20	inoculation
38	plante légumière				

BRAZIL

30	sol	14	nutrition	12	phaseolus vulgaris
26	plante céréalière	14	plante oléagineuse	11	aluminium
24	Amérique du Sud	13	oligoélément	11	glycine max
24	sol tropical	13	plante legumière	11	symbiose
22	sol latéritique	13	Zea mais	10	étude en serre
21	fixation azote	12	bactérie	10	plante fourragère
16	rendement	12	fertilisation azotée		

FRENCH-SPEAKING COUNTRIES

51	sol	7	algérois	5	forêt
23	zone tropicale	6	argile minéral	5	Guyane française
12	agriculture	6	climat	5	inoculation
11	classification	6	donnée MEB	5	karst
11	Sénégal	6	ERTS Landsat	5	milieu aride
10	Afrique	6	morphologie	5	morphologie volcan
10	morphodynamique	6	plante fruitière	5	mycorhize
10	sol tropical	6	symbiose	5	occupation sol
9	cartographie	6	végétation	5	pédogenèse
9	fixation azote	5	analyse image	5	plante céréalière
8	microflore	5	Antilles	5	plante oléagineuse
7	Afrique ouest	5	bactérie	5	sol sableux
7	satellite Landsat	5	classification super-	5	structure sol
7	télédétection		visée	5	végétal
7	télédétection multispectrale	5	écologie		

keyword in a group of words linked to each other at a quantitative level. For details on this methodology see (a) M. Callon, Courtial, W. A. Turner, and S. Bauin, "From Translations to Problematic Networks: An Introduction to Co-Word Analysis," *Social Science Information* 22 (2) (1983): 191–235; and (b) Courtial, *Introduction à la Scientométrie* (Paris: Anthropos, 1990).

Table 4.3. Fourteen thematic co-word clusters (A-N) with keyword associations

Nitrogen Fixation [A]	Environmental Management-Development [J]
Legumes, forage	Pollution [K]
Mycorrhiza [B]	Water-Erosion, Erosion [L]
Nitrogen Fixation [A]	Soil Chemistry [E]
Nitrogen cycle	Cations exchange
Forest Ecosystems [C]	Acidification
Tropical rain forests	Soil and Nutrition/Fertilization [D]
Aforestation	Tropical soil
Agroforestry	Organic Matter [M]
Nutrition-Fertilization [D]	Medicinal Plants [N]
Oligoelements	Nitrogen Fixation [A]
Phosphor	Enzymatic activity
Soil Chemistry [E]	Soils [G]
Complex exchanges	Mineralogical clays
Tropical Crops [F]	Tropical Crops [F]
Cereals	Industrial cultivation
Soils [G]	Agriculture-Development [J]
Alteration	Nitrogen Fixation [A]
Amendments	Rhizosphere
Profiles	Tropical Crops [F]
Morphodynamics [H]	Oil plants, fruits, fibre
Karst	Pastures
Desertification	Fertilization [D]
Soils [G]	Green manure
Cartography	Nitrogen Fixation [A]
Teledetection [I]	Anabaena azolla

clusters (represented by clusters A through N) and the principal keyword associated with them, by decreasing order of appearance. This is a simplified representation that does not reproduce the rich complexity and thematic links. The first clusters can be considered core research (nitrogen fixation, mycorrhiza, forest environments, fertilization), while the last ones are more marginal but sufficiently coherent in the sample to appear as newer research interests (green manure, Anabaena azolla). The main clusters indicate that research is oriented toward biological factors effecting soils and their relation to soil characteristics and agricultural development. It is interesting that classic soil science nestles between the core subjects and the new.

In summary, the following observations about disciplinary interests are deduced. On the whole, northern countries have a rather equally distributed tropical research effort in all subject categories, with some emphasis on geomorphology and pedogenesis. The distribution of research in the south and in the periphery show an emphasis on fertilization and microbiology;

the south accentuates these topics proportionally more than does the periphery. This is a slightly different view of the south doing mostly what their Northern counterparts are doing. These are clear and conscious choices on the part of southern countries. This does not mean there is no influence from the north. Types of work done in the north often appear elsewhere later, set in a southern context. There is a close relationship between the work of northern and southern scientists in some countries. One case is French-speaking African researchers who publish in the same areas as French soil researchers but with a higher proportion of fertilization studies from the African countries. We find proportionally higher citations in the French-speaking world to those subject areas that are less well represented in other southern and periphery countries. France and French-speaking countries seem to have chosen different ways of approaching soil research.[31]

Another factor which obviously plays an important role in publication is language. Although English is essentially the *lingua franca* of science, many researchers evaluate the cost of translation against the need to have access to the literature in English. This is especially true for the non-English-speaking world, which has numerous publication opportunities of its own, even if these are less well distributed than the journals written in English. India has an English-speaking scientific community and publishes regularly in mainstream journals. Brazilians publish more in Portuguese than English. In former French colonies which are independent countries today, scientists publish mainly in French, tend to avoid the English-speaking mainstream and may constitute a French-speaking mainstream.[32] The natural attraction to the linguistic precursor of the former colonial power certainly plays an important role.

In tropical soil science, English represents 75% of the total citations, followed by French (10%). Surprising is the relatively strong position of

31. These results suggest a need to study the international research collaborations as well as the impact of strong scientific countries on smaller and newer ones. Some research is beginning to appear on this topic: (a) N. Narvaez-Berthelemot, L. P. Frigoletto, and J. F. Miquel, "International Scientific Collaboration in Latin America," in *Proceedings of the International Conference on Science Indicators*, ed. Arvanitis; (b) J. El Alami, J. C. Doré, and Miquel, "International Scientific Collaboration in Arab Countires," in ibid.; (c) M. T. Fernández, A. Agis, A. Martín, A. Cabrero, and J. Gómes, "Cooperative Research Between the Spanish National Research Council and Latin American Institutions," in ibid.; and (d) F. W. Lancaster and S. Abdullah, "Science and Politics: Some Bibliometric Analysis," in ibid. Currently, our team at ORSTOM is researching scientific collaborations between European and Third World laboratories forged by the Science Technology and Development Program of the European Communities in Tropical Agriculture and Tropical Medicine.

32. This is suggested by D. Pillot in *Francophone and Anglophone Farming Systems Research: Similarities and Differences*; keynote address to the Fourth National Seminar on Farming Systems (Thailand, Prince of Songkla University, 1987). Rather than a cultural or linguistic difference between the French and English farming systems research, it appears that different approaches are promoted by different types of research institutions.

Portuguese/Brazilian (7%), and the low 5% for Spanish. This may reflect a poor coverage of Spanish-speaking countries in the database, though PAS-CAL has a reasonable coverage and even has a Spanish keywords section. The most probable explanation is the Latin American tendency to publish in very local publications. Velho claims that this is certainly the case with Brazil, but, as our figures show, it affects the Spanish-speaking American countries more than Brazil.[33] French stands at 10%, in part because of work done by institutions like ORSTOM and CIRAD, along with French-speaking African countries.

Inconclusion, the factors affecting the publication pattern of scientists are diverse. The most important appear to be

(1) the existence of a local publishing capacity,
(2) the type of research results that are published,
(3) the subfield of a science that is the subject of research, and
(4) the ability to publish in English (and to a lesser extent French) versus national languages.

E. On the Structure of Scientific Communities

These criteria are reflected in the way scientific communities are structured. This hypothesis needs to be tested by further bibliometric and sociological studies. Some disciplines are more centrally organized, with the center in rich countries; others are more decentralized, with strong regional/national networks. International relationships are a basic element in understanding the scientific literature of a discipline. A good scientific community is one that manages its relations intelligently within the international arena and with the mainstream.[34]

An implicit assumption in bibliometric studies such as this is that science is disseminated mainly through the literature, although there is a more complex configuration. Science is also spread through informal communication channels and oral communication, along with publications. This is true in developing countries where scientists are less pressed by the "publish or perish" dogma, and evaluation of research is based less on publication.[35]

This does not imply that most tropical scientists are less professional in

33. Velho, "Science on the Periphery."

34. R. Waast, *Proceedings of the Seminars on the Emergence of Scientific Communities in Developing Countries: Algeria, Brazil, India, and Venezuela* (Paris: ORSTOM Editions, 1980).

35. (a) Velho, "The Meaning of Citation in the Context of a Scientifically Peripheral Country." (b) Lomnitz, M. W. Rees, and L. Cameo, "Publication and Referencing Patterns in a Mexican Research Institute," *Social Studies of Science* 17 (1) (1987): 115–33.

the world of academia. Careers are based not only upon an ability to teach and subject expertise, but often upon national political abilities, which are as important as professional research capacity.[36] Brazilian scientists are not keen on distinguishing between their roles as research scientist and professional in society,[37] and national politics play a much more important role than for scientists in industrialized countries.[38]

Such factors help to explain the importance of communication at scientific meetings, congresses, and symposia, which represented 15% of the total references. The highest figure for a southern country was 41%. Southern and periphery countries made up 52% of all the papers at congresses. PASCAL indexes the published proceedings of conferences which reflect participation by Third World scientists. Our experience in Africa and Latin America, however, confirms that the diffusion activity of our Third World colleagues is often through participation in scientific meetings. It is also through participation at these meetings that scientists and their countires acquire international standing and visibility. Third World countries are beginning to participate heavily in a diversity of scientific enterprises, which will place them more and more as equal partners in a wide span of research activities. This should result in national scientific communities, while internationalizing the scientific norms of behavior which are common to all researchers.[39] Bibliometrics cannot measure the latter process, but some indicators clearly suggest the process has begun. One of them is the analysis of coauthorship. The studies of D. de Beaver and R. Rosen indicate that collaborative research enhances productivity, and that collaborative research is growing.[40]

In our soil production sample, Third World countries had high numbers of authors per article (2.48 authors for India; 2.79, for Brazil). European countries, by contrast, had a low number, 1.48 to 1.63. With the exception of South Africa, all periphery countries and the United States are in the

36. (a) Argenti, "From Standardization to Relevance and Back Again." (b) Schwartzman, "Coming Full Circle: For a Reappraisal of University Research in Latin America," *Minerva* 24 (4) (1986): 456–75. (c) Vessuri, "El Proceso de Profesionalización de la Ciencia Venezolana: La Facultad de Ciencias de la Universidad Central de Venezuela," *Revista Quipu* (Mexico) 4 (2) (1988): 253–84. (d) Vessuri, "La Formación de la Comunidad Científica en Venezuela," in *Ciencia Acadécmica en la Venezuela Moderna*, ed. Vessuri (Caracas: Fondo Editorial Acta Científica Venezolana, 1984).
37. Schwartzman, *A Space for Science*.
38. Botelho, "Far from Silicon Valley: Give Me a Laboratory and I Will Not Raise the World," presented at the XVth Annual Meeting of the Society for the Social Studies of Science (Minneapolis, Minn., 1990).
39. Fortes, *La Formación del Científico en México*.
40. D. de B. Beaver and R. Rosen, "Studies in Scientific Collaboration, Parts I–III," *Scientometrics* 1 (1–3): 64–84, 133–49, 231–45, 1978–79.

middle, with their number of authors ranging from 2.1 to 2.2. The striking figure is that of the big Third World producers India, Egypt, and Brazil. One can note an historical and somewhat anomalous process: the older countries (Europe) have the lowest number of coauthors, followed by the first peripheral countries (United States, Australia) and the younger countries, which all belong to the Third World.[41] The factors which have created this anomaly are complex and deserve further investigation.

F. Changing Our View of Scientific Production

It appears that new scientific communities are emerging with their own agendas and functions. This is important in the assumption that research will solve development problems. We now know that this is not necessarily the case. There is a need for the strengthening of local scientific communities which are incipient in many countries, especially in Africa.[42] Few simple ways of knowledge or other international technology transfers between north and south exist,[43] and little internal technology transfer, from laboratories to the productive sector.[44] Stronger scientific communities in the south mean stronger involvement in fundamental science. The assumption that developing countries mainly publish in areas of applied science is not wholly accurate. Some countries are dedicated to themes bearing directly on agriculture, but they do not seem to limit themselves to the applied aspects.

Scientific knowledge cannot be truncated in small bits with basic research on the one hand and applied on the other. A research problem is defined as being of interest, and, within that problem, scientists will eventually occupy all types of research. Scientific research is a process, not merely a production activity, which policymakers often overlook when urging scientists to

41. This is also the conclusion of Velho, "The Meaning of Citation in the Context of a Scientifically Peripheral Country," and of other studies on the institutionalization of research in Third World countries. See, for example, the articles in Vessuri, "El Proceso de Profesionalizacón de la Ciencia Venezolana." We disagree with the argument that coauthors in the developing countries are more numerous because every single member of a lab signs an article written by only one scientist. No figures assert such practices. But developed country scientists publish relatively more articles in common with foreign colleagues.

42. Gaillard, "La Recherche Scientifique en Afrique."

43. A. Rath, "Science, Technology, and Policy in the Periphery: A Perspective from the Centre," *World Development* 18 (11) (1990): 1429–43.

44. (a) Argenti, "From Standardization to Relevance and Back Again." (b) A. Pirela, R. Rengifo, and Arvanitis, "Vinculaciones Universidad-Empresa en Venezuela: Fabula de Amores Platonicos y Cicerones," *Acta Científica Venezolana* 42 (1991): 239–46. (c) J. Ruffier, "Pensar la Modernización de la Industria Uruguaya," in *Uruguay: El Debate Sobre la Industrialización Posible*, ed. G. Argenti (Montevideo: Ciesu/Ediciones de la Banda Oriental, 1991), pp. 13–49.

do applied research.[45] Let us give an example which is particularly interesting. In a previous study, a research program of the legume *Canavalia*, we found a large body of literature on the biochemistry of the plant's toxicity. Not until there existed a coordinated research program on the plant did Venezuela begin to study its biochemistry in order to understand the nature of its toxic elements and ways to eliminate or neutralize them. Some dozen years of research were needed to understand all the aspects involved in the management of the plant. Ten years earlier, a United States laboratory had worked on biochemical methods involving the toxins of the plant, particularly one toxic chemical. Of course, the Venezuelan laboratory used the results of the United States laboratory, but can we really talk of a transfer of knowledge? We believe this is not the case, even if the Venezuelan laboratory used some of the methods and results of the United States laboratory, the Venezuelan research was very different: in the case of the United States, the plant was used as a laboratory specimen, in the other it was the transformation of the plant itself that was of interest.[46]

In summation, the question of scientific literary output in developing countries is sometimes obscured by the views of the north and by an obsolete conception of how science evolves.[47] We too often evaluate developing countries' scientific production only by the small share of published research that is indexed and made available in the north.[48] Scientific communities of the south are not necessarily governed by the same criteria as the north, but are characterized by informality and their merger of research with other activities. The dynamics of scientific research, basically so different in the south, are a major factor. A new appraisal of the scientific production in developing countries is needed using valid bibliometric studies.[49] The questions then become: what determines that a topic is "researchable?" What orients the interests of the scientists in the south toward new areas of

45. Arvanitis, *De la Recherche au Développement*.

46. Arvanitis and T. Bardini, "Analyse d'un Programme Pluridisciplinaire par Deux Méthodes d'Analyse de Réseaux: Le cas du Groupe de Recherche sur *Canavalia*," in *Proceedings of the International Conference on Science Indicators*, ed. Arvanitis.

47. Sociology of science has changed our view. See B. Latour, *Science in Action* (Open University Press, 1987). For applications of a new conception of science in the policy and management of research, see D. Vinck, ed., *Gestion de la Recherche. Nouveaux Problémes, Nouveaux Outils*, (Brussels: De Boeck Professional Publishing, 1991).

48. For an example, see S. Arunachalam and U. N. Singh, "Access to Information and the Scientific Output of India," *Journal of Scientific and Industrial Research* 51 (1) (1992): 99–119.

49. (a) S. Thomas, "The Evaluation of Plant Biomass Research: A Case Study of the Problems Inherent to Bibliometric Indicators," in *Proceedings of the International Conference on Science Indicators*, ed. Arvanitis. Thomas has shown that, in the case of biomass research, neither publication, in peer reviewing, is, by itself, an appropriate instrument for the evaluation of the impact of research. (b) Arvanitis and Gaillard, "Pour un Renouveau des Indicateurs de Science pour les Pays en Développement," in *Proceedings of the International Conference on Science Indicators*, ed. Arvanitis.

research? How do northern and southern research teams cooperate? These and other important questions can perhaps be examined by bibliometric analysis. Obviously, more work needs to be done on these topics, with the close participation of divergent disciplines including sociology, soil sciences, economics, and other agricultural sciences.

Appendix. Construct of the Bibliometric Database for Analysis of Tropical Soil Sciences

All documents were extracted from the 1983 *Bulletins Signalétiques* (No. 226 and No. 381) of the PASCAL database in the following categories:

Bulletin Signalétique No. 226:
   Formations superficielles
   Géomorphologie
     Sols
Bulletin Signalétique No. 381:
   Généralités
     Comptes-rendus généraux, rapports d'activité, congrès, bibliographies
     Méthodes et techniques diverses
       Télédétection
   Sols, Agronomie générale
     Généralités
     Techniques et méthodes d'analyse
     Cartographie des sols
     Classification des sols
     Pédogenèse
     Physico-chimie du sol
       Eléments minéraux, oligo-éléments, propriétés ioniques et d'échange
       Matière organique, évolution de la matiére organique, complexe argilohumique, cycle de l'azote et du carbone
     Propriétés physiques
       Structure et texture, densité, comportement mécanique, échanges gazeux et thermiques
       Dynamique de l'eau et des solutés (état et transfert)
   Microbiologie des sols, enzymes du sol, interactions microorganismes-végétaux
   Fertilisation minérale et organique, nutrition
     Généralités
     Diagnostic foliaire
     Fertilisation des différentes cultures
     Fertilisation azotée
     Fertilisation potassique
   Oligo-éléments
     Utilisation des déchets solides et liquides

Maladies de carence, toxicité
Pollution du sol
Amendements et engrais minéraux divers, correction de pH
Amendements et engrais organiques
Substrats artificiels, hydroponie, fertilisation par CO_2
Relations sol-plante
Conservation des sols, érosion
Potentialités, aménagement du territoire

5. Introduction to the Literature of Soil Science in Developing Countries

S. W. BUOL

North Carolina State University

The study of soil as an independent science had its beginning slightly over 100 years ago in northern temperate zones. Tropical soil research lagged by several decades. The observation that soil properties developed in response to climatic factors of temperature and amount of precipitation caused early soil scientists to hold exaggerated expectations of the soil properties to be found in the largely unexplored tropics. Thus, extremes of soil properties were the first explored and reported in tropical soil science literature.

In *Tropical Soils* by Edward Mohr and F. A. Van Baren (1954) there is an early example of reporting extreme soil conditions and thereby formulating a theory and concepts of tropical soil formation, as distinct from temperate soils.[1] The state of knowledge in soil science in tropical areas largely reflected data concerning total elemental analyses and weathering functions with depth in rock types not commonly present in temperate areas. Although rock differences were carefully discussed, there was a concerted effort to conceptualize tropical weathering and tropical soil-forming processes as separate and distinct from implied temperate zone processes. These limitations were recognized by Mohr and Van Baren at the conclusion of their volume: "As no international determination key, based on fundamental features of soils, is yet available for use as a basis for soil denomination, the use of descriptive terms is to be preferred to a geographic indication."[2]

Agronomic aspects of tropical soil science before 1960 focused on plantation crops, primarily cotton, oil palm, and rubber. Native farming, low-input technologies were observed, with their low output, which were passed

---

1. Edward Mohr and F. A. Van Baren, *Tropical Soils: A Critical Study of Soil Genesis as Related to Climate, Rock and Vegetation* (New York: Interscience Pub., Inc., 1954), (New impression of 1954 edition).

2. Ibid., p. 479.

on through the literature as the accepted production potentials of tropical soils. The inadequacy of this practice was best stated by Charles Kellogg: "Certainly none of us would think of estimating the potentialities for production of soils in temperate regions on the basis of non-scientific management. Suppose we looked at native production in Holland, Belgium, Florida, even the Midwest of the United States, without fertilizer, insecticides and plant breeding?"[3]

Kellogg's plea for broader scientific research and management of tropical soils was noted and in 1954, at the International Congress of Soil Science held in the Congo, almost half of the papers dealt with soils of tropical areas. It is noteworthy that nearly all of the papers reporting soil research in the tropics were from past or current colonies, reflecting the former colonizers' support for soil science investigations.

The colonies played yet another important role in the expansion of soil science research in the tropics. Following World War II, students of soil science were able to take advantage of financial support for study in developed countries via three primary routes. Those in former colonies were able to study abroad within the universities of their colonial rulers, i.e., United Kingdom, France, the Netherlands, and Belgium. Those in developing countries under communist forms of government tended to study in the Soviet Union. Those in countries receiving support from the U.S. Agency for International Development (AID) studied in the United States. Some developing countries managed uniquely to embrace all three routes. A substantial number of students were also able to pursue advanced degrees in countries such as Australia and Canada. But where the colonial association was strong, invariably the developing country duplicated the contrasting concepts and methodologies of its academic benefactors. Characteristic of this duplication are the former French colonies in northern Africa and Latin America where Philippe Duchaufour and ORSTOM (the Office de la Recherche Scientifique et Technique Outre-Mer), with its many publications as well as associated French organizations, dominated soil science.[4] These relationships created wide and useful diversity in undergraduate and graduate training in soil science within a broad spectrum of developing countries.

The 1960s saw tremendous change in understanding soils of tropical regions. For the first time soil scientists developed quantitative methods of classifying the properties of soils called for by Mohr and Van Baren rather than relying on outdated pedogenic concepts. Two independent efforts were in the forefront of this change. The FAO/UNESCO *Soil Map of the World*

3. Charles E. Kellogg, "Tropical Soils," *Transactions of the 4th International Congress of Soil Science* 1 (1950): 266–76.
4. Philippe Duchaufour, *Précis de Pédologie*, 2d ed. (Paris: Masson, 1965).

project and the Food and Agriculture Organization (FAO) *Soils Bulletins* series. The former, a collaboration between FAO and UNESCO, developed 104 mapping units, with a new system of diagnostic properties and nomenclature, on 1:5,000,000 scale maps of the world.[5] This effort was marked by several meetings and, more importantly, field studies to characterize soils in developing countries under the guidance of FAO scientists. This World Soil Map project was followed by the first publication of the FAO *Soils Bulletins*.[6] To date, sixty-three such Bulletins have been published, many of them in more than one language. Intended primarily as on-site resources, these Bulletins cover a diversity of subjects. Important titles have included *Survey of Soil Laboratories in 64 FAO Member Countries*, 1965; *Soil Conservation for Developing Countries*, 1976; *Micronutrients and the Nutrient Status of Soils: A Global Study*, 1982; 2nd *Management of Gypsiferous Soils*, 1990. These Bulletins have provided not only data but methodology references for many developing country scientists. Most of the Bulletins report work done in developing countries in cooperation with local soil scientists and utilize the FAO classification of soils to provide a basis of extrapolating the information to other areas of like soils around the world.

At the 1960 International Society of Soil Science meeting, the United States Soil Survey Staff distributed a published version of a totally new system of soil classification that made provision for any soil in the world, although at the time only soils in the United States were extensively included.[7] The most significant aspect of the system was the recognition of soil temperature and soil moisture dynamics as properties of soil rather than climatic factors controlling soil genesis. This initiative and the continuing activities of the USDA Soil Survey Staff (see Chapter 11) were to serve as foci for extensive communication in the soil science community. A bibliography compiled by G. D. Bailey lists 1,310 publications from 1960–79 using this system to identify world soils.[8]

In 1960, P. H. Nye and D. J. Greenland presented to soil scientists and agronomists a quantitative view of the chemistry involved in the major farming practice in the tropics; namely, shifting cultivation.[9] A decade

5. R. Dudal, *Definitions of Soil Units for the Soil Map of the World* (Rome: World Soil Resources Office, Land and Water Development Division, FAO, 1968).

6. Food and Agriculture Organization, *Soils Bulletins (1963 to date)* (Rome: FAO, Via Delle Terme Di Caracalla, 1965).

7. U.S. Soil Survey Staff, *Soil Classification: A Comprehensive System 7th Approximation* (Washington, D.C.: USDA-SCS, U.S. Govt. Print. Off., 1960).

8. G. D. Bailey, *Bibliography of Soil Taxonomy, 1960–1979* (Wallingford, U.K.: CAB International, 1987).

9. P. H. Nye and D. J. Greenland, *The Soil under Shifting Cultivation* (Harpenden, U.K.: Commonwealth Bureau of Soil Science, 1960 [Technical Communication no. 51]).

later, a review of soils research in tropical Latin America summarized the literature from this region and focused attention on the relationship of soil management responses on individual types of soils in the tropics.[10]

The combination of quantitative classification of soils based upon measured properties in the soil and a factual and quantitative understanding of shifting cultivation allowed soil scientists to see clearly that many of the earlier assumptions about unique tropical soils and their behavior under cultivation should be abandoned. This change in approach to tropical soils by most of the influential centers of graduate training in the United States and in Europe, coupled with an influx of young scholars from developing countries, created a basis for obtaining factual evaluation of tropical soil properties and their corresponding production potential. Results of these agronomic studies presented at international meetings were now evaluated and compared on the basis of soil properties, and not by geographic or political boundaries.

With the advent of new, high-yield potential germplasms for food crops in tropical areas during the green revolution of the 1960s, soil fertility limitation of the use of germplasms by farmers became the focus of international research centers. Textbooks, such as Pedro A. Sanchez's *Properties and Management of Soils in the Tropics*,[11] utilized the ability to quantitatively compare soils, via *Soil Taxonomy* and the FAO world soil map legend, to relate research on soils in the tropics to research done on like soils in temperate regions.[12] The extent of this work is revealed in A. C. Orvedal's massive, five-volume bibliography, which cites publications relating to soils in various locations in the tropics and permits researchers and students to avail themselves of research relevant to the kind of soil in their local studies.[13]

Significant contributions to the factual understanding of soils in the tropics were made possible by soil classification workshops, which brought the soil characterization technology of the National Cooperative Soil Survey program of the United States to bear on various areas of the tropics. These characterization data were made available through published proceedings of

10. Pedro A. Sanchez, ed., *A Review of Soils Research in Tropical Latin America* (Raleigh, N.C.: Soil Science Dept., N.C. State Univ., 1973). (Trans. to Spanish, 1973).

11. Sanchez, *Properties and Management of Soils in the Tropics* (New York: Wiley, 1976). (Trans. to Spanish, 1981).

12. U.S. Soil Survey Staff, *Soil Taxonomy* (Washington, D.C.: USDA-SCS, U.S. Govt. Print. Off., 1975 [*USDA Agriculture Handbook* no. 436]).

13. A. C. Orvedal, *Bibliography of Soils of the Tropics* (1975, Africa; 1977, South America; 1978, Middle America and West Indies; 1980, Islands of the Pacific and Indian Oceans; 1983, Mainland Asia, Pakistan, Nepal, and Bhutan), (Washington, D.C.: U.S. Agency Int. Dev., Dev. Support Bur., Off. Agric., 1975–83 [*AID. Technical Bulletin* no. 17]).

International Soil Classification Workshops conducted by Soil Management Support Services (SMSS) (see Chapter 11).

Soil science as a scientific discipline appears to have developed to a degree of maturity where it can encompass all soils of the world and subject them to rigorous scientific standards. Young soil scientists, eager to achieve recognition among their world peers, are now doing research in all developing countries, albeit their numbers are small and their influence as yet unrecognized in many countries.

The scientific efforts of the soil science community in the developing world are now focused on soil as a natural resource, to be studied for the practical aspects of food and fiber production and environmental quality. This is in contrast to the time when soils were studied because they were unknown natural phenomena. Much still needs to be done to integrate a knowledge of tropical soil properties into the multitude of soil management practices common in developed countries, where soil is a vital reagent.

To a large extent the conceptual and methodological differences that were so obvious in the 1970s have been reduced. The particulars of this change in a country vary greatly, which makes the identification of responsible programs or policies difficult to apply at national levels. Nevertheless, among the contributing factors one must consider the number of developing country scientists who have returned to their countries after completing studies abroad. By way of example, in 1980, institutions with either a soil science department or an agronomy department totaled sixty-five in South America, thirty-two in Africa, and ninety-one in Asia and the Middle East.[14] Degree offerings at these institutions were not determined. The opportunities afforded developing country scientists to communicate via training courses, collaborative regional projects, and international conferences greatly enhances their efforts to keep scientifically current. The fear of appearing backward in the eyes of peers from other developing countries is a strong motivating force to do good work and represent their nation in the scientific community. The many successful soil management technologies that have been developed in tropical or developing areas have obviated the necessity to compare tropical and temperate soils. Many tropical soil myths of incompatibility with productive farming practices have been debunked, at least among agronomic scientists, although these same myths linger in the ranks of too many natural scientists. When confronted with evidence that others have solved many of the supposed limitations of tropical soil, developing

14. Burton Swanson, ed., *International Directory of Agricultural Education Institutions*, 3 vols. (Champaign-Urbana, Ill.: Bureau of Education Research, University of Illinois at Champaign-Urbana, 1981).

country scientists are in a better position today to return from such meetings with renewed resolve to show their people what can be done.

Although each developing country has its own evolutionary pathway regarding soils research and teaching, governments and society at large are beginning to recognize the vital role of soil as an essential natural resource that must be used for sustainable development. Organizations of many ideological backgrounds have sought to include some form of soil information dissemination into their programs, but to date few national or international organizations have had soil science as their central focus. Among these few are the International Society of Soil Science and the various national soil science societies. The relatively recent founding of regional soil societies in diverse areas of the developing world reflects the professional development of soil scientists as a whole. The establishment of the International Board for Soil Research and Management (IBSRAM) also reflects this international concern for improved distribution of information about soil. While these efforts currently are small and circulation of publications limited, most of the developing countries rely on more established organizations for information. These include, but are not limited, to the *Soil Science Society of America Journal*, *Australian Journal of Soil Research*, *Canadian Journal of Soil Science*, *Journal of Soil Science* (British), and independent publications such as *Geoderma*, *Catena*, and *Soil Science*. In Central and South America, *Turrialba*, *Bragantia*, *Suelos Ecuatoriales*, *Fitotecnia Latinoamericana*, and other regional and institutional journals became important avenues for publishing soils research. The story is the same in the Arab world, in Africa, and in Asia. An example of developing country readership is reflective of the number of developing country members of the Soil Science Society of America: in 1961, there were 141 members; in 1971, 333; in 1982, 499; and in 1990, 545. A decrease in Soil Science Society of America membership from within India during the last decade perhaps reflects the growing attraction for their own *Journal of the Indian Society of Soil Science*. The Brazilian Soil Science Society may be creating a similar trend in that country with their publication *Revista Brasileira de Cienca do Solo*.

Although access to world literature is good in a few developing countries, many soil science departments lack needed teaching and reference literature. In many library facilities in tropical regions, one finds that the international journals stop with the last personal donation from a senior scientist or the curtailment of expenditures of hard money from the developing country. Since the vast majority of soil science literature is published in English, translations of modern textbooks, if ever available, lag several years, offering beginning students concepts many years out of date. The antiquated nature of many libraries is quite evident in facilities with modern analytical equipment which is underutilized because the soil analytical techniques do not effectively address real soil problems. This is especially disturbing because many of the agricultural problems in developing countries are manifested in the chemistry and fertility of the soil and require accurate and appropriate soil analysis.

A primary challenge of soil research in most developing countries remains bringing soil science information to bear on site-specific problems. A wide range of ameliorating resources are gaining wider currency. Most developing country farms are small, with limited resources to provide on-site expertise. Contrary to earlier beliefs that through intense weathering the soils in the tropics were homogenized in tropical landscapes, abrupt contrast of soil properties in these regions is at least as common as in temperate areas, if not more so. Many of these differences are masked by the small scale of existing soil maps in developing countries (see Chapter 11). Similarly, results of cropping research in many developing countries are contradictory, and on-farm results often bedevil agronomic research and extension efforts. Case studies of contradicting results all too often reveal that the errant site is on a soil with properties quite at odds with those on general soil maps of insufficient scale to illustrate farm-size areas. The need for better utilization of agronomic technology and improved environmental quality is axiomatic, but it awaits the application of the correct technology to the particular soils of specific sites in order to improve. A more rapid exchange of quantitative information about soil use may aid in this endeavor. The evolution of soil science from its former reliance on concepts and qualitative observation to quantitative evaluation of chemical, physical, and mineralogical properties of the soil is basic to this continued development.

6. Contributions to Pre-1960s Soil Science Literature in Third World Countries

ARMAND VAN WAMBEKE

Cornell University

A brief review of the soil science monographs targeted at the Third World and included in the *Core List of Soils Monographs* in Chapter 9 shows that nearly all were published after the 1960s. The graph in Figure 6.1 shows the spread of these Third World core monographs over a forty-year period, clearly demonstrating that the majority of the monographs were published in the last fifteen years. This selection of monographs provides an up-to-date guide to the core literature of soil science in low-latitude countries.

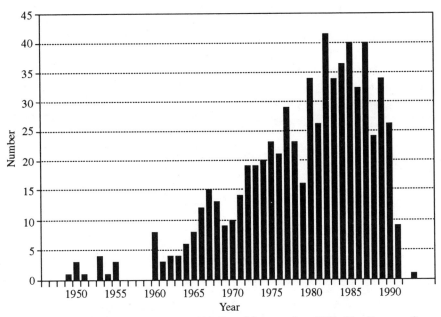

Figure 6.1. Distribution of Third World Core Monographs, 1950–91. (*Source:* Core Agricultural Literature Project analysis, Chapter 9.)

There is no doubt that work before the 1960s contributed significantly to the growth of soil science in the Third World. What was its place in the agricultural literature? What contributions did research carried out in low-latitude countries make to soil science? How effectively did it serve development? It is the purpose of this chapter to give some answers to these questions.

This review has been organized according to time, subject matter, and geographic origin. The synopsis is not comprehensive. Omissions and sample selections may have reduced its objectivity. The sources that have been consulted are not wholly representative, because the number of publications produced in Third World countries seldom reflects the magnitude of their soil research and teaching activities. Most scientists have few opportunities to publish, or the incentives to publish either do not exist or are less rigorous than in industrialized nations. Language barriers also have kept many publications from entering the mainstream; on the other hand, soil science is sitespecific and often only relevant to readers with local interests. We hope that others will have the time, opportunity, and interest to fill the gaps that undoubtedly exist in this review.

A. The Early Phases

Soils have always been a subject of concern to people who depend on their productivity for subsistence. To start this review, it is appropriate to credit the farmers in many Third World countries who, since ancient times, have designated soils by vernacular names. These names have seldom been recorded, although they are often still used; examples would be *regur*,[1] the name that farmers in India give to black cotton soils,[2] and the local names of soils in Senegal and the Sudan cited by G. Aubert and B. Newsky.[3] More recently, the validity of the vernacular names that are used in Mauritania has been investigated by P. N. Bradley.[4]

Historical reviews about the development of branches of science usually start from the time they are commonly accepted as independent disciplines.

1. Indian Council of Agricultural Research, *Final Report of the All India Soil Survey Scheme* (Calcutta: Government of India Press, 1953 [Bulletin no. 73]).
2. T. J. Newboldt, "On the Rigar or Black Cotton Soil of India," *Proceedings of the Royal Society* 4 (1838): 53–54.
3. G. Aubert and B. Newsky, "Note on the Vernacular Names of the Soils of the Sudan and Senegal," *Proceedings of the 1st Commonwealth Conference on Tropical and Sub-Tropical Soils, 1948* (Harpenden, U.K.: Commonwealth Bureau of Soil Science, 1949 [Technical Communication no. 46]).
4. P. N. Bradley, *Peasants, Soils and Classification* (Newcastle upon Tyne, U.K.: University of Newcastle upon Tyne, Dept. of Geography, 1983 [Research Series no. 14]).

For soils, Jean Boulaine (Chapter 2) places this event between 1870 and 1900. This does not mean there were no publications on soils before that time. In most countries, the need to examine soils resulted either from famines or from trade interests in commodities such as coffee, sugar, cotton, or indigo. The constant decline in productivity also was an incentive for soil studies. Soils of Third World countries were, in fact, the object of many significant publications. Two examples are given in the following paragraphs: India and Indonesia.

According to K. K. Guha Roy the scientific study of soils in India began in 1895.[5] However, in 1807, F. Buchanan described, for the first time, soil materials that hardened irreversibly into ironstone and called them *laterite*.[6] Guha Roy's bibliography cites ninety-five publications on the geology of India during the period between 1807 and 1894, all of them written by non-Indians.[7] Most soil literature appeared in agricultural, horticultural, geological, or geographical journals. Guha Roy lists several soil science disciplines.[8] Some of the citations, with their first publication dates given in parentheses, are liming (1823), salinity (1834), crop suitability (1836), soil types (1838), chemistry (1839), alkalinity (1862), black cotton soils (1863), and soil surveys (1883).

In Indonesia, the decline over time in crop production was a major concern to agriculturalists.[9] In soil fertility management, experiments with guano were reported in 1854 by P. F. H. Fromberg.[10] In 1890, J. M. Van Bemmelen (1868–1941) conducted research on the relationships between the chemical and mineralogical composition of soils and the yields of industrial crops.[11] K. F. Holle, in 1866, started investigations on the effects of terracing on the conservation of soils of hill slopes.[12] C. H. Edelman compiled a bibliography of early soil publications on Indonesia.[13]

5. K. K. Guha Roy, *Bibliography of Soil Science and Fertilizers with Reference to India* (Calcutta: Government of India Press, 1954 [Indian Council of Agricultural Research Bulletin no. 74]).

6. F. Buchanan, "A Journey from Madras Through the Countries of Mysore, Canara, and Malabar," *Geological Magazine* 2 (1807): 436–37, 440–41.

7. Guha Roy, *Bibliography of Soil Science and Fertilizers*.

8. Ibid.

9. F. A. Van Baren, "De Ontwikkelingsgeschiedenis van de Bodemkunde in Indonesia," *Chronica Naturae* 106 (6) (1950): 231–45.

10. P. F. H. Fromberg, "Over de Guano, hare Oorsprong, Chemische Samenstelling en Werkzaamheid als Meststof," *Natuurkundig Tijdschrift voor Nederlandsch Indië* 6 (1854): 63–84.

11. J. M. Van Bemmelen, "Die Zusammensetzung des Vulkanischen Bodens in Deli (Sumatra) und in Malang (Java), und des Fluss-Tonbodens in Rembang (Java), Welche für die Tabakskultur benutz Werden," *Landwirtschaft Versuchst* 37 (1890): 257–78.

12. K. F. Holle, "Een Groot Gevaar, dat Sluipend Nadert," *Tijdschrift voor Nijverheid en Landbouw in Nederlandsch Indië* 18 (1866): 261–89.

13. C. H. Edelman, *Studies over de Bodemkunde van Nederlandsch-Indië* (1947 [Publicatie no. 24 van de Stichting Fonds Landbouw Export Bureau 1916–1918]).

B. Twentieth-Century Soil Science Literature

Most reviewers consider the beginning of the twentieth century as a turning point in the development of soil science. What was the place of soil publications in the Third World agricultural sciences literature? An examination of the *1931–1934 Bibliography of Soil Science, Fertilizers and General Agronomy* may help explain the status of soil science before World War II.[14] It lists 965 serial publications as their sources. Among them were 102 agriculture and forestry journals published in Third World countries and twenty-three soil science journals all published in developed nations, except one with two issues.

In Europe, the soil science contributions that specifically related to Third World countries mostly appeared in serial publications of institutions that specialized in overseas activities, for example, ORSTOM (Office de la Recherche Scientifique et Technique Outre-Mer) in France; another French agency, the Institut de Recherches Agronomiques Tropicales et de Cultures Vivières, published *L'Agronomie Tropicale*, a monthly journal founded in 1946. In Portugal, the overseas institution was the Junta de Investigaçoes do Ultramar; in the United Kingdom, the Colonial Office and the Imperial Bureau of Soils; in Belgium, the Institut National pour l'Etude Agronomique du Congo Belge (INEAC); in the Netherlands, the Koloniaal Instituut. Before World War I, Germany also had publications on overseas territories, for example, *Berichte über Land- und Forstwirtschaft in Deutsch-Ostafrika*, published by the Kaiserlichen Gouvernement von Deutsch-Ostafrika. In most developed countries, academies also accepted soil science papers from overseas.

There were no specialized overseas agricultural or natural resources government research institutes in the United States prior to 1950, and, consequently, no centrally published tropical agriculture literature. Soil publications on Third World countries were included in United States professional journals. The same situation prevailed in Germany after World War I. German contributions were usually published in the *Zeitschrift fur Pflanzenernährung, Düngung und Bodenkunde*.

Another concentration of soil science literature resides in proceedings or transactions of international, Asian, interafrican, or interamerican soil conferences. For example, most publications on the soils of Cameroon referred to by the Commonwealth Bureau of Soils Science, between 1938 and 1960 were part of such proceedings. One third of the papers presented in Com-

14. Imperial Bureau of Soil Science, *Bibliography of Soils Science, Fertilizers and General Agronomy 1931–1934* (Harpenden, U.K., 1935).

mission Five at the Seventh International Congress of Soil Science were related to Third World countries.[15]

The twentieth century soil science literature that was published more than twenty-five years ago on low latitude regions had many different origins. For that reason, it presents a wide diversity in subject matters, disciplines, perspectives, and philosophical approaches. Well defined schools, declared or undeclared, followed preferred pathways. Their thinking usually conformed with the theories that best explained the characteristics of their study area or contributed to the understanding of specific soil-plant relationships. The longevity of these schools of thought often corresponded to needs for research to solve specific problems. Isolation of scientists was one of the major drawbacks and contributed, to some extent, to the individuality of the schools.

There is probably no good way to subdivide the subject matter. The approach followed here begins with the soil science literature in a number of countries that are thought to be representative, subsequently discusses some major linkages, and, finally, reviews a number of subdisciplines.

Country and Regional Reviews

Asia

A. B. Ghosh regards the establishment of the Imperial Institute of Agricultural Research at Pusa, in 1906, as the beginning of soil science in India.[16] J. W. Leather was the first agricultural chemist at the institute. His work on the reclamation of sodic soils (called *usar* in India) is still a classic in this field.[17] There were other institutions as well. In 1943, the All India Soil Survey Scheme started paving the way for future resource inventory operations. Some of the major contributors to the diversity of programs were W. H. Harrison, A. N. Puri (physicochemistry and physics), S. P. Raychaudhuri and Bhagavatula Viswanath (soil resource inventories), S. L. Das (phosphorus), J. N. Mukherjee (physical chemistry and clay mineralogy), T. R. Bhaskaran, B. Chatterjee (soil acidity), A. Sreenivasan and V. Subrahmanyan (nitrogen), N. R. Dhar (microbiology), R. M. Gorrie (erosion), P. Ramiah and R. N. Singh. In 1933, V. I. Vaidyanathan summarized the results of more than 5,000 permanent manurial experiments.[18]

15. International Society of Soil Science, *Transactions of the 7th International Congress of Soil Science. Volume IV.* (Madison, Wis., 1960).
16. A. B. Ghosh, "History and Development of Agriculture in India with Special Reference to Soil Science during 1800–1950," *Indian Society of Soil Science Bulletin* 14 (1984): 14–28.
17. Ibid.
18. Ibid.

Indian soil science was closely linked to research and training in agriculture. Education was often part of the agricultural experiment stations' programs; there were six agricultural educational institutions in 1906,[19] with another ten state agricultural universities added during the 1960s.[20] Indian soil science was one of the first to be institutionalized. The Indian Soil Science Society was founded in 1934, the same year as the French society and before all others except the German, Danish, and Japanese societies.[21] The *Bulletin of the Indian Society of Soil Science* started publication in 1938.

In the Philippines, the first soil survey was accomplished in 1903 by C. W. Dorsey, an American soil scientist. Not until 1921 was a Division of Soils and Fertilizers created in the Bureau of Sciences. This unit mainly carried out chemical analyses. After independence, the Bureau of Soil Conservation was established in 1951, employing more than one hundred people.[22] The first issue of the *Philippine Journal of Soil Conservation* was published in 1953.

Mexico

In Mexico interest in soils as a basic science in support of agricultural research began to develop in 1906, at the time that its agricultural experiment stations were created. Two government agencies and the Escuela Nacional de Agricultura, founded in 1884, carried out physical and chemical soil analyses. The areas of major concern were soil fertility and irrigation management. By the end of the 1930s, the agronomy department of the Comisión Nacional de Irrigación, created in 1926, had collected data on 334 soil series that were identified by forty-five projects.[23] The Comisión had its own numbered publication series and was the first government agency in Latin America to engage in soil classification.[24] The Secretaría de Recursos Hidráulicos, in 1964, had on file about 130 soil maps of irrigation

19. T. D. Biswas and G. Narayanasamy, "Documentation and Dissemination of Soil Information," *Indian Society of Soils Science Bulletin* 14 (1984): 55–62.

20. C. P. Streeter, "A Partnership to Improve Food Production in India," a report from the Rockefeller Foundation (New York, 1969).

21. ISSS Secretariat, personal communication, Directory of National Societies, ISRIC, Wageningen, Netherlands, 1992.

22. M. M. Alicante, "The Bureau of Soil Conservation—Its History, Activities and Salient Accomplishments," *Philippine Journal of Soil Conservation* 1 (1953): 5–18.

23. R. J. Laird, "Evolucion de la Ciencia del Suelo en México, Algunos Avances y los Desafíos para el Futuro" (1992, Mimeographed).

24. M. Rodriguez, "Desarollo de la Ciencia del Suelo en los Últimos 25 años en America Latina," in *Progreso y Futuro; Las Ciencias Agrícolas en América Latina* (San José, Costa Rica: IICA, 1967), pp. 523–565.

project areas.[25] Many soil publications appeared in the serial *Irrigación en México* and, later, in *Ingeneria Hidráulica en México*.

K. D. Glinka's soil textbook was translated into Spanish in 1940. The first attempt to classify the soils of Mexico in great groups following European guidelines was made in 1937 and revised several times in 1948, 1959, and 1960. The first national Soil Science Congress of Mexico was held in 1963. The Rockefeller Foundation in 1943 started a cooperative research program with the Oficina de Estudios Especiales. Soil management was among the disciplines that contributed to the increase in the production of many food crops.[26] R. Ortiz Monasterio was one of the major Mexican contributors to this program.

The soil science department at the Escuela Nacional de Agricultura was created in 1958 with M. Villegas Soto as its leader. It marked the beginning in Mexico of increased growth in soil science during the 1960s.

South and Central America

Latin American countries all gained independence in the first half of the nineteenth century. Distance, language and political barriers did not facilitate communications with North American or European colleagues. According to a bibliography on soils of Latin America, the earliest scientific publications go back to 1894 (Argentina), 1903 (Brazil), 1928 (Chile), 1944 (Colombia), 1934 (Costa Rica), 1911 (Cuba), 1945 (Ecuador), 1944 (El Salvador), 1951 (Guatemala), 1929 (Mexico), 1945 (Nicaragua), 1929 (Panama), 1918 (Peru), 1933 (Uruguay), and 1936 (Venezuela).[27]

In spite of political isolation, contacts between professionals continued to contribute to the dissemination of knowledge. Until World War II, many publications on soils appeared in foreign journals: for example, Adolpho Matthei, a Chilean pedologist, published a paper in German on the agriculture of Chile which included sketchy soil maps of parts of that country.[28] In the same year, he published a first soil map of South America, in German.[29]

The flow of information went both ways. In Argentina, P. Lavenir, a French agronomist, was instrumental in the establishment of an Instituto de

25. A. C. Orvedal, *Bibliography of the Soils of the Tropics*, Vol. 3 (Washington, D.C.: Office of Agriculture, Technical Assistance Bureau, AID, 1978 [Technical Series Bulletin no. 17]).

26. Rodriguez, "Desarollo de la Ciencia del Suelo."

27. Food and Agriculture Organization, *Bibliography on Soil and Related Sciences for Latin America* (Rome; FAO, 1966 [World Soil Resources Report no. 23]).

28. Adolpho Matthei, "The Soil Geography of South America," *Soil Research* 5 (1936): 75–98.

29. Matthei, *Agrarwirtschaft und Agrarpolitik der Republik Chile* (Berlin: Verlagsbuchhandlung Paul Parey, 1936 [Berichte über Landwirtschaft. no. 119 Sonderheft]).

Suelos y Agrotecnia in 1943. Previously, he had been actively involved in the study of the soils of Argentina as head of the chemistry and agrology laboratory of the Ministry of Agriculture. His first publication on soils appeared in 1901,[30] followed by a 577 page monograph in 1910.[31] The Instituto de Suelos y Agrotecnia started its own publication series in 1945.

After World War II, soil science benefited from the surge in institutional development assistance. The U.S. Agency for International Development (AID) financed projects that linked seven United States agricultural colleges with seven Latin American universities. AID also contributed to the soil survey of several countries. At the same time, the Food and Agriculture Organization (FAO) undertook a soil resource inventory throughout the continent, with assistance from private organizations. Incentives to initiate the study of soil fertility came, for example, from the Rockefeller Brothers IBEC Research Institute, which sponsored experiments on *cerrado* (savannah) soils of Brazil. Major contributors in 1960 and 1961 were L. M. M. de Freitas, W. L. Lott, A. C. McClung, and D. S. Mikkelsen. Responsibilities for soil resource inventories were often vested in cadastral or military institutes that issued their own publication series. An example is Colombia, where the Instituto Geográfico Militar y Cadastral began to map soils for tax assessment purposes in the 1940s. A book on soil classification and land evaluation was published by its director, J. V. Lafaurie Acosta, in 1946.[32]

Other soil resource agencies were integrated into ministries of agriculture. In Chile, work began in 1942.[33] The Comissão de Solos, in Brazil, started the systematic collection of soil information of this huge country in 1957. Its successors continued the work and are still active. Leading individuals were M. N. Camargo and R. Costa de Lemos in classification and M. V. Vettori in soil analyses. J. Bennema, D. L. Bramão,[34] and R. W. Simonson with FAO were closely associated with this work.[35]

There is no scope in these few pages to detail the exponential growth of soil science publications in the 1950s and 1960s. Interested readers should consult an article by M. Rodriguez on the development of soil science in

30. P. Lavenir, "Análisis de Tierras y Aguas (del Territorio de Santa Cruz)," *Boletín Agrícola y Ganadero* 22 (1901): 29–37.

31. Lavenir, "Contribución al Estudio de los Suelos de la República Argentina," *An. M. Agro. Química* 2 (2) (1910): 1–577.

32. J. V. Lafaurie Acosta, *Clasificación y Valoración de Tierras, Interpretación Ponderal del Suelo* (Bogotá: Editorial Centro, Inst. Gráfico Ltda., 1946).

33. C. Diaz Vial, "Desarollo de los Estudios de Suelos en Chile Durante en Decenio 1948–1958," *Agricultura Técnica* 28 (2) (1958): 59–77.

34. D. L. Bramão and P. Lemos, "Soil Map of South America," *Transactions of the 7th International Congress on Soil Science* 4 (1960): 1–10.

35. Edward C. J. Mohr and Van Baren, *Tropical Soils: A Critical Study of Soil Genesis as Related to Climate, Rock, and Vegetation* (New York: Interscience, 1954).

Latin America during the twenty-five years preceding 1967.[36] Illustrative is the growth of the soils literature of Costa Rica, which according to F. Bertsch expanded from 165 citations in the decade 1940–1949, to 190 in the next, to 400 in the 1960s and 880 in the 1970s, a level roughly maintained in the 1980s. Bertsch divided these fifty years of citations on Costa Rican soils into the following broad subject categories:[37]

Fertilizing and fertilizers	41.5%
Soil fertility and chemistry	25.2
Classification and genesis	15.2
Mineral nutrition	11.7
Physical properties	6.1

A large part of the soil science literature in Latin America resides in reports to government agencies that were produced under international technical assistance programs. These reports are often difficult to access because of limited distribution. The bibliographies by A. C. Orvedal, however, may help in tracing these publications.[38]

Linkages: Soil Resource Inventories, Soil Classification, Genesis

The United States Influence

Most United States pedologists who worked on overseas assignments had gained much experience in the Soil Survey of the United States Soil Conservation Service. Their work in the United States usually dealt with the rather detailed mapping of soils at 1:20,000 scale, based on systematic field observations. There was a strong commitment to serve farmers (see Chapter 11). Phases of soil series were practically the only units that were mapped. Climate was seldom a criterion on detailed maps and for the majority of surveyors, climatic attributes were not a priority concern. There were no reasons to modify the survey practices as they were conducted in the United States, and these were often transferred intact to the Third World countries.

There are many examples of transfers of soil survey techniques. H. H. Bennett and R. V. Allison mapped the soils of Cuba at a scale of 1: 800,000, with a legend that only recognized series.[39] The first criterion to

36. Rodriguez, "Desarollo de la Ciencia del Suelo."

37. F. Bertsch, *Bibliografía de Suelos de Costa Rica* (San José, Costa Rica: Univ. of Costa Rica, Centro de Investigaciones Agronómicas, 1987).

38. A. C. Orvedal, *Bibliography of the Soils of the Tropics*, Vol. 1–5 (Washington, D.C.: Office of Agriculture, Technical Assistance Bureau, AID, 1975–1983). (Technical Series Bulletin no. 17).

39. H. H. Bennett and R. V. Allison, *The Soils of Cuba* (Washington, D.C.: Tropical Plant Research Foundation, 1928).

identify series was color. Bennett visited Venezuela in 1942, where soil studies had started in 1936, and introduced the soil series concept as the basic map unit.[40] C. E. Simmons, J. M. Tarano, and J. H. Pinto made the soil map of Guatemala, distinguishing soil series grouped only by local names.[41] The series descriptions contained important land use information. In many instances, nationals followed the same United States guidelines, for example C. Dondoli and J. A. Torres mapped a part of Costa Rica at a scale of 1:59,000.[42] C. Diaz Vial also used series names to designate soils of Chile.[43]

The U.S. Soil Survey Staff *Soil Survey Manual*[44] and its predecessor by Charles E. Kellogg[45] was the publication that had the strongest impact on soil mapping in the Americas. The manual was completely translated into Spanish in Venezuela. It promoted consistency in the collection of soil survey data. The United States approach differed from French and Portuguese views by placing less emphasis on high level soil classification units and adhering more strongly to homeland guidelines, i.e., the U.S. Soil Survey.

Other United States authors who had a major impact in Third World countries were R. E. Storie;[46] R. D. Hockensmith and J. G. Steele[47] and A. A. Klingebiel[48] on land capability classifications. *USDA Agricultural Handbook* no. 60 on the diagnosis and reclamation of salt-affected soils, edited by L. A. Richards was a regularly consulted publication.[49] Feasibility studies of irrigation projects often followed the guidelines of the U.S. Bureau of Reclamation.

The above comments do not mean that broader approaches to soil classification, e.g., the impact of climate on soil characteristics, did not influence

40. E. Casanova Olivo, *Introducción a la Ciencia del Suelo* (Caracas: Central University of Venezuela, Faculty of Agronomy, 1991).

41. C. E. Simmons, J. M. Tarano T., and J. H. Pinto Z. *Clasificación de Reconocimiento de los Suelos de la República de Guatemala* (Guatemala, City: Inst. Agropecuario Nacional, Ministerio de Agricultura, Editorial del Ministerio de Educación Pública, 1959).

42. C. Dondoli B. and J. A. Torres M., *Estudio Geoagronómico de la Región Oriental de la Meseta Central* (San José, Costa Rica: Ministerio de Agricultura e Indústrias, 1954).

43. C. Diaz Vial, "Desarollo de los Estudios de Suelos en Chile."

44. U.S. Soil Survey Staff, *Soil Survey Manual* (Washington, D.C.: USDA, 1951 [USDA Agriculture Handbook no. 18]).

45. Charles E. Kellogg, *Soil Survey Manual* (Washington, D.C.: USDA, 1938 [USDA Misc. Publication no. 274]).

46. (a) R. E. Storie, "The Classification of Natural Land Divisions and the Application of this Classification to Land Use and Conservation," *Proceedings of the Sixth Pacific Science Congress* 4 (1940): 867–68. (b) Storie, "Rating Soils for Agricultural, Forest and Grazing Use," *Transactions of the 4th International Congress on Soil Science* 1 (1950): 336–39.

47. R. D. Hockensmith and J. G. Steele, *Principles of the Land-Capability Classification* (Washington, D.C.: USDA Soil Conservation Service, 1949).

48. A. A. Klingebiel, "Soil Survey Interpretation-Capability Groupings," *Soil Science Society of America Proceedings* 22 (1958): 160–63.

49. L. A. Richards, ed., *Diagnosis and Improvement of Saline and Alkaline Soils* (Washington, D.C.: U.S. Govt. Print. Off., 1954 [USDA Agriculture Handbook no. 60]).

soil scientists in the United States: Eugene Hilgard (1833–1916) divided soils into arid and humid, on the basis of moisture relationships.[50] C. F. Marbut (1863–1935), in his publication on the *Vegetation and Soils of Africa*, wrote that "the character of the soil at maturity is due mainly to climatic forces and to the character of its native vegetation."[51] Marbut's soil classification system published in 1935[52] was a complete change from earlier versions based on geology and weathering.[53] In general, however, soil maps produced under what may be called the American school placed most emphasis on the properties of soil units (e.g., soil series) that were identified during detailed surveys. More generalized units that are used for small-scale maps were often considered of academic interest and were seldom the object of more than a small section in the soil survey reports.

The European Influence in the Nineteenth and Early Twentieth Centuries

The European influence on the Third World soil science literature is characterized by its diversity. At the turn of the century, the knowledge acquired by soil scientists in Europe was based on experience gained in temperate regions. Tropical soils had no place in the soil science literature. When information on soils of low latitudes became available, European pedologists had already acquired strong views on the structure of their science and had adopted concepts and principles that best conformed with their knowledge.

Third World studies on soil properties in the nineteenth and early twentieth centuries came essentially from geologists, mineralogists, and chemists. They were mostly published in their respective professional journals. New findings on soils of warm or frost-free climates were essentially point data. Before World War II, practically no detailed soil maps were available; it was difficult to assess the geographic extent of the soil features that were described. The properties that attracted most attention among pedologists in Europe were those that showed striking differences from properties of temperate soils. Authors often focused on aspects closely related to their area of specialization: chemistry or mineralogy.

There was a firm belief among European scientists that climate was the primary factor that imparted to soils their major characteristics. Vasilii V.

50. More details on the ideas of Eugene Hilgard are given in Chapter 11.

51. H. L. Shantz and C. F. Marbut, *The Vegetation and Soils of Africa* (New York: National Research Council and American Geographical Society, 1923), p. 138.

52. Marbut, "Soils of the United States," *USDA Atlas of American Agriculture*, Part III (Washington, D.C.: U.S. Govt. Print. Off., 1935), pp. 11–14.

53. Marbut et al., *Soils of the United States* (Washington, D.C.: USDA, 1913 [USDA Bureau Soils Bulletin no. 96]).

Dokuchaev (1846–1903) had been the leading soil scientist in this respect, strongly influencing the thinking of soil geographers during the early 1900s. It was commonly accepted that the present climate was the most important soil forming factor which differentiated soils at the highest level of generalization and gave each climatic region its characteristic soil. J. S. Joffe, for example, wrote that "in the humid tropics and subtropics the reactions involved in the processes of soil formation differ from those in any other climatic zone."[54]

In western Europe, German pedologists such as Emil Ramann (1851–1926) adhered to Dokuchaev's views, as did French soil scientists. For French-speaking Africa, Aubert,[55] at ORSTOM, developed his approach to soil geography along the same lines. At a later stage, Aubert and Philippe Duchaufour similarly published their soil classification system intended to encompass world conditions.[56] There was a strong belief that climatic regions and associated vegetation belts determined the general distribution of major soil groups in the world, and, because of this, there should be a fundamental difference between soils of the tropics and soils of the temperate regions. European views were that tropical and nontropical soils belonged to two different kingdoms, each having their own set of processes, characteristics, and subdivisions.

It is interesting to note that, during that period, many names were coined to designate complex soil forming processes. They were used to explain the presence of soil properties that characterized "zonal" soils. Usually the processes were not well understood in terms of chemistry, mineralogy, or physics. For example, laterization was the process that ultimately would bring all mature soils in the tropics to harden into ironstone; friable red or yellow soils were considered immature.[57]

The attitudes that prevailed in the first half of the twentieth century stimulated the independent thinking of European soil scientists working abroad. As stated before, they were often isolated, with very few opportunities to benefit from contacts with other scientists. The autonomy of the European soil science schools was further enhanced by the diversity of cultures and languages that characterizes the old continent. These conditions all contributed to individuality: first, independence from the rules that governed the classification of temperate soils (at any rate, none of the homelands had

54. J. S. Joffe, *Pedology,* 2d ed. (New Brunswick, N.J.: Pedology Publications, 1949). (1st ed. 1936), p. 101.
55. Aubert, *Les Sols de la France d'Outre-Mer* (Paris: Imprimerie Nationale, 1941).
56. Aubert and Philippe Duchaufour, "Projet de Classification des Sols," *Comptes Rendus VI Congrés Int. des Sols* (Paris) E (1956): 507–604.
57. J. M. Campbell, "Laterite: Its Origin, Structure and Minerals," *Mineralogical Magazine* 17 (1917): 67–77, 120–28, 171–79, 220–29.

tropical soils); and second, soils were to be studied for the sake of an independent soil science, not necessarily at the service of any other discipline. This attitude, however, often varied from one individual to another.

Much of the interest of European overseas pedologists was directed to Africa. Reproduced here is a list of the scientists who contributed to the *Soil Map of Africa* by J. L. D'hoore.[58] It is a good illustration of the strong European influence on tropical soil science before most countries on that continent achieved independence.

The soils map of Africa was constructed with the direct collaboration of the following authors:

GENERAL LEGEND: G. Aubert, H. Greene, J.V. Botelho da Costa. AFRICA OCCIDEN TAL ESPAÑOLA : P. Dutil. ALGÉRIE : J. Boulaine, J. Durand, P. Dutil. ANGOLA : J.V. Botelho da Costa, A.L. Azevedo, R. Pinto Ricardo, E.P. Cardoso Franco, E.M. Silva da Camara. BASUTOLAND : C.R. Van der Merwe, R.F. Loxton. BE CHUANALAND : J.S. De Beer, R.F. Loxton, C.R. Van Straten, C.R. Van der Merwe. BURUNDI : A. Van Wambeke, R. Frankart. CAMEROUN : P. Segalen. CONGO (BRAZZAVILLE) : J.M. Brugière, G. Bocquier. CONGO (LÉOPOLDVILLE) : K. Sys, R. Frankart, P. Jongen, A. Van Wambeke, P. Gilson, J.M. Berce, A. Pecrot, M. Jamagne. CÔTE 'IVOIRE : N. Leneuf, B. Dabin, G. Riou. CÔTE FRAN ÇAISE DES SOMALIS : H. Besairie, J. Boulaine. DAHOMEY : R. Fauck, P. Willaime. ETHIOPIA : R.E.G. Pichi-Sermolli, Anonyme Doc. F.A.O. GABON : Y. Chatelin. GAMBIA : M. Brunt. GHANA : C.F. Charter, H. Brammer. GUINÉE : R. Maignien. GUINEA CONTINENTAL ESPAÑOLA : A. Hoyos de Castro, W. Kubiena. GUINÉ POR TUGUESA : P.J. da Silva Teixeira. HAUTE VOLTA : R. Maignien, M.G. Gaveau. KENYA : G.H. Gethin Jones, R. Scott, C.G. Trapnell, J. Thorp. LIBERIA : W. Reed. LIBYA : P. Dutil. MADAGASCAR : J. Riquier, J. Hervieu. MALI : B. Dabin, B.K. Kaloga. MAROC : J. Boulaine, P.B. Billaux, G.B. Bryssine. MAURITANIE : P.A. Audry, S.P. Perreira Barreto, M.G. Gaveau. MOÇAMBIQUE : G.D.H. Gouveia, A.L. Azevedo. NIGER : G. Bocquier, M.G. Gaveau. NIGERIA : H. Vine, G.M. Higgins, A.W.S. Mould, A.J. Smyth, P.R. Tomlinson, P.D. Jungerius, C.H. Ob-ihara. NYASALAND : G. Jackson, A. Young. RÉPUBLIQUE CENTR 'AFRICAINE : M. Benoit-Janin, P. Quantin. REPUBLIC OF SOUTH AFRICA : C.R. Van der Merwe, R.F. Loxton. RHODESIA (NORTHERN) : R. Webster, C.G. Trapnell. RHODESIA (SOUTHERN) : B.S. Ellis, J.G. Thompson, RWANDA : A. Van Wambeke, R. Frankart. SÉNÉGAL : P.B. Bonfils, C.C. Charreau, J. Maymard, R. Maignien, R. Fauck. SIERRA LEONE : A.R. Stobbs, R. Allbrook. SOMALIA : R.E.G. Pichi-Sermolli, R. Dudal. SOUTH WEST AFRICA : R. Ganssen, C.R. Van der Merwe, R.F. Loxton. SUDAN : G. Aubert, H. Greene, G. Worrall. SWAZILAND : G. Murdoch, C.R. VAn

58. J. L. D'hoore, *Soil Map of Africa* (Lagos, Nigeria: Commission for Technical Cooperation in Africa, 1964).

der Merwe, R.F. Loxton. TANGANYIKA : G.H. Gethin Jones, W.E. Calton, R. Scott. TCHAD : J. Pias. TOGO : M. Lamouroux. TUNISIE : P. Roederer. UGANDA : E.M. Chenery, C. Ollier, J.F. Harrop, J.G. Wilson, S.A. Radwansky.

There were two major trends in the approach to soil geography: one placing emphasis on "high level" or broadly defined units and one favoring "low level" or more narrowly defined units. The position taken by the first group was described by G. W. Robinson: "they seemed to follow the philosophy of neo-Platonists, who accept the existence of metaphysical entities (i.e., soil groups) of which all examples observed in the field were just occasional manifestations."[59] Most soil scientists at that time referred their low-level units to "great soil gropus" which were not precisely defined or mutually exclusive. It was generally accepted that there was a gap, or hiatus, between high-level and low-level soil classes. More recent classifications, as the *7th Approximation*[60] and *Soil Taxonomy*,[61] closed the hiatus by creating one comprehensive, multicategoric system with mutually exclusive classes.

Of course, there were marked differences among pedological schools in Europe, even within the same homeland. Whether the emphasis was placed on high-level or low-level units varied from one country to another. High-level units prevailed in France and Portugal, who administered large territories in Africa; British approaches often favored units that had more local practical values. French and Portuguese pedologists strongly adhered to the principle that soils cannot be identified unless their genesis is understood. Consequently, considerable research on soil formation processes was carried out.

Soil scientists of the United Kingdom apparently placed less emphasis on global classification schemes. One of the major concerns expressed at the First Commonwealth Conference on Tropical and Sub-tropical Soils, in 1948, was that "workers in tropical areas should concentrate on the collection of information without unduly concerning themselves with the problems of soil classification."[62] Most publications of the Commonwealth Bureau of Soil Science were either site specific or problem oriented. For

59. (a) G. W. Robinson, "Some Considerations on Soil Classification," *Journal of Soil Science* 1 (2) (1950): 150–55. (b) Commonwealth Bureau of Soil Science, *Proceedings of the 1st Commonwealth Conference on Tropical and Sub-Tropical Soils, 1948* (Harpenden, U.K., 1949 [Technical Communication no. 46]), p. 132.

60. USDA, Soil Survey Staff, *Soil Classification, A Comprehensive System, 7th Approximation* (Washington, D.C.: USDA Soil Cons. Serv., U.S. Govt. Print. Off., 1960).

61. USDA, Soil Survey Staff, *Soil Taxonomy: A Basic System of Soil Classification for Making and Interpreting Soil Surveys* (Washington, D.C.: U.S. Govt. Print. Off., 1975 [USDA Agriculture Handbook no. 436]).

62. Commonwealth Bureau of Soil Science, p. 104.

example, J. B. Harrison discussed the weathering of igneous rocks,[63] J. A. Prescott and R. L. Pendleton, laterites.[64] British soil scientists seldom ascribed to systems, because they were thought to be incomplete, rigid, and potentially prejudicial toward future growth. However, serious efforts were made to achieve uniformity in the descriptive terms that were used to collect data.

Typical of British soil resource inventories were those produced by the Regional Research Centre of the British Caribbean and published by the Imperial College of Tropical Agriculture in Trinidad between 1958 and 1966. Each survey area was mapped in splendid isolation by more than thirty scientists, who set up their own legends on each island and had not reference to the work of their colleagues.[65] The inventories, however, contained valuable information important for land use. The Colonial Office also published books on the land resources of some of their colonies, for example, in British Honduras (now Belize), by A. C. S. Wright et al.[66] More than ten years elapsed between the decision to conduct the land survey (1948) and the publication of the book.

Tropical Soils

The first books on tropical soils as a whole were probably those written by Paul Vageler[67] and Edward Mohr.[68] Early descriptions of tropical podzols with their highly acidic raw humus were those of H. J. Hardon[69] and P. W. Richards.[70]

Three attributes of tropical soils were of major interest to European pedologists in the early 1900s: (1) extreme soil depths that were considered the product of very intense weathering processes; (2) the occurrence of soil

63. J. B. Harrison, *The Katamorphism of Igneous Rocks under Humid Tropical Conditions* (Harpenden, U.K.: Imperial Bureau of Soil Science, 1933).

64. J. A. Prescott and R. L. Pendleton, *Laterite and Lateritic Soils* (1952 [Commonwealth Bureau of Soil Science Technical Communication no. 47]).

65. N. Ahmad, "Foreword," in *Soil and Land Use Surveys No. 27*, G. D. Smith (St. Augustine, Trinidad: Dept. of Soil Science, Faculty of Agriculture, Univ. of the West Indies, 1982).

66. A. C. S. Wright, D. H. Romney, R. H. Arbuckle, and V. E. Vial, *Land in British Honduras* (London: Colonial Office, 1959 [Colonial Research Publication no. 24]).

67. (a) Paul Vageler, *Grundriss der Tropischen und Subtropischen Bodenkunde* (Berlin: Ferlargsgesellsch. für Ackerbau M. B. H., 1930). (b) Vageler, *An Introduction to Tropical Soils*, trans. H. Greene (London: Macmillan, 1933).

68. (a) Mohr, *Tropical Soils*. (b) Mohr, *The Soils of Equatorial Regions with Special Reference to the Netherlands East Indies*, trans. Robert Pendleton (Ann Arbor, Mich.: Edwards Bros., 1944) (Translation of *De Bodem der Tropen in het Algemeen, en die van Nederlandsch Indie in het Bijzonder*, Amsterdam, 1933).

69. H. J. Hardon, "An Example of Podsol in Tropical Lowlands," *Pedology* 3 (1938): 325–31.

70. P. W. Richards, "Lowland Tropical Podzols and Their Vegetation," *Nature* 148 (1941): 129–31.

materials that indurated irreversibly into ironstone upon drying, which Buchanan had described much earlier and called laterites;[71] and (3) the bright yellow or reddish colors generally attributed to iron oxides forming at high temperatures.

Investigations on ironstone crusts and on the deep weathering profiles associated with them were published by Harrison,[72] Lacroix,[73] and, subsequently, many others (114 between 1936 and 1962).[74] Molar SiO_2 to Al_2O_3 ratios were proposed by H. Harrassowitz as diagnostic criteria of soil development or weathering intensities.[75] The low ratios observed in soils of warm climates were considered to be the result of removal of silica by large quantities of rainwater percolating through the profiles. The molar silica to aluminum or silica to sesquioxide ratios that were proposed during this initial phase of tropical soil science are still the basis of some classifications, although many critical chemical levels have since been replaced by clay-mineralogical data that were not available at the beginning of the twentieth century.

Other contributions to soil science that originated in Third World countries should be mentioned. For example, G. Milne created the concept of soil catenas used for naming map units in East Africa.[76] Silica pans were first recognized in Zambia.[77] Hans Jenny et al. strongly influenced research on the dynamics of soil organic matter in tropical soils after publishing the results of their investigations in Colombia.[78]

This section on soils of the tropics would be incomplete without mentioning contributions from Soviet soil scientists which began approximately in the 1960s. B. G. Rozanov and I. M. Rozanova described soils of Burma.[79] S. V. Zonn studied the genesis of soils of southern China.[80]

71. Buchanan, "A Journey from Madras."

72. Harrison, "The Residual Earths of British Guiana Commonly Called Laterite," *Geological Magazine* 5 (7) (1910): 439–52, 488–95, 553–62.

73. A. Lacroix, *Les Laterites de la Guiné et les Produits d'Altération qui leurs sont Associés,* ser. 5, V (Paris: Nouv. Arch. Mus. Hist. Nat., 1914), pp. 255–356.

74. Commonwealth Bureau of Soil Science, *Bibliography of Laterite (1962–1963)* (Harpenden, U.K.: 1963).

75. H. Harrassowitz, "Laterit," *Forschritte der Geologie und Paleontologie* 4 (1926): 253–566.

76. (a) G. Milne, "Some Suggested Units of Classification and Mapping Particularly for East African Soils," *Soil Research* 4 (2) (1934): 183–98. (b) Milne, "Composite Units for the Mapping of Complex Soil Associations," *Transactions of the 3d International Congress on Soil Science* 1 (1935): 266–70.

77. G. W. Lamplugh, *Geological Magazine* 9 (1902): 575.

78. Hans Jenny, S. P. Gessel, and F. T. Bingham, "Comparative Study of Decomposition Rates of Organic Matter in Temperate and Tropical Regions," *Soil Science* 68 (1949): 419–32.

79. B. G. Rozanov and I. M. Rozanova, "Soils of the Mountainous Subtropics and of the High Mountains of Burma," *Vestnik Moskovskogo Universiteta, Seriia, Biologiia Pochvovedenie* 3 (1963): 71–78.

80. S. V. Zonn, "Some Problems of the Development of Soils on the Red Erosion Crust in Georgia and Southern China," *Acta Pedol. Sin.* 6 (1958): 52–53.

Mediterranean Soils

The red soils of the Mediterranean basin, often referred to as *terra rossa* and found in southern Europe, northern Africa, and the Middle East, attracted much debate and controversy. Many soil scientists collected data on these soils and on properties associated with them, such as horizons of calcium carbonate accumulation. A few examples give some important dates in the development of Mediterranean soil science: thesis research was conducted at the University of Algiers;[81] J. Agafonoff published a review of the soils of Tunisia in the *Annals of the Botanical and Agricultural Service of Tunisia*;[82] G. Bryssine published most of his findings on the soils of Morocco in an agronomic research publication in Rabat in 1959;[83] and the soil map of Lebanon was published by the ministry of agriculture in Beirut.[84]

Soil Management

The soil management literature is as diverse as the literature of soil geography. It is generally commodity oriented and often published in journals that specialize in crop sciences. A large part of it deals with industrial commodities. Pioneering work in various disciplines of soil science was achieved in Africa by research on shifting cultivation. Major contributions were published by P. H. Nye and D. J. Greenland[85] and H. Laudelout.[86] F. Jurion and J. Henry summarized the Belgian research experience with intensifying shifting cultivation in Zaire.[87] An important part of the investigations dealt with the inclusion of cash crops in the rotations and the maintenance of soil fertility, essentially with minimum inputs.

Soil management studies included more aggressive development practices which attracted much attention. One of them was the Sudan Gezira scheme; its soil problems were described in the 1930s by H. Greene.

81. (a) J. Durand, "Etude Géologique, Hydrologique et Pédologique des Crôutes en Algérie," Thesis, Univ. of Algiers, SES, 1953. (b) Jean Boulaine, "Lessols des Plaines du Chélif," Thesis, Univ. of Algiers, SES, 1957.

82. (a) V. Agafonoff, *Les Sols de France, au Point de Vue Pédologique* (Paris: Dunod, 1936). (b) Agafonoff, "Les Sols Types de Tunisie," *Annales du Service Botanique et Agricole de Tunisie* 12–13 (1936): 319–51.

83. G. Bryssine, "Les Facteurs Climatiques de la Pédologie au Maroc," *Cahiers de Recherches Agronomiques* 2 (1949): 43–70.

84. B. Geze, *Carte de Reconnaissance des Sols du Liban au 1/200.000,* Notice Explicative (Beirut: Ministry of Agriculture, 1959).

85. P. H. Nye and D. J. Greenland, *The Soil under Shifting Cultivation* (1960 [Commonwealth Bureau of Soils Technical Communication no. 51]).

86. H. Laudelout, *Dynamics of Tropical Soils in Relation to Their Fallowing Techniques* (Rome: FAO, 1961).

87. F. Jurion and J. Henry, *Can Primitive Farming be Modernised?* (Brussels: INEAC Publication, Hors Serie, 1969).

In the Caribbean area and Central America, the Imperial College of Agriculture, founded in 1921 in Trinidad, and the Puerto Rico Agricultural Experiment Station generated a considerable body of knowledge on the management properties of the soils of the area. J. A. Bonnet, M. A. Lugo Lopez, F. Abruña, and J. Vicente-Chandler published most of their work in the Puerto Rico Agricultural Experiment Station Bulletins and the *Journal of Agriculture of the University of Puerto Rico*. The Imperial College, with F. Hardy as its major contributor, published most of its soil science papers in *Tropical Agriculture*, a journal founded in 1924. In Brazil, the Instituto Agronomico, in Campinas, and the Escola Superior de Agricultura 'Luiz de Queiroz,' in Piracicaba, published most of their soil papers in the journal *Bragantia*, founded in 1941.

Erosion and soil conservation were major areas of concern. A. F. Suarez de Castro, in Colombia;[88] J. C. Bhattacharjee and K. M. Mehta, in India; C. W. Rose, in East Africa; and F. Fournier started work on these subjects before 1964. The influence of soil structure and seed bed preparation on soil and water conservation was investigated by H. C. Pereira and co-workers in East Africa.

Land-use planning and land evaluation are part of soil management. J. P. Mamisao published a paper on this subject in the first issue of the *Journal of the Soil Science Society of the Philippines* in 1949.[89]

Soil Biology

Soil biology is the most site specific among the soil disciplines, and, for that reason, tropical soil microbiology contributes substantially to the Third World soil science literature. Termites and their influence on soil properties are an example. T. E. Snyder lists 5,809 publications from 1350 B.C. to 1965 on termites; of course, not all are related to soils.[90] W. V. Harris during 1949–70, and J. P. Watson, in the 1960s, in East Africa; P. Boyer (1948–59), in West Africa; and Pendleton, in Thailand, authored papers on soils influenced by termite activities.

Biological nitrogen fixation is a much researched topic for soil fertility maintenance. The free-living, tropical, nitrogen-fixing genus *Beijerinckia* was isolated in 1936 from Malayan soils by R. A. Altson.[91] In Brazil, it was

88. F. Suarez de Castro, *Conservación de Suelos*, 1st ed. (1956).

89. J. P. Mamisao, "Land-Use Planning for Soil Conservation Farming," *Journal of the Soil Science Society of the Philippines* 1 (1949): 47–54.

90. T. E. Snyder, *Second Supplement to the Annotated Subject-Heading Bibliography of Termites, 1961–1965* (Washington, D.C.: Smithsonian Institution, 1968).

91. R. A. Altson, "Studies on Azotobacter in Malayan Soils," *Journal of Agricultural Science* 26 (1936): 268–80.

first described by J. Döbereiner and A. F. de Castro in 1955.[92] Symbiotic nitrogen fixation with legumes, first referenced in 1930, became a major area of interest in tropical agriculture only in the 1960s, with research aimed at the conditions under which *Rhizobium* strains operate most effectively in tropical soils.[93]

Soil microbiology research fluctuates in time according to needs. The diagram by P. S. Nutman (Figure 6.2) shows the cumulative number of reports and research papers on various aspects of soil microbiology published in *Tropical Agriculture* from 1924 to 1981.[94] The horizontal lines in the figure denote lack of interest: an example is the absence of reports on organic manures and nitrogen cycles between 1940 and 1968 that coincides with the increasing use of chemical fertilizers.

Studies on nitrogen dynamics in east African soils were summarized by J. Meiklejohn.[95] The effects of wetting and drying on the release of nitrogen by decomposing organic matter was described by H. F. Birch for East African conditions.[96] The nitrogen flush at the beginning of wet seasons in tropical areas is still often referred to as the "Birch effect." Research conducted on the biology of tropical soils in central Africa was summarized by J. Meyer and Laudelout in 1960.[97] Y. Dommergues was particularly interested in interactions between the soil microflora and environmental conditions in West African soils.

C. Closing Remarks

There is no doubt that a significant part of the Third World soil science literature is missing in this pre-1965 review. Some significant initiatives taken at the end of that period are not accounted for. The *World Soil Resources Reports* of the FAO started its series in 1961 and has, at present, published sixty-six monographs. The preparation of the FAO-UNESCO *Soil Map of the World* also began in 1961 and generated, during twenty years,

92. J. Döbereiner and A. F. de Castro, "Ocorrencia e Capacidade de Fixação de Nitrogenio de Bacterias do Genero Beijericnkia nas Series de Solos da Area Territorial do Centro Nacional de Ensino e Pesquisas Agronomicas," *Bol. Inst. Ecol. Expl. Agr.* 16 (1955).

93. P. S. Nutman, "Microbiology of Tropical Soils—A Historical Review," *Tropical Agriculture* (Trinidad) 63 (2) (1986): 90–94.

94. Ibid.

95. J. Meiklejohn, "Nitrogen Problems in Tropical Soils," *Soils and Fertilizers* 18 (1955): 459–63.

96. (a) H. F. Birch, "Pattern of Humus Decomposition in East African Soils," *Nature* (London) 181 (1958): 788. (b) Birch, "Further Observations on Humus Decomposition and Nitrification," *Plant and Soil* 11 (1959): 262–86.

97. J. Meyer and Laudelout, "The Biology of Tropical Soils," *Agricultura, Louvain* 8 (1960): 567–94.

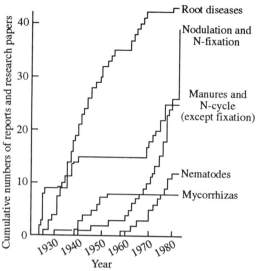

Figure 6.2. Cumulative number of papers published in *Tropical Agriculture* on various aspects of soil microbiology. (*Source:* P. S. Nutman, "Microbiology of Tropical Soils— A Historical Review," *Tropical Agriculture* [Trinidad] 63 [2] [1986]: 90–94. By permission of author and publisher.)

an abundance of literature. In the 1950s, the Soil Conservation Service of the United States Department of Agriculture started work on *Soil Taxonomy* which stimulated a large interest in soil systematics and which still continues. There were parallel initiatives at the national level in Third World countries.

There has been little room to acknowledge the activities of teaching and research institutions that had a considerable impact on the training of national soil scientists from Third World countries. In addition to the United States universities, one should mention the Soil Museum in Wageningen, the Netherlands, created on January 1, 1966; the International Institute for Aerospace Survey and Earth Sciences in Delft, later in Enschede, the Netherlands; ORSTOM in France; and the International Training Centre for Postgraduate Soil Scientists in Ghent, Belgium.

The 1960s witnessed the beginning of an increased growth in soils literature in the Third World. Chapter 5, by S. W. Buol, reports on these developments.

D. Acknowledgments

The assistance of Hans Van Baren, at the International Soil Reference and Information Centre (ISRIC), in Wageningen, is gratefully acknowledged for providing quick references on early contributions to soil science in Indonesia. I. P. Abrol, of the India Council of Agricultural Research, helped to identify sources of information on Indian soil science literature. The history of Mexican soil science was taken from an unpublished paper by R. J. Laird, made available by N. Aguilera Herrera.

7. Soil Science Societies and Their Publishing Influence

W. E. LARSON

University of Minnesota

Research scientists, by nature and training, take satisfaction in developing new knowledge. To be successful, they must remain aware of the current information through reading and oral communication. Exchange of ideas and thoughts among scientists has always been stimulating.

Progress in science occurs when one new discovery builds on another. In this way, the findings from one laboratory can be incorporated with the findings from other laboratories to create an increasing bank of knowledge. Combining the information requires a carefully written account of methodologies and discoveries that is available in a library for all to read. In addition, documentation of the activities of a laboratory or scientist helps prevent duplication of effort.

In the early nineteenth century, scientists could communicate directly among themselves, since their numbers were few. In these early days, however, duplicate research findings by widely separated scientists sometimes went undetected for significant periods of time. As scientists became more numerous, regular meetings were held where they could present their findings and exchange ideas. Even in the early stages of science, it was desirable to provide written accounts of research for others to reference and critique.

A. Scientific Societies

Early scientific societies were founded to satisfy the need for oral discussions and generate written accounts of scientists' work. These early societies were broad in scope, often including all of science or natural science. While the scope has narrowed over the past century and the organization of professional societies has become more complex, the basic objective has remained the same—to provide a forum for the exchange of information, oral and written.

G. M. Browning, writing in the *Journal of Soil and Water Conservation*, listed ten major objectives common to all professional societies, with various societies concentrating on different emphases.[1] These objectives are

(1) promote basic and applied research,
(2) foster high standards in education,
(3) provide information to members and others,
(4) encourage professional improvement,
(5) provide and stimulate a friendly forum,
(6) recognize meritorious contributions,
(7) develop and maintain high ethical standards,
(8) protect the rights and interests of members,
(9) provide information to aid public policy formulation, and
(10) cooperate with other organizations.

Other roles for societies include establishing standards, uniform nomenclature, units of measurement, provisions for informal communications during meetings, and recruitment services for employees and employers.

Soil science emerged as a separate entity early in the twentieth century.[2] While formal and informal meetings and discussions were held dating back into the nineteenth century, formally organized and recognized societies dedicated to soil science or agronomy were not formed until the early 1900s. The first International Conference of Soil Scientists met in Budapest, April 14–24, 1909, with later meetings at Stockholm, in 1910; St. Petersburg, Russia, in 1914; and Rome in 1924.[3] However, a formal society, the International Society of Soil Science (ISSS) was not organized until 1924, with J. G. Lipman of Rutgers University as the first president. The first congress of the ISSS was held in Washington, D.C., in 1927. The ISSS usually held its congresses every four years, with some irregularity during World Wars I and II. The president of the society has usually come from the country hosting the meeting.

The American Society of Agronomy (ASA), which included soils and crops, was organized in 1907.[4] However, as thesociety grew and scientists became more specialized, ASA divided its annual program into soils and

1. G. M. Browning, "The Role of Professional Societies in Resource Development," *Journal of Soil and Water Conservation* 17 (1962): 59–61.
2. A. J. Moffat, "William Cobbett: Politician and Soil Scientist," *Geography Journal* 151 (1985): 351–55.
3. J. F. Lutz, "History of the Soil Science Society of America," *Soil Science Society of America Journal* 41 (1977): 152–73.
4. Ibid.

crops in 1924. In 1931, the ASA reorganized and formally recognized soils and crops sections. The first meeting of what eventually was called the American Soil Survey Association (ASSA) was held on February 7–8, 1920. The ASSA's formation reflects the early emphasis on soil survey and the strong leadership for soil survey in the United States and Europe. In 1924, ASA made provisions for affiliated societies and extended an invitation to the ASSA. This invitation was accepted, and the affiliation continued until 1936, when the ASSA and the soils section of the ASA were combined under the ASA to form the Soil Science Society of America (SSSA). In 1937, the SSSA took over the activities of the American section of the ISSS.

The combining of the soils section of the SSSA, the ASA, and the American section of the ISSS into the SSSA resulted from a need for better communication among all soil scientists and a desire for improved services for members at reasonable costs. The first president of the SSSA was Richard Bradfield, originally from Ohio State University, who later went to Cornell University, where a building has been named for him.

The Great Plains dust bowl in the United States in the 1930s, as well as severe water erosion in many parts of the country, forcibly brought to the attention of the U.S. Congress the need for a national effort in soil erosion control. Congress established the Soil Erosion Service in 1933 as a temporary agency in the U.S. Department of Interior. It was made a permanent agency in 1935, moved to the USDA, and renamed the Soil Conservation Service (SCS). While the primary activity of the SCS was providing technical information to farmers, the agency also included soil survey and research divisions. H. H. Bennett, a soil scientist, is usually credited with providing the leadership for the establishment of the Soil Erosion Service and the SCS, and was chief of both from 1933 to 1951. The strong United States movement in soil erosion control during the 1930s and early 1940s resulted in the formation, in 1945, of the Soil Conservation Society (now the Soil and Water Conservation Society, SWCS). The interest in erosion control quickly spread to other countries.

The ASA and the SSSA grew rapidly in the 1940s, particularly after World War II. Rapidly expanded use of fertilizers and increasing crop yields occurred during this period. The need for food in war-ravaged countries made agriculture a highly visible industry worldwide, with a great need for new knowledge and increased numbers of scientists. To meet this growth and the need for expanded services to members, a headquarters for the ASA and the SSSA was established in Madison, Wisconsin, in 1945, with a full-time executive secretary. The staff worked in rental quarters until 1960, when a society-owned headquarters was built.

To coordinate soil erosion control and management, research, and education on a world basis, the World Association of Soil and Water Conservation (WASWC) was formed in 1983 in Ankeny, Iowa, and now has over 500 members. The WASWC is led by an international president, an executive secretary, and a number of regional vice presidents.

Several national societies of importance are worthy of mention. The Sociedade Brasileira de Ciência do Solo was founded in 1947 in Brazil. The society holds regular meetings for its members and publishes a journal which is organized into subject matter commissions. The French Association Française pour l'Etude du Sol, founded in 1934, is organized into Dojohiryo Gakkai was organized in Japan in 1927 and publishes a journal. A Soil Science Society of Sri Lanka was organized in 1979 and has published some literature. Polskie Towarzystwo Gleborznawcze, organized in 1934, holds meetings in Poland for its members and publishes a soil science annual, *Roczhiki Gleborznawcze* (since 1970, four numbers per year), and a review of scientific literature in agriculture and forestry (four numbers per year). The scientific committees resemble the commissions of the ISSS. The Indian Society of Soil Science was founded in 1934 in Calcutta. B. C. Burt was the first president and J. N. Mukherjee, the first secretary. The objectives of the Indian Society are (a) to cultivate, promote, and disseminate knowledge of soil science, (b) to publish journals and reports fulfilling the objectives of the society, (c) to work with the ISSS and other organizations having similar objectives, and (d) to assist in the fulfillment of the previous objectives. The *Journal of the Indian Society of Soil Science* was started in 1953, with issuance every 6 months, but, in 1956, was made a quarterly. The society has published a large number of bulletins, proceedings, and reviews. To help fulfill the objectives of the society and to increase membership participation, a large number of chapters at various locations have been formed (twenty-four as of 1984). An annual convention is held for sharing members' papers, giving special topic presentations, presenting awards, and a holding business meeting (Indian Society of Soil Science, 1984). The International Society of Soil Science held its twelfth congress in New Delhi in 1982. Table 7.1 lists the national soil science societies of the world and the dates of their formations.

From inception, the ISSS, ASA, and SSSA were societies formed to exchange research information and foster academic education. Somewhat in contrast, SWCS was initially formed to allow conservationists to improve their efficiency, effectiveness, and vocational competencies and to utilize their combined talents and influence in solving the problems facing conservation programs.[5] The goals and objectives of all of these societies have

5. H. W. Pritchard, "Glimpses from SCSA History," *Journal of Soil and Water Conservation* 20 (1965): 123–27.

Table 7.1. National soil science societies

Country	Society	Year of formation
Algeria	Association Algerienne de la Science du Sol	1989
Argentina	Associacion Argentina de la Ciência del Suelo	1960
Australia	Australian Society of Soil Science	1956
Austria	Österreichische Bodenkundliche Gesellschaft	1955
Bangladesh	Soil Science Society of Bangladesh	1958
Belgium	Belgische Bodemdundige Vereniging/Société de Pédologie	1950
Brazil	Sociedade Brasileira de Ciência do Solo	1947
Bulgaria	Bulgarian Society of Soil Science	1958
Canada	Canadian Society of Soil Science/Société Canadienne de la Science du Sol	1932/1954
Chile	Sociedad Chilena de la Ciência del Suelo	1973
China	Soil Science Society of China	1945
Colombia	Sociedad Colombiana de la Ciência del Suelo	1955
Costa Rica	Associación Costa-Ricana de la Ciência del Suelo	1979
Cuba	Soil Science Society of Cuba	1985
Czechoslovakia	Ceskoslovenska Pedólogicka Sekcia	1968
Denmark	Dansk Forening for Jordbundsvidenskab	1928
East Africa	Soil Science Society of East Africa	1975
Ecuador	Sociedad Ecuatoriana de la Ciência del Suelo	1977
Egypt	Soil Science Society of Egypt	1950
Ethiopia	Ethiopian Society of Soil Science	1990
Finland	Kansainvälisen Maaperäseuran Suomen Osasto	1971
France	Association Française pour l'Etude du Sol	1934
Germany	Deutsche Bodenkundliche Gesellschaft	1926
Ghana	Soil Science Society of Ghana	1964
Greece	Elliniki Edfologiki Eteria	1983
Hungary	MAE Talajtani Tarsasag	1957
India	Indian Society Soil Science	1934
Indonesia	Hiti Himpunan Ilmu Tanah Indonesia	1972
Iran	Soil Science Society of Iran	1975
Ireland	Soil Science Society of Ireland	1977
Israel	Israel Society of Soil Science	1950
Italy	Societá Italiana della Scienza del Suolo	1952
Japan	Nippon Dojohiryo Gakkai	1927
Korea, North	Korean Soil Science Association	
Korea, South	Hangug Toyang Biryo Haghoe	1968
Malaysia	Persatuan Sains Tanah Malaysia	1971
Mexico	Sociedad Mexicana de la Ciência del Suelo	1962
Netherlands	Nederlandse Bodemdundige Vereniging	1935
New Zealand	New Zealand Society of Soil Science	1952
Nigeria	Soil Science Society of Nigeria	1968
Norway	Norsk Forning for Jordfurskning	1981
Pakistan	Soil Science Society of Pakistan	1984
Philippines	Soil Science Society of the Philippines	1948
Poland	Polskie Towarzystwo Gleborznawcze	1937
Portugal	Sociedade Portuguesa de Ciéncia do Solo	1955
Romania	Societatea Nationala Romana Pentru Stinta Solului	1961
Senegal	Association Sénégalaise de la Science du Sol	1988

Table 7.1. (Continued)

Country	Society	Year of formation
South Africa	Grundkundevereniging van Suid-Afrika	1954
Spain	Sociedad Española de la Ciência del Suelo	1947
Sri Lanka	Soil Science Society of Sri Lanka	1979
Sudan	Sudanese Society of Soil Science	
Sweden	Svenska Markläresällskapet	1943
Switzerland	Bodenkundiliche Gesellschaft der Schweiz	1975
Tunisia	Association Tunisienne de la Science du Sol	1979
Turkey	Turkish Society of Soil Science	
United Kingdom	British Society of Soil Science	1947
U.S.A.	Soil Science Society of America	1936
U.S.S.R.	All-Union Society of Soil Scientists of the USSR	1939
Venezuela	Sociedad Venzolana de la Ciência del Suelo	1954
Yugoslavia	Jugoslovensko Drustvo za Proucavanje Zemljista	1953
Zambia	Soil Science Society of Zambia	1990
Zimbabwe	Soil Science Society of Zimbabwe	

Source: ISRIC, the Netherlands.

changed over the years, but, in broad terms, societies continue to foster the creation and dissemination of knowledge concerning soil science.

The ISSS now has about 7,000 members; the ASA, about 12,600; the SSSA, about 6,100 (also included as ASA members); and the SWCS, about 13,000 members. The tri-societies of ASA, CSSA (Crop Science Society of America), and SSSA now have about thirty full-time employees in their headquarters in Madison, Wis. The *American Registry of Certified Professionals in Agronomy, Crops, and Soils* (ARCPACS) was fostered by the tri-societies and now has about 1,900 certified professionals. Scientists can be certified in the following fields: Certified Professional Agronomist; Certified Professional Agronomist/Crop Scientist/Soil Scientist; Certified Professional Agronomist/Soil Scientist; Certified Professional Soil Scientist; Certified Professional Soil Scientist/Soil Classifier; and Certified Professional Soil Classifier. New categories recently initiated are Certified Professional Horticulturist, Certified Professional Plant Pathologist, and Certified Professional Weed Scientist. In addition, a Certified Crop Advisors Program will be implemented in 1993.

Certification by ARCPACS provides evidence of professional competence for those whose activities affect the well-being of the general public. Certification (a) encourages professional development, growth, and renewal; (b) enhances the visibility of the profession; (c) promotes high standards of performance; and (d) publicizes the Code of Ethics. Scholarly preparation, knowledge, and experience are necessary requirements for certification. Requirements for eligibility are (a) at least a baccalaureate degree in agronomy, crops, soils, a closely allied field of science, or equivalent, (b) completion of the minimum core requirements, and (c) five years of professional experience in the area of certification. If the baccalaureate degree does not meet the core requirements, passing a written exam may be necessary. Evidence of continued competence is required every five years.

In 1976, the SWCS and ARCPACS combined in a certification program for specialists in erosion control. Under this arrangement, SWCS has the responsibility for determining eligibility for certification, while the ARCPACS office handles the records and financial services. Professionals are certified as Erosion and Sediment Control Specialists.

Regional societies are organized in East and West Africa, while an East and Southeast Asia Federation of Soil Science Societies was established recently. Since 1940, the number of soil science societies has increased approximately linearly, at 1.0 per year (Figure 7.1).

B. Soil Science Journals

Currently, over 975 journals are given in the CABI *Soil and Fertilizers* monthly abstract journal, of which only seventy-eight are soils specific.

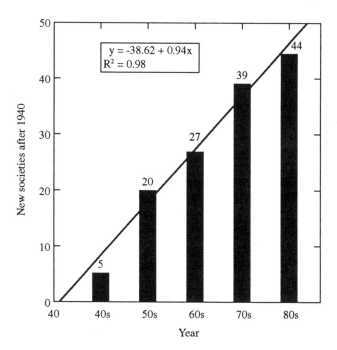

$$y = -38.62 + 0.94x$$
$$R^2 = 0.98$$

Figure 7.1. Soil science societies formed since 1940. The decades run, e.g., 1940–49.

Soil science societies publish a significant portion of the literature in soil science. Table 7.2 lists the fifteen most highly cited society journals in CABI databases. Eleven of these fifteen periodicals are issued by United States scientific societies, suggesting that publication of periodicals by scientific societies is most prevalent in the United States. About 18% of all citations from over 750 journals are from society-sponsored journals. Thus, about fifteen society journals (about 2% of all journals) account for almost one-fifth of the citations. According to Garfield the *Soil Science Society of America Journal* was the most cited of all soil science journals in 1989 (*SCI* database).[6]

From their initiation, most of the soils societies (e.g., ISSS, ASA, SSSA, SWCS) have published proceedings of their meetings. In some cases, e.g., SSSA, proceedings evolved into technical journals. In Table 7.2, all of the titles were first issued in the twentieth century, except for *Science*, first issued in 1883. All the journals in Table 7.2 that began prior to World War II were American, except the *Annals of Applied Biology* from the United Kingdom. The dates of first issue of a society journal often lag behind the date of establishment of the society, but may indicate when the society became strong. *Agronomy Journal* (initially, the *Journal of the American Society of Agronomy*) dates back to 1913, and the *Soil Science Society of America Journal* (initially, the *Proceedings of the Soil Science Society of America*) was first published in 1936. The ISSS continues to publish the technical papers and business activities of its congresses under the title *Transactions of the International Soil Science Society*. Initially, all of the papers presented at the congress were published in one or more volumes, called the *Transactions of the ISSS*. However, as the ISSS grew, there was a trend toward publication of only the invitational and key papers, along with abstracts of the unsolicited papers. The sheer volume of papers and cost of printing have been the primary restrictions for limiting material in the transactions.

C. Content of Society Periodicals

The organization and content of the journals of the scientific soil societies reflect the interest of the membership and, perhaps, their response to their information needs. The early interest, and indeed the demands of society,

6. Eugene Garfield, "The Effectiveness of American Society of Agronomy Journals: A Citationist's Perspective," in *Research Ethics, Manuscript Review and Journal Quality*, ed. H. F. Maryland and R. E. Sojka (Madison, Wis.: American Society of Agronomy, Crop Science Society of America, Soil Science Society of America, 1992).

Table 7.2. Fifteen leading science society journals by soil citation counts using *CAB Abstracts*

Journal	First published	Publisher and location
1. Agronomy Journal	1907–09	American Society of Agronomy, Madison, Wis.
2. Journal of Environmental Quality	1972	American Society of Agronomy, Madison, Wis.
3. Plant Physiology	1926	American Society of Plant Physiologists
4. Soil Science Society of America Journal	1936	Soil Science Society of America, Madison, Wis.
5. Journal of the Indian Society of Soil Science	1953	Indian Society of Soil Science, New Delhi
6. Ecology	1920	Ecological Society of America and the Duke University Press, Durham, N.C.
7. Journal of Soil and Water Conservation	1946	Soil Conservation Society of America, Ankeny, Iowa
8. Phytopathology	1911	American Phytopathological Society, St. Paul, Minn.
9. Science	1883	American Association for the Advancement of Science, Washington, D.C.
10. Transactions of the ASAE	1958	American Society of Agricultural Engineers, St. Joseph, Mich.
11. Annals of Applied Biology	1914	Association of Economic Biologists, London
12. Netherlands Journal of Agricultural Science	1953	Royal Netherlands Society for Agricultural Science, Wageningen
13. New Zealand Agricultural Science	1965	New Zealand Institute of Agricultural Science, Wellington
14. Crop Science	1961	Crop Science Society of America, Madison, Wis.
15. Acta Agriculturae Scandinavica	1950	Scandinavian Association of Agricultural Scientists and the Royal Swedish Academy of Agriculture and Forestry, Stockholm

Source: CAB Abstracts data, 1984–present.

was primarily concerned with soils as a medium for plant growth, as well as in soil classification. This emphasis continued until the last few decades, when soil scientists became concerned with soils for a wide variety of uses. The serial publications of twenty-four soil science societies are listed in Table 7.3. The society and publication names give an indication of the interests of society members.

Table 7.3. Soil science societies and their serial publications

Society	Place of publication	Publications
Agronomy Society of New Zealand	Christchurch, N.Z.	*Special Publications*
American Society of Agronomy	Madison, Wis.	*Agronomy Abstracts* *Agronomy Journal* *Agronomy News* *ASA Special Publications* *Journal of Agronomic Education* *Journal of Environmental Quality* *Journal of Production Agriculture*
Association Française pour l'Etude du Sol	Paris, France	*Bulletin*
Australian Society of Soil Science	Cowra, New South Wales	*Soils News*
British Society of Soil Science	Oxford, U.K.	*Soil Use and Management*
Canadian Society of Soil Science	Ottawa, Canada	*Canadian Journal of Soil Science*
Conservation Society of India	Hazarbagh, India	*Journal of Soil and Water*
Deutsche Bodenkunliche Gesellschaft	Weinheim, Germany	*Mitterlungen der Deutsche Bodenkundliche Gesellschaft* *Zeitschrift fur Planzenernahrung und Bodenkunde*
Egyptian Society of Soil Science	Alexandria, Egypt	*Egyptian Journal of Soil Science*
Fertilizer Society of London	London, U.K.	*Proceedings*
Indian Society of Soil Science	New Delhi, India	*Bulletin* *Journal of the Indian Society of Soil Science*
International Peat Society	Helsinki, Finland	*Bulletin*
International Society for Soilless Culture	Wageningen, Netherlands	*Soilless Culture*
International Society of Soil Science	Wageningen, Netherlands	*Bulletin*
Japanese Society of Science & Plant Nutrition	Tokyo, Japan	*Soil Science and Plant Nutrition*
New Zealand Society of Soil Science	Wellington, New Zealand	*New Zealand Soil News*
Sociedad Chilena de la Ciência del Suelo	Santiago, Chile	*Boletin*
Sociedad Brasileira de Ciência do Solo	Campinas, Brazil	*Revista Brasileira de Ciênca do Societe Solo*
Société Belge de Pédologie	Ghent	*Pédologie*

Society	Place of publication	Publications
Soil and Crop Science Society of Florida	Gainesville, Fla.	*Proceedings*
Soil and Water Conservation Society of America	Ankeny, Iowa	*Journal of Soil and Water Conservation*
Soil Science Society of America	Madison, Wis.	*Soil Science Society of America Journal*
		Soil Science Society of America Special Publication Series
Soil Science Society of Bangladesh	Dacca, Bangladesh	*Bangladesh Journal of Soil Science*
Soil Science Society of Ceylon (Sri Lanka)	Colombo, Sri Lanka	*Handbook of the Soils of Sri Lanka*
World Association of Soil and Water Conservation	Ankeny, Iowa	*Newsletter*

Source: *International Union List of Agricultural Serials* (pub., date).

The original constitution of the SSSA provided for six subject matter sections (later divisions): (1) Soil Physics, (2) Soil Chemistry, (3) Soil Microbiology, (4) Soil Fertility, (5) Soil Genesis, Morphology and Cartography, and (6) Soil Technology. These subject matter divisions followed the pattern of the ISSS.[7] Almost from the start, unofficial subdivisions were recognized. By 1970, the divisions had been renamed and added to: (S-1) Soil Physics, (S-2) Soil Chemistry, (S-3) Soil Microbiology and Biochemistry, (S-4) Soil Fertility and Plant Nutrition, (S-5) Soil Genesis, Morphology, and Classification, (S-6) Soil and Water Management and Conservation, (S-7) Forest and Range Soils, (S-8) Fertilizer Technology and Use, and (S-9) Soil Mineralogy.

The ISSS has six divisions (commissions), the Brazilian Sociedade Brasileira de Ciência do Solo has six, while the Polish Polskie Towarzystwo Gleborznawcze has nine divisions, including those listed above, divisions on soil data banks, and the history of soil science in Poland. The material published in society journals is often listed under the subject matter of one of the divisions. The smaller societies often do not list divisions by title, although they may informally recognize differences in subject matter.

As the field of soil science has grown, members of developing subspecialties desired an organizational structure within societies where they could present papers and discuss their findings with colleagues of similar

7. Lutz, "History of the Soil Science Society of America."

interest. In some cases, the specialty groups have wanted to be a recognized subgroup so as to have more influence in society affairs. The Agronomy Society of America divisions are: (A-1) Resident Education, (A-2) Military Land Use and Management, (A-3) Agroclimatology and Agronomic Modeling, (A-4) Extension Education, (A-5) Environmental Quality, (A-6) International Agriculture, (A-7) Agricultural Research Station Management, and (A-8) Soil and Plant Science Applications (Provisional). As indicated by the names, soil science and agroclimatology are highly involved in many of these divisions.

From their beginnings, or soon thereafter, the ASA, SSSA, and CSSA have each published a journal which contained the technical information of the respective societies. However, as the interests of the societies have grown, so have the number and size of the journals. Six journals are now published by the three societies. The journals, the sponsoring society, and date of initiation are given in Table 7.4.

The *Journal of Soil and Water Conservation* (JSWC) publishes a mix of research-oriented articles and less technical material. About half of the *JSWC* material is original research.

In the United States, the SSSA (along with the ASA) has been the dominant publisher of refereed research papers. A listing of the total number of papers and the number, by division, at ten-year intervals, is given in Table 7.5. From 1950 to 1990, the number of papers published in the *SSSAJ* has tripled (from 94 to 282). These numbers do not reflect the total growth in refereed papers in the United States, since other journals exist, and several journals have been established which carry soil science papers. Notably, the *Journal of Environmental Quality* (established 1972), *Journal of Production Agriculture* (1988), *Soil Tillage Research* (1980), *Journal of Soil and Water Conservation* (1946), *Transactions of the American Society of Agricultural Engineers* (1958), and *Journal of Canadian Soil Science* (1957) have carried many American papers. *Soil Science* (established 1916) has long carried a significant number of American papers.

The number of papers published in *SSSAJ* may reflect the emphasis on different subdisciplines within soil science in the United States. The papers in Soil Physics (Division 1), Soil Chemistry (Division 2), and Soil Microbiology and Biochemistry (Division 3) have roughly tripled over the past forty years (Table 7.5). Division 5 (Soil Genesis, Morphology and Classification) has shown modest growth. Papers in Division 6 (Soil and Water Management and Conservation) have increased dramatically. Division 8 (Fertilizer Technology and Use) split from Division 4 (Soil Fertility and Plant Nutrition), and Division 9 was an outgrowth from Division 2.[8]

8. Division 7 (Forest and Range Soils) will be dealt with in a future volume on forestry in *The Literature of Agricultural Sciences* series.

Table 7.4. Journals published by the ASA, CSSA, and SSSA and their content

Journal	Year of establishment	Sponsoring society	Content
Agronomy Journal	1913	ASA	All aspects of crop and soil sciences, including crop physiology, production, and management, along with their relationship to soil fertility and climatic conditions.
Crop Science	1961	CSSA	Recent developments in crop breeding and genetics, crop physiology and biochemistry, ecology, cytology, crop and seed production, statistics, and weed control.
Soil Science Society of America Journal	1936	SSSA	New developments in soil physics, mineralogy, chemistry, microbiology, fertility and plant nutrition, soil genesis and classification, soil and water management, forest and range soils, and fertilizer use technology.
Journal of Agronomic Education	1972	ASA	Innovative concepts and techniques for improving education programs in universities, extension, and industry, as related to crops and soils.
Journal of Environmental Quality	1972	ASA	Information concerned with reviews and technical reports on protection and improvement of environmental quality in natural and agricultural systems.
Journal of Production Agriculture	1988	ASA	Production-oriented information from agronomy, crop science, soil science, agricultural economics, forages, pastures, animal science, range management, weed science, entomology, plant pathology, horticulture, and forestry.

Most journals have a defined area of subject matter about which they will consider unsolicited papers. The definition is not always rigid, and judgment is required as to whether the subject matter of the manuscript fits the

Table 7.5. Number of unsolicited papers published in the *Soil Science Society of America Journal* by division in ten-year intervals

	Division	Vol. 14 1950	Vol. 24 1960	Vol. 34 1970	Vol. 44 1980	Vol. 54 1990
S-1	Physics	12	14	25	35	40
S-2	Chemistry	20	37	40	54	54
S-3	Microbiology and Biochemistry	10	21	20	39	28
S-4	Fertility	25	26	33	27	30
S-5	Genesis and Classification	19	18	28	26	24
S-6	Soil and Water Management	8	10	15	19	47
S-7	Forest Soils	—	—	13	19	32
S-8	Fertilizer Management	—	—	9	4	17
S-9	Mineralogy	—	—	12	23	10
TOTAL		94	126	195	246	282

Numbers include papers by division. Notes and invitational papers not included.
—indicates the division was not in existence at that time.

journal's area of interest. The prime consideration by authors as to which journal they submit their manuscript is usually where they believe the manuscript will be read and studied most. Obviously, authors want to publish in journals which carry papers central to the theme of their manuscript, because their peers are most likely to read these journals. Other considerations are page charges, rigidity of the review process, timeliness of publication after manuscript receipt, and quality of the journal.

D. Monographs, Books, and Reports

In addition to refereed journals, many scientific societies publish monographs, books, proceedings of conferences, and other material. For example, the SSSA has published twenty-two monographs, four volumes in the book series, and eighteen special publications. The SSSA monographs are numbered and concentrate on taking a relatively narrow scientific subject and dealing with that subject in depth. The book is broader in scope than the monographs and often includes literature reviews and/or state-of-the-art treatments. Special publications have varied in subject matter and length, and usually result from special symposia given at the annual meetings of the society. The SWCS has published about thirty titles since 1980. The *Transaction of the ISSS Congress* is usually published usually by an organization in the congress host country, in cooperation with ISSS. Likewise, ISSS participates in sponsoring periodic conferences, the proceedings of which are published by other organizations, in cooperation with ISSS. The material includes state-of-the-science reviews and material related to a current problem or, to a lesser extent, research needs.

The most comprehensive service for abstracting soil scientific literature is that of *Soils and Fertilizers*, which publishes about 6,000 abstracts per year.[9]

E. Publishers

In the United States in the early part of the twentieth century, much of the original research in agriculture was published by the sponsoring institutions, the state agricultural experiment stations within universities, and the United States Department of Agriculture. All of these published bulletin series and special reports. The Government Printing Office in Washington, D.C., issued (under the auspices of USDA and the Association of Land-Grant Universities) the *Journal of Agricultural Research* from 1913 to 1949.

The bulletins and reports of USDA and land-grant universities often resulted in relatively lengthy reports of a comprehensive research study. With the growth of scientific societies, most of the original research is now published the society or private publishers, in journals with relatively little printed by the sponsoring institutions. The societies have encouraged publication of the original research in journals, while the reviews and interpretive articles have gone into books and monographs. For financial reasons, societies usually put a page limit on journal articles which encourages reports of limited length. For example, the *SSSAJ* allows members to publish four printed pages at a minimal cost. Pages beyond four bring a heavier publication charge. Some have said that page limits encourage fragmented, rather than comprehensive, research reporting. Societies often publish material at relatively low cost, partly because most of the editorial responsibilities are undertaken by unpaid members. Experiment stations and the USDA often give financial reasons for their dwindling publication of more lengthy reports. In some countries, societies are linked to government organizations and the publications are jointly supervised by the society and a government agency. In these cases, the journals may be financially subsidized by the government.

In the last decade or two, refereed research journals and monographs published by private publishing houses have proliferated in Europe. Unlike the United States, scientific societies in many countries have not taken the

---

9. *Soils and Fertilizers* (Harpenden, U.K.: Commonwealth Bureau of Soils). Now published by CAB International, Wallingford, U.K.

lead in publishing their scientific material, thus encouraging the private or commercial publishers. In many cases, the national societies have been slow to react to the need for journals outside the traditional fields of scientific investigation, partly for financial reasons. Starting a new journal cannot be taken lightly, because costs usually outweigh income for the first few years. Publication of a journal should be a long-term commitment. Societies in developing countries are most often limited by financial constraints in publishing journals. In the United States, publication costs, such as page charges, are usually paid by the sponsoring research institution, which is not true in many countries.

Competition from recently established journals and books by a few multinational, private-sector publishers has encouraged societies to shorten the time between receipt of a manuscript and hard copy publication. The private-sector trend in publications may also have resulted in a broadening of subject matter publication by scientific societies (e.g., establishment of the *Journal of Production Agriculture* by ASA), and encouraged low-cost printing procedures.

The trend in internationalization of scientific literature may result in a few journals concentrating on publication of the more basic work, while journals from individual societies within a country concentrate on the papers concerned more with applied research unique to that country. That trend is already apparent.

F. Quality of Publications

Many of the scientific societies have had a major influence on improving the quality of writing. Scientists must write in clear, concise terms. The material must be clear so that misinterpretation of the information does not occur. Cost of publication demands brevity whenever possible. The content of a society's published material is usually reviewed by the authors' scientific peers, while editing for style is done by society employees. Most societies have style manuals and promotional material which encourage improved presentation. Reviews and technical editing by peers put pressure on the writer to provide high-quality material.

In some cases, the societies, through their style manuals and review procedures, have encouraged rather rigid scientific procedures and interpretations. Suppositions and interpretations not based on the usual research procedures have been discouraged. Speculative interpretations are often criticized by peer reviewers as not being substantiated by direct evidence. Currently, a need exists for high-quality journals that provide an outlet for fresh ideas and interpretations on broad issues where evidence may come from a number of sources. Commenting on the results of a survey, R. E. Sojka et al. concluded that the members of ASA supported greater open-mindness in publishing unusual results where the analysis procedures were acceptable.[10]

G. Changing Emphasis

The membership in early soil science societies was very concerned with classification of soils and with soil as a medium for plant growth. Of these concerns, soils for agricultural crop production predominated, with some literature devoted to forestry, ecology, and range. With the advent of cheap and plentiful supplies of mineral fertilizer after World War II, interest in soil fertility and plant nutrition increased dramatically.

During the last several decades, scientists and the public have become increasingly worried about degradation of the world's natural resources. While erosion was recognized early as a severe threat, movement of agricultural chemicals and toxicants into both surface-and groundwaters has only recently come to the fore as an international problem. Another major international problem is compaction of the soil from modern machinery. Utilization of domestic wastes on land as an effective plant nutrient, without contamination, has received the attention of many soil scientists in the last two decades. Whether it be for agriculture, forestry, urban and industry expansion, recreation, or other uses, matching land qualities with land use is an area where soil scientists can contribute much. The influence of soil management on the hydrologic cycle is of vital importance. Since most precipitation reaches the soil surface, the condition of the soil determines, to a large extent, whether the water infiltrates and is stored, moves laterally to appear later in springs or seeps, moves into groundwater, or is transported across the land surface to streams and water bodies. Because of the growing awareness of the finiteness of quantity and the potential for contamination of water resources, soil scientists are increasingly called upon to help solve hydrologic problems. The establishment of the *Journal of Environmental Quality* by the American Society of Agronomy, in 1972, reflects this interest. Currently, a host of journals in environment, natural resource, sustainable agriculture, farming systems, and related subjects have been initiated. Most of these are privately sponsored. For soil scientists to contribute their potential in the wide array of current world problems, it is

10. R. E. Sojka, H. F. Maryland, and E. E. Gbur, "American Society of Agronomy Members' Experiences and Perceptions of the Peer Reviewing-Editing Process," in *Research Ethics, Manuscript Review and Journal Quality*, ed. Maryland and Sojka.

extremely important to have journals which will publish and recognize their work. Likewise, it is important for societies to provide programs at their meetings where all facets of soils research can be presented. Chapter 14 gives the name and date of first issue of some journals initiated since 1980 which reflect the current broader use and management of soils. A listing of recently published books would probably give an equally broad array of soil science titles.

H. Breadth of Research

As scientific societies have proliferated and narrowed their subject matter, scientists have tended to associate themselves with only one or a few societies in their discipline. Scientists often attend only one scientific society meeting per year, being limited by time, travel costs, and administrative restrictions. This has tended to confine research reporting to their discipline and may have discouraged interdisciplinary research.

According to a survey by L. Busch and W. B. Lacey, 86% of soil scientists in the United States subscribe to a journal in their field, and the mean number of fields in which soil scientists have at least one subscription is 2.7.[11] This compares with 2.4 for all agricultural disciplines. In addition to soil science journals, soil scientists most often subscribe to journals of agronomy (general and crop science) and general science. Soil scientists publish in an average of 2.3 journals (number of journals with at least one publication), which is at the mean for all of the disciplines. Busch and Lacey have suggested that, indeed, the narrowness of scientific societies has been a major stumbling block in problem-solving, interdisciplinary, agricultural research, and, thus, has hampered progress.[12] Busch and Lacey stated in 1983, "There is a need for a high quality interdisciplinary United States agricultural journal."[13] The *Journal of Production Agriculture* was initiated in 1988 by ASA to provide an outlet for a variety of interdisciplinary research and interpretive papers, and may have been partially motivated by Busch and Lacey's statement. In a survey of ASA members by Sojka et al., nearly one-half of the respondents felt there was a need to publish more articles of a philosophical, speculative, and socially analytical nature in Tri-Societies' journals.[14]

11. L. Busch and W. B. Lacey, *Science, Agriculture, and the Politics of Research* (Boulder, Colo.: Westview Press, 1983).
12. Ibid.
13. Busch, *Science, Agriculture, and the Politics of Research*, p. 238.
14. Sojka, "American Society of Agronomy Members' Experiences."

I. What Role for Societies?

The scientific societies have, up to now, concentrated on hard-copy periodicals, as have the commercial presses. It is often said that, in the future, scientific communication will be facilitated by electronic devices which will not necessarily require the current hard-copy journals. One scenario might be that scientific manuscripts would be submitted to the society electronically, transmitted to and from reviewers electronically, and, when approved, put on a menu from which society members could select articles for printing on their own office equipment. In addition, networking for communicating scientific information more informally will probably expand dramatically. The ASA and SSSA now accepts approved manuscripts electronically for inclusion in all six of their journals. The societies are investigating the feasibility of making their journals available on CD-ROM.

Modern electronic transmission and copying procedures for written material have often resulted in communication of research using both informal and formal procedures. Currently, when an author submits a manuscript to a journal, that author may also send copies of the manuscript to a number of key peers for their information. In turn, the peers may copy the manuscript and give it to other interested individuals. Thus, by the time of publication, many scientists may have seen the manuscript in unpublished form. While informal and perhaps inefficient, this procedure does, to some extent, provide timeliness and a written account from which formal credit is obtained.

The scientific societies are in a good position to foster and develop electronic scientific communication. In addition to the hardware mechanisms and financing, procedures for scientist and institution credits need to be worked out. The scientific societies could provide a tremendous service by providing leadership in this exciting field.

In 1991, F. P. Miller, president of the Soil Science Society of America, made a strong case for soil science taking a broader role in the years to come.[15] He has clearly enumerated areas of public concern in agriculture, environment, and natural resources, where soil scientists could contribute more. Miller, in his presidential report to the society, also stressed the need for soil scientists to communicate more actively and work with scientists from other societies, emphasizing the need for joint technical meetings. A symposium at the 1991 annual meeting of the SSSA was entitled "Opportunities in Soil Science—Thinking Outside the Box." The concern in this symposium was to identify areas where soil scientists could contribute more

15. F. P. Miller, "Soil Science: Should We Change Our Paradigm?" *Agronomy News* (Oct. 1991): 8–9.

outside their specialized fields of expertise. The SSSA published a book in 1992 which discussed the *Opportunities in Basic Soil Science Research.*[16]

Of the ten professional society objectives given by Browning and listed earlier in this chapter, the last two need further discussion and attention.[17] The study of Busch and Lacey[18] and the written comments by Miller[19] both emphasize the need for improved interdisciplinary interaction in research and education. The interaction must be at all levels—among individual scientists, among science administrators, and among professional societies. The need for interdisciplinary cooperation will become greater as the role of soil scientists in agriculture, natural resources, and environment broadens.[20] The scientific societies can offer leadership in encouraging this interdisciplinary work.

One area where this interdisciplinary work may be effective is in providing technical information to aid public policy formulation. The scientific societies that focus on research and education have not given this objective adequate attention, although there is increased awareness of the need. Scientists must be willing to provide technical information about public policy matters in a language decisionmakers understand. Societies can assist in this process by determining which members are best qualified to coordinate the activites. Some societies have fulfilled rather well the role of providing information to policymakers, others have been less successful, usually because this has not been seen as an important society function.

The ASA, CSSA, and SSSA recently have made several successful attempts at providing technical input into government actions. In 1985, they established a Congressional Fellows Program, which paid young scientists to work on the staffs of United States representatives and senators providing input at the time policy matters are being formulated. The scientists are there to provide technical information, not to represent a political view. They also help keep members of the societies informed of relevant considerations by Congress. A second example was a major workshop held in 1981, with ten societies cooperating in developing statements on *Soil and Water Resources: Research Priorities for the Nation.* A proceedings and an executive summary were published and distributed widely.[21] Personal contacts and follow-ups were extensive.

16. R. J. Luxmoore, S. H. Mickelson, and Pamm Kasper, *Opportunities in Basic Soil Science Research* (Madison, Wis.: Soil Science Society of America, 1992).

17. Browning, "The Role of Professional Societies in Resource Development."

18. Busch, *Science, Agriculture, and the Politics of Research.*

19. Miller, "Soil Science."

20. Ibid.

21. (a) W. E. Larson, L. M. Walsh, B. A. Stewart, and D. H. Boelter, *Soil and Water Resources: Research Priorities for the Nation, Executive Summary* (Madison, Wis.: Soil Science Society of America, 1981a). (b) Larson, Walsh, Stewart, and Boelter, *Soil and Water Resources: Research Priorities for the Nation* (Madison, Wis.: Soil Science Society of America, 1981b).

8. Present-Day Soil Information Systems

DAVID L. ANDERSON
U.S. Soil Conservation Service

J. DUMANSKI
Land Resource Research Centre, Ottawa, Canada

Managing soil information is critical in supporting environmental assessments at global, national, and local levels. Modern needs for natural resource information require access to large, integrated, computerized databases in which soil data are critical components.

The U.S. Department of Agriculture (USDA) began systematically mapping county soils early in this century. Matched with the United States soil classification scheme, these surveys and maps now number in the thousands and comprise a massive archive of data. With the advent of computers in the 1970s, these surveys and their point data served as the basis for modeling computer-based soil databanks, which today form a soil base for sophisticated, interactive information systems.

Initial computerized soil information came from these databanks, which later changed to incorporate the concepts of an information system. A databank is a repository of data, generally alphanumeric, with controlled input. The concept of an information system includes the databank but extends to include interactive manipulation of the data into user-designed output models which can be applied to specific purposes.[1] David L. Anderson defines an information system as a group of elements, including both people and machines, organized to maximize this transformation of data into purposeful information whose characteristics include user-manipulation, organization, and user input/output.[2] These information systems exist to support decision making.

---

1. J. Dumanski, B. Kloosterman, and S. E. Brandon, "Concepts, Objectives and Structures of the Canada Soil Information System," *Canadian Journal of Soil Science* 55 (1975): 181–87.

2. David L. Anderson, "Criteria for Developing a Comprehensive Soil Resource Information System," Master's thesis, Colorado State University, 1982.

Soil information systems are configured man-machine systems, designed to manage and provide the information necessary to answer questions about soil resources and physical interactions. This information can take the form of

(1) a map showing soil bodies as polygons;
(2) ungeneralized data points or lines;
(3) data tables (tabular or attribute information) that describe the characteristics of the spatial areas; and;
(4) interpretations (both statistical and algorithmic) from data stored in the system.

A soil information system must be sufficiently generic and flexible to produce these and other outputs quickly and cheaply.

Many scientists have worked on aspects of creating these systems, notably in Canada,[3] France,[4] Australia,[5] the Netherlands,[6] and the United States.[7] A group of authors has written on the philosophy and requirements for soil information systems, emphasizing such fundamentals as orderliness in data collection and storage, accessibility, editing and updating, simplicity, flexibility, and feedback, all of which are important guidelines for information systems at all levels.[8] P. Dumas et al. discussed in detail important considerations in implementing data systems.[9] J. Beresford wrote of the need to predefine the users of the system, as well as their needs and projected expectations.[10] M. Rousselot extended this by stressing the importance of establishing a permanent relationship between users and specialists, including those involved in data gathering and data processing.[11]

---

3. Dumanski, "Concepts, Objectives and Structures of the Canada Soil Information System."

4. R. Van den Driessche, "Current Work on Computerized Soil Data Processing in OR-STOM," in *Report of the FAO/UNESCO Ad Hoc Consultation on Computerized Soil Data Interpretation for Development Purposes* (Rome: FAO, 1972), p. 26.

5. A. W. Moore, "Projects Involving Storage and Retrieval of Soils Data in Australia," in ibid, p. 22.

6. J. Schelling, "Earth Science Information Systems for the Netherlands," in ibid, pp. 29–30.

7. U.S. Soil Conservation System, *Pedon Coding System for the National Cooperative Soil Survey* (USDA, Soil Conservation Service, 1973).

8. M. K. John, C. J. Van Laerhoven, and P. N. Sprout, "A System of Soils Information Retrieval," *Canadian Journal of Soil Science* 52 (1972): 351–56.

9. P. Dumas, J. Salmona, O. Salomonson, R. Tomlinson, and M. Usui, "Summary Report," in *Data Banks for Development Proceedings of the UNIDO/INSEE Expert-Meeting* (Marseilles, France: Institut National de la Statistique et des Etudes Economiques, Observatoire Economique Mediterraneenne, 1971), pp. 229–66.

10. J. Beresford, "Relationships Between Users and Data Banks," in ibid, pp. 194–229.

11. M. Rousselot, "Methods for Use of Data Banks as a Basis for Decision-Making," in ibid, pp. 136–66.

A. Evolution of Soil Information Systems

In 1975, representatives from eighteen countries met in Wageningen, the Netherlands, for the first, formal, international scientific meeting on soil information systems. The gathering was organized by the Working Group on Soil Information Systems, then part of Commission V of the International Society of Soil Science (ISSS). The working group made provisions for the largest number of ISSS members to participate in laying the foundations of a collective effort for development of information systems which would serve a variety of soil-related applications. In the foreword to the proceedings of this meeting, M. Ciric, chairman of Commission V wrote: "The use of information systems in the development of [soil] classification . . . creates a basis for soil mapping and acts as an instrument for transfer of scientific knowledge and practical experience to ecologically similar conditions. Our information systems will become particularly valuable when they begin to include data . . . leading up to evaluations of world resources of food and raw materials."[12]

This meeting was followed by one in Canberra, Australia, in 1976; thereafter, the group met biennially, until 1983, and published papers of their proceedings. Given the cumbersomeness of computer-generated data in the mid-seventies, these working group meetings were both farsighted and unique in agriculture, seeking to map a future course for soil information systems.

One of the clearest insights into the uses of soil information systems to come from these productive meetings was by V. Linkes, at the meeting in Sofia, Bulgaria (1977): "It is sometimes overlooked that the major problem in the design of soil information systems originates from the soil data themselves. What data are, how and where they are collected and their interrelations and data structure are factors as much associated with the users of soil information as being inherent in the data. The stringent logical requirements of information systems in these fields exclude the use of unclear and undefined conception."[13]

If the working group meetings can be viewed as the first phase in the evolution of modern soil information systems, the second phase logically began as soil scientists sought to apply the technology to rigorous methods

---

12. M. Ciric, Foreword in *Soil Information Systems, Proceedings of the meeting of the ISSS Working Group on Soil Information Systems, Wageningen, Sept. 1975* (Wageningen, the Netherlands: PUDOC, 1975), p. 2.

13. V. Linkes, "Some Problems of Designing Soil Information Systems," in *Development in Soil Information Systems, Proceedings of the Second Meeting of the ISSS Working Group, Sofia, Bulgaria, June 1977* (Wageningen, the Netherlands: PUDOC, 1978), p. 101.

for mapping. As the microcomputer became a standard tool in the work-place, applications in the use of soil information technologies became increasingly sophisticated, including the ability to manipulate the spatial variation of soil properties within delineated landscape units. From 1983 on, this was a subject of keen discussion at international soil survey meetings.

The use of microcomputers and standard, off-the-shelf software has resulted in soil information systems shifting from development to applications. Soil scientists could again focus their attention on the purpose and quality of data collected, rather than having to concern themselves with how to organize the data on a machine using software that was not originally intended for this purpose. Consequently, by the mid-1980s, many systems were consolidating their activities and diverting their attention to developing links with models such as crop growth and degradation, soil interpretation schemes, and other such processes.[14] This was recognized at the 13th International Congress of Soil Science in Hamburg (1986), where the ISSS amalgamated the Soils Information Working Group with the Land Evaluation Group to form the Land Evaluation Information Systems Working Group. This amalgamation has further fostered the evolution of soils information systems as tools for applied modeling and analysis.

Today, use and application of soil information systems have diverged. Peter Burrough identifies four largely separate paths.[15]

(1) *Consolidation and implementation.* A prominent challenge facing national and international soil information organizations relates to the problem of configuring, coherently recording, and storing data from different areas and across international boundaries, where differing systems are in use. This is primarily a question of data standardization and quality.

(2) *Analysis and modeling.* This path relates to the use of digital soil information for analysis and modeling and how the information on the newly created maps should be derived in a logical, systematic way. Standard Geographic Information System (GIS) software which can reclassify soil polygons and pixels according to locally held attributes, in order to make a new map, has made this challenge less complex.

(3) *Basic soil survey data models.* Criteria for data models is only now receiving the attention it deserves. Data models imply that all soil information and all interpretations of those data need to be related to actual survey resolutions and the uses to which those data will be put. Unfortunately, data units of soil surveys are messy, and soil scientists for decades have pondered how these units can better fit into precisely defined classes.

14. A. A. Klingbiel, "Development of Soil Survey Interpretations," *Soil Survey Horizons* 32:53–66.

15. Peter Burrough, "Soil Information Systems," in *Geographical Information Systems: Principles and Applications*, ed. David Maguire et al. (New York: Longman and Wiley, 1991), pp. 153–69.

(4) *The use of expert systems*. One of the best analyses of the potential of expert systems in soil information management is the exploratory work done by M. B. Dale, A. B. McBratney and J. S. Russell.[16] Numerical soil classification, they contend, has been useful in the analysis and organization of small-scale soil data, but has been largely untried at the higher global levels of soil classification because of a lack of suitable data and a corresponding lack of institutional commitment. Numerical classification using expert systems has a potentially useful part to play in establishing soil classes. The user interface and inference procedures of expert systems should allow more integration and realistic assignment of available data and increase the predictability and usefulness of soil information, particularly if presented in a user-friendly mode. The current work in this arena looks promising.

Concurrent with the intellectual challenge of organizing soil information systems has been the evolution of the technology itself. Until the 1980s, the primary medium for managing information and for providing that information to others was written material. Undoubtedly, written text will continue to be important, but today more and more users of soils and natural resource data require information in automated format. Coupled with the availability of new technology, these demands are fueling a growth of automated soil databases.

Prior to the 1980s, small-scale computing technology (microcomputers and software) was not available, or lacked the power necessary to assist soil scientists in managing large volumes of data. This placed considerable procedural constraints on managing the information and, as a consequence, soil scientists aggregated their information into model profiles and/or sites and developed interpretations to reflect this level of information. However, investigators were unable to manage all the data available and, therefore, much valuable and detailed information collected by field soil scientists was underutilized or lost. Many of these constraints currently are being removed with computer technology and more sophisticated software.

Database Management Technology

Recent improvements in computing technology, such as database management systems, geographic information systems, and increased speed and disk storage capacity, allow soil scientists to manage their data in ways that were economically impossible a few years ago.

Database management systems (DBMS) and flexible, fast, user-friendly microcomputers were the most important technologies in the evolution of

16. M. B. Dale, A. B. McBratney, and J. S. Russell, "On the Role of Expert Systems and Numerical Taxonomy in Soil Classification," *Journal of Soil Science* 40 (1989): 223–34.

soils information systems. DBMS, coupled with greatly enhanced speed and storage of microcomputers, have moved information management from the computer processing professional into the hands of the professional user. These combined technologies have greatly expanded the use of information systems.

DBMS, in their most general form, are software capable of supporting and managing an integrated database. These services generally are aimed at organization, access, and control of data.[17] DBMS evolved from the file management systems of the late 1950s and early 1960s and have provided the impetus for the outgrowth of computerized information systems.[18]

Early database management systems utilized hierarchical data models to manage data files.[19] In the mid-1980s, DBMS became available which implemented the relational data model that is used almost exclusively in soil information systems today.

Geographic Information Systems

Geographic Information Systems (GIS) evolved in the 1970s from the computer-assisted design/computer-assisted mapping (CAD/CAM) software.[20] By the mid-1980s, CAD/CAM technology merged with DBMS technology to create the modern GIS. Basically, a GIS stores and manipulates data which are geographically referenced.

A GIS is "a computer-based methodology including hardware, software, and graphics that encodes, analyzes, and displays multiple data layers derived from various sources. GIS provides an efficient means for joining spatial or mapped data to descriptive attribute data for use by planners, resource managers, and a host of other professionals. The application of GIS technology provides users with a tool to solve problems and make decisions. GIS are emerging as the spatial data-handling tools of choice for solving complex geographical problems."[21] "Analysis can be expressed in

17. A. F. Cardenas, *Data Base Management Systems* (Boston: Allyn and Bacon, 1979), pp. 2–171.

18. R. H. Bonczek, *Foundations of Decision Support Systems* (New York: Academic Press, 1981).

19. Anderson, "Criteria for Developing a Comprehensive Soil Resource Information Systems."

20. (a) David J. Maquire, M. Goodchild, and D. Rhind, *Geographic Systems: Principles and Applications* (New York: Wiley, 1991). (b) John Antenucci, *Geographic Information Systems: A Guide to the Technology* (Florence, Ky.: Van Norstrand Reinhold, 1991). (c) Lowell Star and J. Estes, *Geographic Information Systems: An Introduction* (Englewood Cliffs, N.J.: Prentice Hall, 1991). (d) Dana Tomlon, *Geographic Information Systems and Cartographic Modeling* (Englewood Cliffs, N.J.: Prentice Hall, 1990). (e) Donna Peuquet and D. Marble, *Introductory Readings in Geographic Information Systems* (Bristol, Pa.: Taylor and Francis, 1990). (f) David F. Sinton, *Reflections on 25 Years of GIS* (Ft. Collins, Colo.: GIS World, 1992).

21. USDA Geographic Information Systems Work Group, *Report to the Secretary* (Washington, D.C., Feb. 1991), p. 4.

tabular, graphic, and most importantly, in geographically coordinated mapping format."[22] GIS can

(1) retrieve all attributes at a location;
(2) identify all locations with defined attributes;
(3) calculate distances between points and estimate the area within a geographic boundary;
(4) analyze data with interpolation routines;
(5) calculate terrain information from elevation data;
(6) perform union and multiple exclusion operations of areas with intersecting locations; and
(7) draw map overlays.[23]

Modern GIS contain

(1) spatial information represented as polygons, lines or points and
(2) attribute or tabular data which describe the characteristic of the spatial area or point information.

The growth of GIS software is expanding, rapid, and continuing. While there were approximately thirty-five commercial and public-domain GIS software systems in 1988, almost 100 exist today.[24] With the availability of inexpensive microcomputers, GIS and DBMS technology are becoming available in developing countries. This will vastly improve the capability of these countries to manage information effectively. GIS technology represents another important advance in the evolution of soils information systems.[25]

The U.S. Government is moving forward in establishing GIS technology in the Federal community. In 1983, the U.S. Office of Management and Budget created the Federal Interagency Coordinating Committee for Digital Cartography (FICCDC). FICCDC membership is composed of twelve government departments and independent agencies. The purpose of FICCDC is to coordinate the digital cartographic activities of federal agencies, thereby

22. American Farmland Trust, *Survey of Geographic Information Systems for Natural Resource Decision Making at the Local Level* (Washington, D.C.: American Farmland Trust, 1985), pp. 16–21.

23. P. T. Dyke, "Compiling and Managing Data Bases and Geographic Information Systems," in *Agriculture Environments: Characterization, Classification, and Mapping*, ed. A. Bunting (Wallingford, U.K.: CAB International, 1987).

24. USDA, *Report to the Secretary* (Washington, D.C., Feb. 1991).

25. R. F. Tomlinson, "Opening address," in *Environment Information Systems, Proceedings of the UNESCO/IGU 1st Symposium on Geographic Information Systems, IGU Commission on Geographic Data Sensing and Processing, Ottawa, 1970*, ed. R. F. Tomlinson (Ottawa: International Geographical Union, Commission of Geographical Sensing and Processing, 1971), pp. 22–26.

avoiding duplication of effort and waste of resources.[26] FICCDC is taking an active role in establishing digital and attribute data standards. The U.S. Soil Conservation Service has been assigned the responsibility for developing the standards for soils data layers.[27]

Current Data Analysis and Modeling Technology

Modern information systems are becoming more complex, but costs per unit of data have decreased tremendously. To be successful, information systems must be based on modern principles of data administration and management. Modern procedures are highly structured to specific uses and user requirements, and this is reflected in the organization of the technology and the kind of data model implemented in the system.[28]

A data model is a diagram which presents an image of the kind of data stored in the system, and the relationships among the data elements. Relationships among the data are as important as the values of the data attributes themselves. Interrelationships among the data entities are diagrammed to define the various rules for association between two or more data entities. A relationship on a data model represents a specific, data-oriented action that the system must accommodate.[29] The data model is especially important in the development of soils information systems, because the model is a mirror of the level of understanding that exists in the science, becoming the foundation on which the system is built. The design of the data model is critical for information technology to be effective.

A data model is a convenient means of communication, particularly with the complexities that are inherent in the content and organization of soils data. As noted, one of the most difficult aspects of defining the requirements for an information system is developing a view understood and accepted by a majority of the users—a "common view" of a model that represents the data—and this is especially difficult for a natural science such as soils. Soils, by nature, are dynamic, complex, and continuous over the landscape. Characteristics of soil vary over short geographic distances and

26. USDA, *Report to the Secretary*.

27. "A National Geographic Information Resource, The Spatial Foundation of the Information-Bases Society," 1st Annual Report to the Director of the Office of Management and Budget by the Federal Geographic Data Committee, 1991.

28. James Martin, *Information Engineering Book, I—Introduction, II—Planning and Analysis, and III—Design and Construction* (Englewood Cliffs, N.J.: Prentice Hall, 1990).

29. D. L. Anderson, D. J. Ernstrom, and R. M. Stahler, "Application of New Technology in Development of a Soil Survey Information Delivery System," *Proceedings, Resource Technology 90, Second International Symposium on Advanced Technology in Natural Resource Management, Nov. 1990* (Washington, D.C., 1990).

change in response to temporal environmental conditions. To deal with this complexity, soil scientists develop mental concepts or models that they use to represent their idea of soil. The data model reduces these concepts to basic organizational units, or constructs, that define the system in manageable terms. When fuzzy concepts are implemented, the physical design of the database will be as ambiguous as the mental concepts they seek to represent. When this happens, the results obtained from the system will be inconsistent, and the structure of the database will be in a constant state of change.

The data model is used in close conjunction with data flow diagrams and a data dictionary. A data flow diagram is a graphical representation of how data moves and how it is processed within the system. The data dictionary is the repository of the details of the data stored in the system, and the processes which transform the data. Future soils information systems, must utilize these technologies to meet the data management standards required to ensure the integrity of the system.

B. Soils Information Systems in the United States

The USDA Soil Conservation Service (SCS) began to implement computerized soil information in 1975.[30] Early technology consisted of fixed-length data records processed by conventional computer languages. In 1981, DBMS technology was demonstrated as an effective tool for managing soil information.[31] Within the last two years, national policy has been initiated to ensure that all progressive soil surveys are digitized, as a step in implementing a national soils GIS.

The SCS is currently developing a National Soils Information System (NASIS) and has developed a *National Soils Information System Interpretations and Information Dissemination—Draft Requirements* statement which establishes the general concepts and user requirements for the design. NASIS will support soil survey in three important ways:

(1) support field operations to gather new information efficiently in compliance with standards,
(2) apply expert knowledge to make information usable for an increasing variety of purposes, and
(3) make information readily available to meet the needs of a wide variety of users.

30. Maurice Mausbach, D. L. Anderson, and R. W. Arnold, "Soil Survey Databases and Their Uses," *Proceedings of the 1989 Summer Computer Simulation Conference, Austin, Texas, July 1989,* sponsored by the Society for Computer Simulation, ed. Joe K. Clema, pp. 659–64.
31. Anderson, "Criteria for Developing a Comprehensive Soil Resource Information System."

Requirements of NASIS have been established as[32]

(1) *Enable the collectors of soil information to record efficiently their actual obser-vations.* The information system must preserve the integrity, continuity, and entirety of soils information, and must convey to the user accurate information that is consistent with the observations and scientific judgment of the collector/ recorder.

(2) *Flexible input.* The capability must be provided for users to tailor the system to meet local data input needs. Not every location will collect the same informa-tion, however, all information must be collected in an organized way and within national standards.

(3) *Availability of detailed primary soil property data.* Primary data is defined as site-specific data collected, such as pedon descriptions, transects, laboratory analysis, and field notes. It is an objective of NASIS to make maximum use of primary data. The organization and storage of these data will allow their aggre-gation to form more generalized data records.

(4) *Dynamic update.* One of the primary user requirements for NASIS is flexibility in the definition of the system and format of the output. The system must be adaptable as new requirements evolve. There are four areas where the system must be dynamic: (a) addition or modification of new soil attributes (data ele-ments), (b) addition or modification of interpretation criteria, (c) addition or modification of soil records, and (d) customization of data queries and reports.

(5) *Integrated system.* The system must be integrated. This means that beginning with data collection and continuing through information dissemination, all the information that has been recorded, pertaining to a particular soil, should be accessible through the system. In addition, the system is designed to accommo-date integration with other natural resource databases and applications.

(6) *Statements or measures of reliability of the data.* NASIS will contain informa-tion based upon varying degrees of knowledge and supporting documentation. NASIS should provide a way to communicate the methods used to collect the information and indicators of its reliability.

The information system has three soil geographic databases:[33]

(1) Soil Survey Geographic Database (SSURGO) is designed to be used primarily for natural resource planning and management at spatial scales of landowner/ user, township, county, and parish. Soil maps in the SSURGO database are made by field methods that require direct observations along traverses and field

32. Anderson, D. Ernstrom, and R. Stahler, "National Soils Information System (NASIS)—An Overview" (unpublished).

33. (a) Norman B. Bliss and W. U. Reybold, "Small-Scale Digital Soil Maps for Interpreting Natural Resources," *Journal of Soil and Water Conservation* (1988): 30–34. (b) Reybold and G. TeSelle, "Soil Geographic Databases," *Journal of Soil and Water Conservation* 44 (1) (1989): 28–29.

transects. Map unit composition is determined by these observations, using national standards. Maps are made at scales usually ranging from 1:23,000 to 1:31,680.

(2) State Soil Geographic Database (STATSGO) is designed to be used primarily for multi- county state and regional natural resource planning, management, and monitoring. Soils maps for STATSGO are made by generalizing the more detailed SSURGO maps. Map unit composition for STATSGO is determined by transecting or sampling areas on the more detailed maps, according to national standards. They are made at 1:250,000 scale, using USGS 1:250,000 scale topographic quadrangles as the map base.

(3) National Soil Geographic Database (NATSGO) is used for national and regional resource appraisal, planning, and monitoring. The boundaries of the Major Land Resource Areas are used for the NATSGO spatial layer. The NATSGO map is compiled and digitized at a scale of 1:200,000.

The SCS has developed interface computer programs that link the map data with the relational attribute data. These interface programs allow menu-driven access to both the map data and tabular data and remove much of the complexity of interpreting soils information.[34]

C. Opportunities in Developing Systems

Most early work in the development of automated information systems was done in the fields of business and engineering. Data from these fields are relatively easily automated, compared to soils data, because the data and the rules which act upon it are well defined and understood by the community of users. The data models which describe these systems were usually fairly simple and hierarchical.

The automation of natural systems, of which soils are a part, is more difficult. Natural systems require relational data models that can accommodate the description of numerous relationships and interactions. The data models for soil information must correctly represent the true complexity of soils data, and, thus, they can also be extremely complex. Soils are three-dimensional natural bodies with continual variability over the landscape. This variability cannot be easily accommodated with software and hardware configurations that were originally designed for much more simple models in business and engineering.

Soil science is still comparatively young, and concepts and knowledge are still evolving. The data, and the relationships and interdependencies

34. D. J. Lytle and M. J. Mausbach, "Interpreting Soil Geographic Databases," 1991, unpublished.

among the data, are not clearly understood by the user community, regardless of the numerous standards that are available for collecting soils data. This is exacerbated because concepts that govern the collection of soils data are abstract (this is normal for all sciences that deal with continual variables). This fact must be taken into account in designing systems. The design of the system must be flexible and adaptable to change in the content of data and in the explanation of standards for defining data.

Concepts used in collecting and managing soil information are still burdened with limitations developed in preautomation times, when soil information was processed manually. Under these constraints, soil scientists grouped soils with like characteristics, and then recorded information about the soil group. Information technology now allows storage and manipulation of all data, without the necessity of preclassification or aggregation. The management of natural resource data in the information age is causing scientists to rethink some of the paradigms that have governed the management of resource data in the past. As stated by J. Dumanski and B. Kloosterman; "Some changes in the fundamental approach to soil science will be necessary before the full capacity of soil information systems can be achieved."[35]

Difficulty in Sharing Data

One of the major factors limiting the effectiveness of large soil information systems is the difficulty in sharing data between systems. There are two aspects to this problem. One deals with the problem of proprietary rights to the data collected, whereas the other deals with the compatibility of soils data with other natural resource information. Proprietary rights concerns the rights of scientists to their own data prior to publication and can be handled easily through conventions. Compatibility of data with other natural resource information is a far more complex and important worry. For effective decision making the decision-maker must be able to compare soil information with other natural resource data. Seldom are resource problems solved using soil information alone. Problems of sharing information are due to incompatible hardware and software, which manage the data, and incompatibility of the standards that define the data.

Problems of sharing data due to incompatibility of hardware and software are rapidly being solved. Industry standards are evolving which make the portability of information systems much easier if the systems are developed

---

35. Dumanski and Kloosterman, "Soil Information Systems in Perspective," in *Developments in Soil Information Systems*, p. 18.

under these standards. One of these standards is the Structured Query Language (SQL) for which an international standard recently has been established.[36] SQL has become the standard computer database language. There are currently over 100 database management systems products which support SQL.

The issue of data incompatibility due to the definition of the data will require considerable work at national and international levels. Considerable efforts have been directed to establishing standards for the nonautomated soils information.[37] These standards must be defined in much more detail to be applicable to automated systems. Standards for automated soils information are beginning to be addressed in the United States, and efforts are underway to coordinate these standards within the National Cooperative Soil Survey, but much needs to be done internationally.

Selected National Soil Information Systems

There are several sources which give selected lists of important soil information systems now operating in developed countries.[38] In the Third World, application of SIS and soil databanks have been sporadic, generally on a project-by-project basis.[39] However, several important soil information sys-

36. James R. Groff and P. N. Weinberg, *Using SQL* (Berkeley, Calif.: Osborne McGraw-Hill, 1990).

37. (a) Soil Survey Staff, *Soil Survey Manual* (Lincoln, Neb.: USDA Soil Conservation Service, National Soil Survey Center, 1991). (b) Soil Survey Staff, *National Soil Handbook* (Draft) (Washington, D.C.: USDA Soil Conservation Services, 1992). (c) Soil Survey Staff, *Soil Taxonomy: A Basic System of Soil Classification for Making and Interpreting Soil Surveys* (Washington, D.C.: U.S. Govt. Print. Off., 1975 [USDA Agricultural Handbook no. 436]).

38. (a) Jean-Paul Legros, "Computerized Data Sets for Thematic Maps," *Geographical Information Technology in the Field of Environment, Proceedings of the UNEP/UNITAR and EPFL Training Programme in GIS, Lausanne, Switzerland, April 1991* (Lausanne: Swiss Federal Institute of Technology, Dept. of Agricultural Engineering, 1991), Section 14, pp. 59–62. (b) R. Jones and B. Biagi, eds., *Computerization of Land Use Data*, Proceedings of a Symposium in the Community Programme for Coordination of Agricultural Research, May 1987, Pisa, Italy (Brussels: Commission of the European Communities, 1989). (c) J. Bouma and A. K. Bregt, eds., *Land Qualities in Space and Time, Proceedings of a Symposium, Wageningen, August, 1988, organized by the ISSI* (Wageningen, Netherlands: PUDOC, 1989). (d) J. M. Hodgson, ed., *Soil Survey—A Basis for European Protection*, Proceedings of the Meeting of European Heads of Soil Survey, Silsoe, U.K., Dec. 1989 (Brussels: Commission of the European Communities, 1991).

39. (a) J. Alfred Zinck and C. Valenzuela, "Soil Geographic Database: Soil and Application Examples," *ITC Journal* (3) (1990): 270–94. (b) J. C. Day and M. Aillery, "Soil and Moisture Management in Mali: A Case Study Analysis for West Africa," *Agricultural Economics* 2 (3) (1988): 208–22. (c) T. T. Cochrane, *Land in Tropical America*, vol. 3 (Cali, Colombia: Centro Internacional de Agricultura Tropical [CIAT], 1985). (d) A. Ward and B. Middleton, "Sediment Yield Estimation and the Design of Sediment Control Structures in Rural Africa," in *Challenges in African Hydrology and Water Resources*, ed. D. E. Walling et al., Proceedings of the Harare Symposium, July 1984 (Wallingford, U.K.: International Association of Hydrological Science, 1984), pp. 415–25. (e) K. S. Rao et al., "Computer-Aided Brightness Temperature Map of Indian Subcontinent: Inference on Soil Moisture Variations," *Remote Sensing of the Environment* 20 (2)

tem projects at the national level deserve mention. Prominent among them is the Canadian Soil Information System (CanSIS), which was operational by 1984, and today is widely used by universities and research agencies in several provinces. Developed and maintained by the Land Resource Research Institute of Agriculture, in Ottawa, it now serves as a national, computerized information system, facilitating the storage, display, and manipulation of data related to soil science and natural resource studies.[40]

The Food and Agriculture Organization (FAO) and the International Soil Resources and Information Centre (ISRIC), in Wageningen, have collaborated to design the FAO-ISRIC Soil Database, which is in the final stages of development. This is a flexible, user-friendly storage and retrieval system of soil profile data, primarily for use in field projects. ISRIC has added to this system by creating SOTER, their world soil and terrain fixed numeric coding system, with the addition of climatic data. Both systems can be run on IBM-compatibles and have user manuals available in English.[41]

An interactive information system for the storage, retrieval, and manipulation of land resource data called the Western Arid Resource Information System, or WARIS, is now in full operation in Australia and is particularly suitable for point and profile soil survey data.[42] WARIS, which dates back to the mid-seventies Western Arid Region Land Use Studies, has developed over the years into a general resource information system particularly useful for manipulating data recorded in accordance with the *Australian Soil and Land Survey Handbook*.[43] The system is in wide use in Australia today.

In China, work on soil information systems began in 1983, at the Nanking Soil Institute, in Honan Province, and, by 1990, had expanded to the Soil Fertilizer Institute of the Chinese Agricultural Academy, the Water Resource Institute of Shansi Province, and, finally, to the Department of

(1986): 195–207. (f) "International Benchmark Sites Network for Agrotechnology Transfer," *Proceedings of the International Symposium on Miniumum Data Sets for Agrotechnology Transfer*, ICRISAT, Patancheru, India, March 1983 (Patancheru: ICRISAT, 1984). (g) P. Jones, "Current Availability and Deficiencies in Data Relevant to Agro-Ecological Studies in the Geographic Area Covered by the IARCS," in *Agricultural Environments*, ed. A. H. Bunting, pp. 69–84.

40. (a) Dumanski, "Concepts, Objectives and Structures" (b) K. B. MacDonald and Kloosterman, *The Canadian Soil Information System (Cansis): General User's Manual* (Ottawa: Land Resource Research Institute, 1984). (c) MacDonald and L. R. Russell, "A Method to Determine the Computer System Requirements for a Large-Scale Geographic Information System," *Computers and Electronics in Agriculture* 3 (4) (1989): 305–15.

41. Food and Agriculture Organization, *FAO-ISRIC Database (SDB): World Soil Resources Reports No. 64* (Rome: FAO, 1989).

42. K. M. Rosenthal et al., "WARIS: A Computer-Based Storage and Retrieval System for Soils and Related Data," *Australian Journal of Soil Research* 24 (4) (1987): 441–56.

43. R. C. McDonald et al., *Australian Soil and Land Survey Field Handbook* (Melbourne: Inkata Press, 1984).

Land Resource Science, Beijing Agricultural University, in the capital. However, there is little coordination between institutions in China at present, and no national standards have been implemented to form a system.[44]

Although the concept of the modern soil survey began in Russia, SIS efforts in the former Soviet Republics have been slow to evolve. As early as the Tenth International Congress of Soil Scientists in Moscow (1974), Russian soil scientists began working on a soil computer system called SIS-PODZOL, at the Dokuchaev Soil Institute. But not until 1985 did the system became fully operational using the new edition of the USSR soils classification system.[45] SIS-PODZOL itself is divided into seventeen subsystems, including aerial photo interpretation (PICTUR), agricultural zoning (PSEHORA), soil fertility models (PLOMOD), operative diagnostics of agricultural crop mineral nutrition (ISOD), diagnostics using ion-selective etectrode techniques (UDOD), and land suitability categories (LAND-INDEX), among others. Current work is focused on soil conservation (ERSUC) and the evaluation of the erosion hazard within the frame of the international soils degradation program GLASOD.

In Japan, the primary impetus for creating a soil information system has come from the National Institute of Agro-Environmental Sciences in Ibaraki. The soils subsystem of Japan's Agricultural Land Resource Information System (ALRIS) is a Fortran language, SIS, known as JAPSIS. Three main objectives to this system have been described: (1) an inventory of soil resources; (2) the creation of automatic interpretation maps; and (3) the statistical manipulation of large amounts of soil survey data in order to develop a taxonomic classification of soil generic theory and soil geography. JAPSIS is itself divided into several subfiles, including a Soil Name File, the Soil Survey Data Index File, and a Soil Map Index File, and incorporates the data from two major programs, namely, the Fertilizer Application Improvement Program (FAI), which ran from 1953 to 1961, and the data gathered from the Soil Survey Program of the Maintenance of Farm Land Fertility (1959–78).[46]

Many European Community countries also have extensive soil information systems, with France's Système de Transfert de l'Information Pé-

44. Yanchun Shi, President of Beijing Agricultural University, Personal correspondence, November 1991.

45. V. A. Rozhkov, Deputy Director, Dokuchaev Soil Science Institute, Personal communication, January 1992.

46. (a) Yoshitake Kato, "A Computerized Soil Information System for Arable Land in Japan (JAPSIS)," *JARQ* (Japan Agricultural Research Quarterly) 21 (1) (1987): 15–22. (b) Kato, "A Computerized Soil Information System for Arable Land in Japan. I. Concept, Objective, Process and Structure," *Soil Science and Plant Nutrition* 30 (1984): 287–97.

dologique et Agronomique (STIPA)[47] and the Belgian AARDEWERK[48] being particularly strong examples. Poland, Hungary and the Czech and Slovak Republics lead among former Eastern Bloc countries.[49]

D. Future Trends

Traditionally, the structure and configuration of soil information systems have been driven by changes in technologies in the computer industry, and this is expected to continue into the foreseeable future. Soil information systems were first designed as large, centralized facilities, such as those already mentioned, but, increasingly, there will be a rapid decentralization into small private systems, as microcomputers and GIS become widely available. Examples of local systems are numerous in the United States and Europe, especially at county levels, as well as at academic institutions where computer applications of remote sensing, land survey, and natural resource inventories are being developed with a variety of software, such as GIS.[50] However, no concise directory of these efforts has been compiled. Since these soil systems are local and oriented toward specific field projects, little hard data has been gathered on their numbers and scope. A search in the National Agricultural Library's bibliographic database AGRICOLA revealed that for the five-year period 1986–91, 468 records were retrieved using the keywords "soil" and "computer," forty-six using "soil" and "database," and twenty-two using "soil" and "GIS." This is a twofold increase in aggregate of counts of similar citations for the period 1980–85. The 468 records in the first search are not all soil specific, since some deal with allied disciplines such as hydrology and remote sensing, however, only forty-eight, or 10.2%, were to national or international systems. The rest described local or project-specific computer applications.

This trend towards decentralization and the use of local systems is expected to continue, but, in the future, the personalized systems will be working with much larger sets of data, using compact disk technology and

47. J. P. Legros and N. S. Nortcliff, "Conception d'un Vocabulaire pour la Description du Milieu Naturel et des Sols," *Pédologie* 40 (2) (1990): 195–213.

48. (a) Jos Van Orshoven, Director, Belgian Committee for the Establishment of a Soil Information System, Personal communication, January 1992. (b) Van Orshoven et al., "A Structured Database of Belgian Soil Profile Data," *Pédologie* 38 (2) (1988): 191–206.

49. G. Varallyay, "A View From Eastern Europe," Paper presented at the Joint Annual Meeting of the ASA, SSSA, and CSSA, Denver, Colo., October 1991.

50. (a) L. L. Boersma et al., eds., *Future Developments in Soil Science Research*, SSSA Golden Anniversary Annual Meeting, New Orleans, Nov.-Dec. 1986 (Madison, Wis.: Soil Science Society of America, 1987). (b) R. Webster and M. Oliver, *Statistical Methods in Soil and Land Resource Survey* (Oxford: Oxford University Press, 1990).

other media. Conceivably, a desktop system will be capable of processing all the soils information in the world on a digitized world map. Pending some further developments in software, the soils information will be integrated with vastly greater amounts of other related data, such as long-term climate, vegetation, and land use. These data will be used to generate input variables for models on agriculture, environment, and related areas.

Coincident to this evolution in speed, storage, data management, and integration, there will be an evolution in how soil data will be perceived and how they will be handled. Our understanding and descriptions of soils will change from current practices of describing mostly static (or assumed static) attributes and storing these in a soil database to describing dynamic soil systems. Soils are living systems, and many of their most important properties, such as soil water, active organic matter, and so forth, are dynamic. Soil data of the future will describe the soil less as a static set of properties and more as a changing, dynamic, microenvironment, which is sensitive to past and present use and treatment. More attention will be given to rates of change of soil properties relative to major influences, such as climate changes, land use, and other changes, and to natural relationships between attributes. Much more use will be made of techniques such as pedotransfer functions to describe relationships among variables and for data compression and storage.[51]

There is a movement in systems science toward describing the world in the form of objects. Object-oriented programming[52] and object-oriented databases are becoming more common.[53] Process simulation is very important to this technique of describing objects and interactions. When systems are described in this way, the distinction between data as static values and relationships among data become less defined. When these techniques are adapted to the process of understanding soil, this will reduce the use of static attributes like bulk density, and increase the use of coefficients for defined relationships (e.g., bulk density = f [clay mineralogy, compaction, season, and sodium]). Also, there will be more efforts to describe relationships in three dimensions, using simulation models. Examples may be plant, water, and nutrient behavior in the rooting zone and the transport of nutrients and pollutants into the vadose and groundwater zones. The advent

51. J. Bouma and H. A. J. Van Lanen, "Transfer Functions and Threshold Values: From Soil Characteristics to Land Qualities," in *Quantified Land Evaluation*, Proceedings of a Workshop by ISSS/SSSA [ITC Pub. no. 6], (1986): pp. 106–11.

52. (a) James Martin and J. J. Odell, *Object Oriented Analysis and Design* (Englewood Cliffs, N.J.: Prentice Hall, 1992). (b) David A. Taylor, *Object Oriented Technology: A Manager's Guide* (Reading, Mass.: Addison-Wesley Pub. Co., 1990).

53. Alan W. Brown, *Object Oriented Databases—Applications in Software Engineering* (Maidenhead, U.K.: McGraw Hill International, 1991).

of three-dimensional modeling will require that more attention be given to describing landscape position in both surface and subsurface states. Also, the near-surface soil characteristics and management-sensitive attributes will have to be better described.[54]

These changes will be motivated by development of three dimensional geographic informational systems based on digital terrain models, or deviations thereof. Relationship attribute descriptors will be linked to these systems in an attempt to manage and display complex and dynamic entities like soil systems. These developments will foster much greater understanding of soils and their role in the environment.

54. Dyke, personal communication.

9. Core Monographs in Soil Science

PETER MCDONALD

Cornell University

A. Purpose and Methods

This chapter examines the core monographs of the past forty years which are still valuable in academic teaching and research today. In order to arrive at a core monographic listing, a twin process of citation analysis and scholarly review was used. Monographs were extracted from a selected number of broad-based source documents. Citations in source documents were subjected to further analysis; data on a variety of citation elements were tabulated, including type of publisher, place of publication and date. The aim was to identify the prominent soils literature of greatest value to advanced students and beginning researchers worldwide representing a continuum from undergraduate literature to post-doctoral research literature.

The Core Agricultural Litertaure Project is described more fully in Chapter 3. One of the Project's main objectives was indepth analysis of the literature, including careful examination of the major agricultural databases such as CAB Abstracts (CABI) and AGRICOLA. Investigations also revealed that broad literature reviews, which might have assisted literature analysis, were not commonly published in soils journals. Most reviews published in journals were limited geographically or were subject specific, and dealt with current research; these were deemed too narrowly focused for this study which sought broader subject coverage and publications applicable to instruction. Instead, the emphasis of the citation analysis rested on monograph source documents which cover these criteria more fully. They were selected to meet other criteria as well, namely to establish a broad subject-divided base, to cover both temperate and tropical soils, to help identify literature use-patterns, to insure geographic distribution of authors, to cover the requisite time period, and to eliminate possible skews. As noted in Chapter 3, the Core Agricultural Literature Project has endeavored in all its analyses to determine if literature from developed countries is oriented differently than that for developing countries in order to make comparisons between these two groups.

To insure guidance and counsel in these efforts, the Core Agricultural Literature Project created a Soil Science Steering Committee comprised of eminent scholars from a variety of soil subject areas. One of the first tasks for this group was to aid in determining the source documents to be used for analysis. The Steering Committee members were:

Francis Broadbent
    University of California, Davis
William E. Larson
    University of Minnesota, St. Paul
Douglas Lathwell
    Cornell University

Parker Pratt
    University of California, Riverside
Larry Wilding
    Texas A & M University

Twenty-one monographs for citation analysis were identified and approved by the Steering Committee along with fourteen other titles from which monographs only were extracted. Four literature reviews were also included. All source documents are listed in Table 9.1.

B. Compilation and Citation Analysis

Analysis of the source documents revealed that of the 19,642 citations tabulated, 5,306 or 27% were to monographs. From these citations the following data were gathered:

Title of publication;
Date of publication;
Format of publication (e.g., journal or monograph);
Category of publisher (of monographs; e.g., commercial press, university, organization or government).

During analysis, titles of monographs were noted and entered into computerized lists. Each time a monograph or a chapter in it was cited, a tally was made for that title. The same was done for journals. Throughout the process, additional data were gathered providing the basis for the analysis in this and the following chapter.

Data were gathered on two major categories of material, serially published material and monographs. Serially published material included journals, serials, annuals, popular magazines and many conferences, in short

Table 9.1. Sources of citations in soil science analysis

Monograph sources

Bolt, G. H. and M. Bruggenwert. *Soil Chemistry: Basic Elements.* Amsterdam and New York: Elsevier, 1978. 277p.*

Brady, Nyle C. *The Nature and Properties of Soils.* New York: Macmillan Pub. Co., 1974. 639p.#

Buringh, P. *Introduction to the Study of Soils in Tropical and Subtropical Regions.* Wageningen: Centre for Agricultural Publishing (PUDOC), 1979. 124p. (3rdW)#

Coleman, David C. et al. eds. *Dynamics of Soil Organic Matter in Tropical Ecosystems.* (NifTAL Project). Honolulu: University of Hawaii, 1989. 249p. (3rdW)#

Dommergues, Y. R. and H. G. Diem. *Microbiology of Tropical Soils and Plant Productivity.* The Hague and London: M. Nijhoff, 1982. 328p. (3rdW)*

Drosdoff, Matthew et. al., eds. *Diversity of Soils in the Tropics.* Madison, Wis.: American Society of Agronomy, 1978. 219p. (ASA Spec. Pub. #34) (3rdW)#

Duchaufour, P. *Pedology: Pedogenesis and Classification.* London: Allen and Unwin, 1982. 448p.*

Engelstad, O. P. *Fertilizer Technology and Use.* Madison, Wis.: Soil Science Society of America, 1985. 633p.*

Feike, Leij and J. H. Dane, *A Review of Physical and Chemical Processes Pertaining to Solute Transport.* Auburn, Ala.: Alabama Ag. Exp. St., Auburn University, 1989. 155p.#

Fitzpatrick, E. A. *Micromorphology of Soils.* London and New York: Chapman and Hall, 1984. 433p.#

Glinski, Jan. *Soil: Physical Conditions and Plant Roots.* Boca Raton, Fla.: CRC Inc., 1990. 250p.*

Greenland, D. J. and R. Lal, eds. *Soil Conservation and Management in the Humid Tropics. Proceedings of a conference sponsored by ARC, Ibadan, Nigeria, 1975.* New York: J. Wiley, 1977. 283p. (3rdW)*

Hargrove, W. L. ed. *Cropping Strategies for Efficient Use of Water and Nitrogen.* Madison, Wis.: American Society of Agronomy, 1988. 218p. (ASA Spec. Pub. #51)#

Hillel, Daniel. *Applications of Soil Physics.* New York: Academic Press, 1980. 385p.*

International Board for Soil Research and Management (IBSRAM). *Management of Acid Tropical Soils for Sustainable Agriculture. Proceedings of an IBSRAM Inaugural Workshop.* Brazil: IBSRAM, 1986. 299p. (3rdW)*

International Rice Research Institute. *Priorities for Alleviating Soil-Related Constraints to Food Production in the Tropics.* Los Banos, Philippines: IRRI, 1980. 468p. (3rdW)#

Kabata-Pendias, Alina and Henryk Pendias. *Trace Elements in Soils and Plants.* Boca Raton, Fla.: CRC Press, 1983. 315p.*

Kuhnelt, Wilhelm. *Soil Biology.* London: Faber and Faber, 1976. 483p.*

Lal, R. and B. A. Stewart. *Soil Degradation: Advances in Soil Science. Vol. 11.* New York and Berlin: Springer-Verlag, 1989. 346p. (3rdW)*

Landon, J. R., ed. *Booker Tropical Soil Manual.* New York: Longman, 1984. 450p. (3rdW)*

McRae, Stuart G. *Practical Pedology: Studying Soils in the Field.* Chichester, U.K.: Ellis Horwood, 1988. 239p.#

Mortvedt, J. J. et al., eds. *Micronutrients in Agriculture. Proceedings of a symposium, Muscle Shoals, Ala., 1971.* Madison, Wis.: Soil Science Society of America, 1972. 666p.#

Orlov, D. S. *Humus Acids of Soils.* Moscow: Moscow University Publishers, 1974. (Translated by USDA/NSF in 1985).*

Rosenberg, Norman J. et al. *Microclimate: The Biological Environment.* New York: J. Wiley, 1987. 496p.*

Table 9.1. (Continued)

Monograph sources

Sanchez, Pedro A. *Suelos del Tropico: Caracteristicas y Manejo.* San Jose, Costa Rica: IICA, 1976. 618p. (3rdW)*

Soils and Rice: A symposium. Los Banos, Philippines: International Rice Research Institute, 1978. 825p. (3rdW)*

Soon, Y. K. ed. *Soil Nutrient Availability.* New York: Van Nostrand Reinhold Co., 1985. 352p. (Selected chapters analyzed)#

Stewart, G. A., ed. *Land Evaluation: CSIRO Symposium.* Melbourne, Australia: Macmillan of Australia, 1968. 392p.#

Tan, Kim H. *Principles of Soil Chemistry.* New York: Dekker, 1982. 267p.#

Tan, Kim H. *Andosols.* New York: Van Norstrand Reinhold Co., 1984. 418p. (3rdW)*

Trudgill, Stephen T. *Soil and Vegetation Systems.* Oxford, U.K.: Clarendon Press, 1977. 180p.#

Unger, Paul W. *Tillage Systems for Soil and Water Conservation*, (Bulletin 54). Rome: FAO, 1984. 278p. (3rdW)*

Walsh, Leo M. et al., eds. *Soil Testing and Plant Analysis.* Madison, Wis.: Soil Science Society of America, 1973. 491p.*

Wilding, L. P. et. al., eds. *Pedogenesis and Soil Taxonomy.* Amsterdam: Elsevier, 1983. 2 vols.*

Young, Anthony. *Tropical Soils and Soil Survey.* Cambridge, U.K.: Cambridge University Press, 1976. 468p. (3rdW)*

Journal sources

Bergkvist, Bo et al., "Fluxes of Cu, Zn, Pb . . . in Temperate Forest Ecosystems: A Literature Review," *Water, Air and Soil Pollution* 47 (4) (1989): 217–286.*

Burns, I. G., "Influence of the Spatial Distribution of Nitrate on the Uptake of N by Plants: A Review . . . ," *Journal of Soil Science* 31 (1980): 155–173.*

Karlen, D. L. et al., "Soil Tilth: A Review of Past Preconceptions and Future Needs," *Soil Science Society of America Journal* 54 (1) (1990): 153–161.*

Rice, James A. and Patrick MacCarthy, "Comments on the Literature of the Humin Fraction of Humus," *Geoderma* 43 (1) (1988): 65–73.*

* Works analyzed
3rdW = Third World soils monographs
# Works from which monographs only were extracted for evaluation

anything published in a numbered series which lacks a unique title per issue (see Chapter 10). Thus journals follow the pattern of publishing several articles on different and specific subjects with a cover title which remains the same and only the volume number and the issue number change. More problematic were monographs since these, on occasion, are serially published but were counted here as unique monographic titles. The distinguishing characteristics for monographs in this study were several, including: that they were cited as distinct works with an author or editor, that the title itself was specific to the work, and that the item was complete in itself. Good examples of serially published works counted as monographs are the series of *FAO Soils Bulletins*, where each of the bulletins has a seperate author, a unique subject related to soils and a distinct title. The same held true for proceedings if they met these criteria and had a seperate editor and a unique title; otherwise they were counted in the journal category. Chapters in books were counted as monographic titles. These definitions generally followed the citation pattern of soil science authors.

Some material was excluded from the lists.

(1) Pamphlet-like material of fifty pages or less which were cited only once or twice.
(2) Local and national government documents of brief pagination on highly specialized and site-specific topics.
(3) Select titles on subjects well outside the discipline of soil science. (Some of these titles were placed in the monographic lists of other disciplines.)
(4) Specialized geographic material when not in a national or international context; only of use for a restricted site.
(5) Publications prior to 1950, except in special highly cited cases.
(6) Early editions were combined with the latest edition although data were kept on all printings, reprints and revised editions.
(7) Works on forest soils are included, but only those highly cited in the counts and given high marks by reviewers because this will be dealt with seperately in a volume on forestry and agroforestry.

In order to assist the reviewers in ranking the literature, the discipline was divided into seven subject categories. These were:

(1) Soil genesis, morphology and classification
(2) Soil physics and agro-climatology
(3) Soil conservation and water management, erosion
(4) Soil chemistry and mineralogy
(5) Soil fertility and fertilizers
(6) Soil biology and biochemistry
(7) General soil science, with applicable titles in sustainable agriculture, agronomy, agricultural engineering, international development and other allied fields.

The Third World list of monographs for evaluation was not divided by subject in part because soil research in these regions relies less rigidly on subject specialties. Source documents were chosen to reflect the full range of this subject and geographical diversity.

C. Results of Source Document Analysis

Monographic analysis of the citation data clearly revealed that the dominant country of publication was the United States with 47.2% of the monograph citations. Europe as a whole had 31.5%. The Third World, as a whole, received 10.2%. Table 9.2 lists the major countries and their percentage of the citation counts.

Table 9.2 indicates that the Philippines was far and away the most commonly cited Third World county of publication, at 3.7%, and fifth of all countries, primarily because of the International Rice Research Institute (IRRI). Nigeria was second, predominantly because of its International Institute of Tropical Agriculture (IITA), with 0.8%. Brazil was third, with 0.5% of the citations, in part, because Empressa Brasileira de Pesquisa Agropecuaria (EMBRAPA) has such a strong publishing record. The Ministry of Agriculture of Indonesia was responsible for most of its count of 0.3%, the same percentage as Colombia. Here, the count is attributed to work carried out at the Centro Internacional de Agricultura Tropical (CIAT), in the town of Cali in the Andean foothills.

There is an overriding correlation between country of publication in the Third World and a single dominant publisher, for example, IRRI in the Philippines, which accounted for 98% of the country's cited publications. Publication trends in the developed world are not so easily categorized. The higher counts for developed world publications mean a greater array of data can be analyzed, particularly on types of publishers.

Analysis of the monograph citations in the source documents gathered data on four major types of publishers. These were commercial publishers;

Table 9.2. Citation counts of publishing country for monographs

Country	% of Total
United States	47.2
United Kingdom	12.2
Italy	6.7
USSR	5.2
Philippines	3.7
Netherlands	3.6
Australia	3.2
France	3.2
Germany	3.1
Japan	1.0
India	0.7
Other	10.2

independent organizations (such as societies and associations); governments, including the United Nations; and universities, their presses, departments, and institutes. Table 9.3 lists the publisher types and their percent of the total cited literature.

Seventy-two separate commercial publishing houses were identified, which were located in twenty countries. The latter statistic is based on first city or country cited in the imprint of the publication, and may not always be representative, because many publishers are multinational and publish in many countries.

The most highly cited commercial publications were

Schlichting, Ernest, and Udo Schwertmann. *Pseudogley and Gley: Genesis and Use of Hydromorphic Soils.* Weinheim, Germany: Verlag Chemie, 1973. 771p.
Sanchez, Pedro A. *Properties and Management of Soils in the Tropics.* New York: Wiley, 1976. 627p.

Independent organizations were the next most commonly cited publishers. This category is made up primarily of professional societies, but included national trade associations, international research institutes, and some corporately funded groups with ties to agriculture. Two American professional societies stood out, both for the developed world and the Third World. The American Society of Agronomy was first, with 18.9% of all organizational citations, followed by the Soil Science Society of America, with 16.7%. Third, with 15%, was the International Rice Research Institute. The top two monographs cited in both the Third World and the developed world list were published by organizations. These were

Mortvedt, J. J., editorial chairman. *Micronutrients in Agriculture.* 2d ed. Madison, Wis.: Soil Science Society of America, 1991. 670p.
Page, A. L., and R. H. Miller, eds. *Methods of Soil Analysis.* Part 2. Madison, Wis.: American Society of Agronomy, Soil Science Society of America, 1982. 2 vols. (*Agronomy Series*, no. 9).

Table 9.3. Publisher types by citation analysis

Type	Percentage
Commercial	30.7
Independent organ.	24.8
Government	23.8
University	20.7

Government publications included both national agencies, such as the British Ministry of Agriculture, Fisheries and Food (MAFF) and its American counterpart, the U.S. Department of Agriculture, as well as international bodies, such as the United Nations and the Food and Agriculture Organization (FAO). The most commonly cited government material in aggregate was published by the FAO, primarily, specific trade statistics and their *Soils Bulletins*. The highest cited single government monograph in source documents was published by the USDA: U.S. Soil Conservation Service. Soil Survey Staff. *Soil Taxonomy: A Basic System of Soil Classification for Making and Interpreting Soil Surveys*. Washington, D.C.: Soil Conservation Service, USDA, 1976. 233p.

For FAO, it was *Shifting Cultivation and Soil Conservation in Africa*. Rome: FAO, 1974. 245p. (*FAO Soils Bulletin* no. 24).

One other FAO monograph not in a series received may citations: FAO/UNESCO *Soil Map of the World* published in ten volumes in Paris in the early 1970s. This contains the only national soils map available for many countries. Soil maps are dealt with more fully in Chapter 12.

Universities, which accounted for only 20.7% of the total monograph citations analyzed, do not have a strong publishing record in soil science. United States universities had the highest count, with 637 citations, of which, 257 (40.3%) were to United States agricultural experiment station (AES) Bulletins at land-grant universities. Although the influence of these publications has waned since the Second World War, they still remain an important source of research. This is demonstrated by the fact that between 1920 and 1940, they accounted for over 95% of university citation counts (see Chapter 14), whereas today, that figure is well below 50%. Top ranked among the experiment stations were Texas A & M and North Carolina State University. The preeminence of the latter rests primarily upon its work in tropical soils, notably, *A Review of Soils Research in Tropical Latin America*, by Pedro Sanchez (North Carolina Agr. Exp. Sta. *Technical Bulletin 219*, 1973). Also highly cited were AES publications from Auburn, Louisiana State, Nebraska, and Washington State universities.

Oxford and Cambridge were responsible for 96% of the United Kingdom's 124 university citations. Australian universities were next, with thirty-three, only one citation less than all the Third World universities combined, at thirty-four.

Other Measurements and Data

It is often useful in citation analysis such as this to test the duration of relevancy of the tabulated material, and the commonest measure is known as half-life. Cited half-life is defined as the number of publication years, going back from the current year, which account for 50% of the total citations recorded. Knowing the half-life of cited material gives a rough measurement of its relevancy over time; the older an article or monograph, the less likely it is to be cited. In aggregate, tabulations of the 5,306 monograph citations in this study show that their half-life is 11.2 years. Citations taper off precipitously prior to this half-life. By the 1940s, they are few and far between. Data were kept from 1929 to the present; for 1929, there were no citations, and prior to that date, only twelve, primarily to the works of Eugene Hilgard,[1] Justus Liebig,[2] and Charles Darwin.[3]

A cited half-life of over eleven years is a long time for scientific literature to remain current. In soil science, we may attribute this to several factors, but especially to the high counts received by seminal works in several editions. Of the five top-ranked monographs, four were in their second or third edition, and the fifth title was published over twenty years ago in 1968. It is well to remember the exceptionally high count the FAO/ UNESCO *Soil Map* received for the period 1966–70.

Two categories of monographic material were tabulated separately—international conferences and dissertations. No place of publication data was kept on conferences in this analysis. In many ways, the papers presented at conferences are unique publications, commonly of hybrid imprint, and it is not uncommon for a conference to be held in one country, sponsored by a society from another, and, finally, its proceedings printed in a third. This is especially true of conferences held in the developing world. To circumvent this problem, conferences were tabulated separately, along with their date.

Conferences, broadly including congresses, symposia, and workshops, accounted for 11% of all monographic citations. Forty-three conferences were listed, although only a few were highly cited, and many have been deleted from the lists at the suggestion of reviewers. The series most commonly cited was the International Congress of Soil Science, sponsored by the International Society of Soil Science (ISSS). Indeed, the ISSS's *9th International Congress of Soil Science*, held in Adelaide, Australia, in 1968 (4 vols. New York: Elsevier), consistently ranked in the top ten monographs on both the developed and developing world lists. It is worth noting that the greatest disparity in ranking by reviewers was with important conferences, with about a third marking them of little value, a third, of greater but secondary importance, and a final third, as very important.

1. Eugene W. Hilgard, *Soils: Their Formation, Properties, Composition, and Relations to Climate and Plant Growth in the Humid and Arid Regions* (New York and London: Macmillan, 1906).
    2. Justus Liebig, *Organic Chemistry in its Applications to Agriculture and Physiology = Organische Chemie in ihrer Anwendung auf Agricultur und Physiologie*, ed. Lyon Playfair, 1st Am. ed. (Cambridge, Mass.: J. Owen, 1841); 1st Ger. ed. (Brunswick Vieweg, 1841).
    3. Charles Darwin, *The Formation of Vegetable Mould, Through the Action of Worms: With Observations on Their Habits* (London: J. Murray, 1881).

Dissertations, which accounted for 5.1% of all citations, obviously have only a marginal impact on the literature of soil science, especially in North America. Eight of the source documents did not cite a single dissertation, and five cited only one. Indeed, over half of the 274 dissertation citations came from one source document, *Pedology: Pedogenesis and Classification*, by the Frenchman Philippe Duchaufour. Only one source document from France was analyzed, so this does not provide enough evidence to decide whether this is a peculiarly French phenomenon or the author's. This is a classic example of a citation skew, where a single document inflates the data in a single category. It is true to say, however, that United States and United Kingdom dissertations accounted for fewer than 10% of the total, and from only a few documents, mostly dealing with the tropics.

D. Weighting the Monograph Lists

In order to rank the titles, several elements were used.

(1) Each time a monograph or a chapter of a monograph was cited in one of the source documents, the title was given a "hit." This element was given the weight of one per citation.
(2) Rankings by reviewers were given a weighted prominence in the calculations. For the developed world, since the lists were divided by subject category, each subject list was weighted separately according to the number of list replies per category. Category 7, the general category, which was sent to all reviewers, was weighted to match category lists with fewer replies.
(3) If a title was reprinted or translated, it was given a score of one.
(4) If a title went through more than one edition, it was given a score of two.

The equation for the ranking computations is

$$\text{(no. of counts)} + \text{(top rank} \times 2) + \text{(2nd rank} \times 1) + \text{(edition} \times 2) + \text{(reprint} \times 1) + \text{(translation} \times 1)$$

This formula was used for both the developed and the Third World lists, which were sent to reviewers separately. The scope of analysis reflected temperate zone soil science, as well as that pertaining to the tropics, which translated into literature of the developed world and developing world, respectively. By this equation, the peer evaluation acounted for 67.7% and 82.3% of the final tabulations for the developed and Third World lists, respectively.

Figure 9.1 shows the scoring curve for both lists.

E. Comparison of the Core Monographic Lists

The Third World and developed world lists combined have 941 distinct titles; 323 are unique to the Third World, and 270, to the developed, with

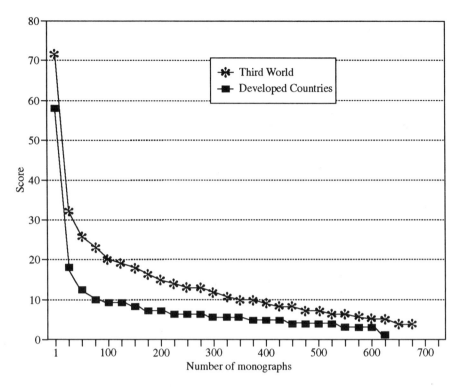

Figure 9.1. Scoring of monographs for developed and Third World countries.

348 common to both. The fact that almost half the titles should be unique to the developed and developing world lists, respectively, is not surprising. Soil science is unlike other agricultural disciplines for a number of reasons. For one, while general principles will apply worldwide, soils themselves respect no boundaries and are variable from site to site. As a consequence, research in one location may not serve well in another. Indeed, the major subject divisions, such as soil genesis, biology, fertility, chemistry, physics, and management, can be further divided by locale. Geographical division of these subdisciplines will create further branching of research priorities. In short, in no other agricultural discipline, except perhaps agricultural economics, will geographic principles play such a large part in defining research. These geographic principles, in turn, are reflected in the literature.

Given this caveat, this effort was careful to include, for the Third World monograph list, only those titles which would have wide application, such as regional overviews. Many titles which were locally published and site specific were dropped from the list as being of marginal value when ranked by our reviewers. However, analysis of these titles reveal that there is in the

Third World an extensive soils literature published and distributed at the national level, often in the language of the local populace, with restricted, local relevance (see Chapter 4). This seemed especially true of Asian countries, such as Malaysia, Myanmar, Indonesia, and the Philippines, where government agricultural publications are prolific. Tracts of a similar nature were also common in South and Central America, but since Spanish and Portugese were the common language of publication, these stood a better chance of being accepted by our reviewers, and some remain on the list.

The original size of the developed world list before the review process began was 1,463 titles, of which 845, or 57.7%, have been removed, creating the core list of 618 titles. For the Third World, the original size was 1,127 titles, of which 456 (40.4%) dropped out, leaving 671 titles as core. Deletions common to both lists were 297, or 26.3%, of the original Third World count and 20.3% of the developed. Instructions to reviewers carefully delineated the aim of identifying monographs of academic and instructional value for university-level work. With this, many titles were dropped from the lists as the review process proceeded. The most common cause of deletion was a combination of low citation count from one source document and that the title was unknown to reviewers. Obscure titles tended to drop out in the early iterations of the review process. Also, many titles were dated and had been superseded by better works. Unless of seminal importance, works from the 1950s and 1960s generally fell out by the second round of review. It should be noted that many, if not most, non-English language monographs were deleted by the reviewers during this process, and that titles in non-Romance languages generally fell out, being difficult for scientists to handle. Finally, some titles were universally considered of poor quality and were widely rejected by reviewers.

Thirty-six reviewers responded to rank the developed world list, and forty-one reviewed the Third World list. They were instructed to evaluate the listings and rank titles deemed important in advanced soil science education and research. Evaluations were requested for only those titles of which they had some knowledge. They were encouraged to recommend titles for inclusion to be reviewed by others scholars. The process was iterative, and four mailings of revised lists were sent out at two-month intervals. The names and affiliations of the reviewers are listed below.

Developed Countries Reviewers

E.G. Beauchamp
  University of Guelph
  Guelph, Canada
Hans-Peter Blume
  Institüt für Pflanzener-
  nährung und Boden-

kunde
  Kiel, Germany
Larry Boersma
  Oregon State University
  Corvallis, Oregon

D. Bouldin
  Cornell University
  Ithaca, N.Y.
E.M. Bridges
  University College of
  Swansea, U.K.

E.J. Dent
Regional Office for Asia and the Pacific, F.A.O.
Bangkok, Thailand

H.E. Dregne
International Center for Arid & Semiarid Land Studies
Texas Tech University
Lubbock, Tex.

S.A. El-Swaify
University of Hawaii, Manoa
Honolulu, Hawaii

M.I. Gerasimova
Moscow State University
Moscow, Russia

R.J. Gilkes
The University of Western Australia
Nedlands, Australia

Wenceslau Goedert
EMBRAPA
Brasília, Brazil

Horst Grimme
Hannover, Germany

Roger Hanson
North Carolina State University
Raleigh, N.C.

K.H. Hartge
Institut für Bodenkunde
Hannover, Germany

Lloyd R. Hossner
Texas A & M University
College Station, Tex.

M. Kutílek
Soil Science Laboratory
Prague, Czechoslovakia

Williard L. Lindsay
Colorado State University
Ft. Collins, Colo.

R.H. Loeppert
Texas A & M University
College Station, Tex.

John Mortvedt
Tennessee Valley Authority
Muscle Shoals, Ala.

Donald R. Nielsen
University of California, Davis
Davis, Calif.

P.H. Nye
University of Oxford
Oxford, U.K.

Georges Pédro
ORSTOM
Paris, France

W.S. Reid
Cornell University
Ithaca, N.Y.

Guennadi Schepaschenko
Dokuchaev Soil Science Institute
Moscow, Russia

Udo Schwertmann
Institut für Bodenkunde
Technische Universität München in Weihenstephan
Germany

Carlos O. Scoppa
INTA-CIRN
Buenos Aires, Argentina

P.B. Tinker
Terrestrial and Freshwater Sciences
Swindon, U.K.

José Torrent
Universidad de Córdoba
Córdoba, Spain

Hans Van Baren
International Soil Reference and Information Centre
Wageningen, Netherlands

Armand Van Wambeke
Cornell University
Ithaca, N.Y.

B.P. Warkentin
Oregon State University
Corvallis, Oregon

Peter J. Wierenga
University of Arizona
Tucson, Ariz.

Duane C. Wolf
University of Arkansas
Fayetteville, Ark.

Carlos Zamora Jimeno
Lima, Peru

S.V. Zonn
Institut Geografii
Moscow, Russia

D.A. Zuberer
Texas A & M University
College Station, Tex.

Third World Countries Reviewers

I.P. Abrol
Indian Council of Agricultural Research
New Delhi, India

Andrés Aguilar S.
Centro de Edafologia
Colegio de Postgraduados
Montecillo, Mexico

Gabriel Alcántar G.
Centro de Edafologia
Colegio de Postgraduados
Montecillo, Mexico

M.S. Bajwa
Punjab Agric. University
Ludhiana, India

A.M. Balba
University of Alexandria
Alexandria, Egypt

Elemer Bornemisza
Center for Agricultural Research
University of Costa Rica
San Pedro, Costa Rica

Elke J.B.N. Cardoso
ESALQ
Universidad de São Paulo
Piracicaba, Brazil

José L.I. Demattê
ESALQ
Universidad de São Paulo
Piracicaba, Brazil

F.J. Dent
Regional Office for Asia

and the Pacific, F.A.O.
Bangkok, Thailand
El-Sayed Elkhatib
University of Alexandria
Alexandria, Egypt
Ramón Fernández G.
Centro de Edafologia
Colegio de Postgradu-
ados
Montecillo, Mexico
Olatunji A. Folorunso
University of Maiduguri
Maiduguri, Borno State,
Nigeria
Manuel Anaya Garduño
Centro de Edafologia
Montecillo, Mexico
G.P. Gillman
L.W. Lambourn & Col.,
Ltd.
Croydon, U.K.
M. El-Haris
University of Alexan-
dria,
Alexandria, Egypt
Marcel Jamagne
Institut National de la
Recherche Agronomique
Centre de Recherches
d'Orléans
Orléans, France
Jorge de C. Kiehl
ESALO
Universidad de São Paulo
Piracicaba, Brazil
Toshiaki Kinjo
Universidad de São Paulo
São Paulo, Brazil
Luis Alfredo Leon
IFDC/CIAT
Cali, Colombia
Walter Luzio-Leighton
Universidad de Chile
Santiago, Chile

Arary Marconi
ESALQ
Universidad de São Paulo
Piracicaba, Brazil
Zilmar Z. Marcos
ESALQ
Universidad de São Paulo
Piracicaba, Brazil
Shah Muhammed
University of Agricul-
ture
Faisalabad, Pakistan
André M.L. Neptune
ESALQ
Universidad de São Paulo
Piracicaba, Brazil
C.S. Ofori
Food and Agriculture
Organization
Rome, Italy
José Luis Oropeza M.
Centro de Edafologia
Colegio de Postgradu-
ados
Montecillo, Mexico
Inés Pino
Comisión Chilena de
Energía Nuclear
Santiago, Chile
S.S. Prihar
PAFPRI
Punjab Agric. Univer-
sity
Ludhiana, India
Maurice F. Purnell
Food and Agriculture
Organization
Rome, Italy
Carlos Ramirez
Center for Agricultural
Research
University of Costa Rica
San Pedro, Costa Rica

U.S. Sree Ramulu
Water Technology Cen-
tre
Tamil Nadu Agric. Uni-
versity
Coimbatore, India
M. Saleh
University of Alexandria
Alexandria, Egypt
R. Sant' Anna
Food and Agriculture
Organization
Accra, Ghana
B.S. Sidhu
Punjab Agric. Univer-
sity
Ludhiana, India
P.S. Sidhu
Punjab Agric. Univer-
sity
Ludhiana, India
Hugo da Silva
Facultad de Agronomia
La Plata, Argentia
M.W. Thenabadu
University of Sri Lanka
Kandy, Sri Lanka
Armand Van Wambeke
Cornell University
Ithaca, N.Y.
J.Y. Yayock
Ahmadu Bello Univer-
sity
Zaria, Nigeria
Huang Zhiwu
South China Agri-
cultural University
Guangzhou, China
Zhao-liang Zhu
Institute of Soil Science
Chinese Academy of
Sciences
Nanking, China

At first glance it may seem unusual that the Third World monograph list is larger than its counterpart by fifty-three titles. Analysis of the responses given by Third World reviewers revealed that they were keener on giving high rankings to a broader scope of material than were the reviewers of the developed world lists. Indeed, a comparison of the top-cited material in both lists reveals a similarity not matched by analysis of the lists in other agricultural disciplines. The higher number of titles on the Third World list may also reflect the nature of tropical soils themselves, which have unique properties. In order to include this pertinent literature, as well as important monographs from the developed world, it would appear Third World reviewers chose more titles for inclusion.

The division by subject category for the developed list is compared below in Table 9.4 to the 348 titles common to both lists. Because developed world titles were divided by subject and Third World titles were not, titles common to both lists afford a tacit way to analyze Third World subject emphasis. Table 9.4 reveals the similarity between the lists. Soil genesis and classification came out on top on both lists, followed by soil biology, soil chemistry, soil management and erosion and last soil physics, with the category of general soil science posting sixty-three of the 348 titles common to both lists.

Certainly the overall similarity in subject emphasis between developing world patterns and those in the Third World deserves further analysis. It suggests there is still a close link between the two despite obvious differences, and this correlating emphasis may be due in part to the influence outside experts still exert in tropical soils research and to the educational influence of developed world universities on scientists returning to their home countries. Both are likely to have measurable impact.[4] Finally, when we look at the discipline as a whole, these figures suggest that soil science has matured, becoming in the process a more inter-connected worldwide enterprise. Expanding international access to a common body of literature may also play a part.

Comparsion of the numbers in Table 9.5, reveals that commercial presses publish the bulk of the core titles in both categories, with governments,

Table 9.4. Subject division of core lists

Subject category	Developed Countries only	348 titles common to both lists
Soil Genesis and Morphology	21.9	20.9
Soil Physics	10.4	8.3
Soil Management and Erosion	11.9	12.1
Soil Chemistry	13.1	12.6
Soil Fertility and Fertilizers	9.8	12.1
Soil Biology and Microbiology	17.9	15.8
General Category	15.0	18.2

4. Some of these topics are dealt with by Rigas Arvanitis and Yvon Chatelin in Chapter 4.

Table 9.5. Comparison of publishers

Type	Developed list		Third World list	
Commercial	358	(57.9%)	312	(46.4%)
Government	116	(18.7%)	171	(25.4%)
Organization	92	(14.8%)	120	(19.4%)
University	52	(8.6%)	68	(8.8%)

organizations and universities following in descending order. It would appear that both government publishers and organizations have a stronger publishing record in the Third World than commercial houses. One is tempted to suggest an economic basis for this, that commercial publishers will publish where the economic market is greatest. Monographs on regional or tropical agriculture which may not have such a market, will be published more readily by local governments or international organizations.

As Table 9.5 shows, the dominant type of publisher was commercial. Eight major commercial publishers stood out as pre-eminent in the core lists. Table 9.6 ranks the top publishers of all types for the two lists. Among the top eleven, eight were commerical, two organizational and one governmental, namely the FAO in Rome. The United States was the country of publication for six of the top eleven, Germany second with two, with the United Kingdom, the Netherlands and Italy with one apiece.

It is not surprising that FAO stands so high in the Third World counts.

Table 9.6. Top-ranked publishers in soil science by numbers of titles published

Publisher	Developed list	Third World list
Wiley (U.S./c)*	55	49
Academic Press (U.S./c)	35	23
Elsevier (Netherlands/c)	29	26
Soil Science Society of America (U.S./o)	29	22
American Society of Agronomy (U.S./o)	27	27
Springer-Verlag (Germany/c)	20	16
Dekker (Germany/c)	13	7
McGraw-Hill (U.S./c)	11	4
Clarendon (U.K./u)	11	12
Van Nostrand (U.S./c)	10	12
Food and Agriculture Organization (Italy/g)	9	59

* Publishers followed by country and type: c = commercial; o = organization; g = government; u = university.

Their series *FAO Soils Bulletins* consistently came up in the analysis of tropical soils source documents and many titles were highly ranked by reviewers. These *Bulletins*, while part of a numbered series, each have a unique title and usually cover applied aspects of the discipline, serving as manuals on specific topics.

Analysis of country of publication is problematic, because so many publishing houses have offices around the globe. In the comparative counts ranked in Table 9.7, the country of publication listed first on the title page was considered the prominent place and tabulated. In both categories, the United States was the primary country of publication worldwide.

The high count of Third World publishers in the column on the right reflects the importance of two major countries, namely India (thirty-five titles) and the Philippines (eighteen). Thailand and Brazil both had six titles, with Colombia five. The FAO in Rome accounts for all the publications from Italy in the column on the right.

Table 9.7. Country of publication

	Developed Countries titles	Third World
United States	295	254
United Kingdom	128	109
Netherlands	51	54
Germany	39	32
France	23	29
Australia	13	10
Italy	9	57
USSR	8	4
Israel	8	8
Austria	5	7
China	3	7
Japan	3	6
Third World	25	69
Other	8	25

F. The Top Twenty Monographs

The top-ranked monographs for both the developed countries and the Third World countries are displayed in Table 9.8. Nine of the titles are in both lists, which demonstrates that there is strong agreement on which books are most valuable. Indeed, the top three in both categories are the

Table 9.8. Top twenty monographs in soil science

Developed Countries	Third World	
14	18	Adams, Fred, ed. Soil Acidity and Liming. 2d ed. Madison, Wis.: American Society of Agronomy, 1984. 380p. (Advances in Agronomy no. 12) (1st ed. 1967)
	8	Alexander, Martin. Introduction to Soil Microbiology. 2d ed. New York: J. Wiley, 1977. 467p.
	16	Allison, Franklin E. Soil Organic Matter and its Role in Crop Production. Amsterdam and New York: Elsevier, 1973. 639p.
	19	Bartholomew, William V., and Francis E. Clark, eds. Soil Nitrogen. Madison, Wis.: American Society of Agronomy, 1965. 615p. (Agronomy no. 10)
17	6	Black, C. A. Soil-Plant Relationships. 2d ed., reprint. Malabar, Fla.: R. E. Krieger, 1984. 792p. (Reprint of 2d ed., New York: Wiley, 1968.) (1st ed. 1957)
7	12	Brady, Nyle C. The Nature and Properties of Soils. 10th ed. New York: Macmillan, 1990. 621p. (1st ed. by Thomas Lyon, New York: Macmillan, 1922)
20		Brewer, Roy. Fabric and Mineral Analysis of Soils. Huntington, N.Y.: R. E. Krieger, 1976. 482p. (Reprint of the ed. published by J. Wiley with supplementary material.)
4	5	Dixon, Joe B., S. B. Weed, and Richard C. Dinauer, eds. Minerals in Soil Environments. 2d ed. Madison, WI; Soil Science Society of America, 1989. 1244p. (1st ed 1977)
	14	Dregne, H. E. Soils of Arid Regions. Amsterdam and New York: Elsevier, 1976. 237p.
18		Duchaufour, Philippe. Pédogenese et Classificati 2d ed., rev. and enl. Paris and New York; Masson, 1983. 491p. (Pedologie no. 1) (1st ed., 1977. 477p. English ed. titled: Pedology: Pedogenesis and Classification. London: Allen & Unwin, 1982. 448p.)
13		Engelstad, Orvis P., ed. Fertilizer Technology and Use. 3d ed. Madison, Wis.: Soil Science Society of America, 1985. 633p.
	15	Greenland, D. J., and M. H. B. Hayes. The Chemistry of Soil Constituents. Chichester, U.K. and New York: J. Wiley, 1978. 469p.
5	9	International Congress of Soil Science, 9th, Adelaide, Australia, 1968. Proceedings . . . New York; Elsevier, 1968. 4 vols. (English, French, or German, summaries in all three languages.)
	13	Jackson, Marion L. Soil Chemical Analysis: Advanced Course. A Manual of Methods Useful for Instruction and Research in Soil Chemistry, Physical Chemistry of Soils, Soil Fertility, and Soil Genesis. 2d ed. Madison, Wis.: Dept. of Science, University of Wisconsin, 1974. 895p. (1st ed. 1956)
	10	Khasawneh, F. E., E. C. Sample, and E. J. Kamprath, eds. The Role of Phosphorus in Agriculture. Proceedings of a Symposium, National Fertlizer Development Center, Tennessee Valley Authority, Muscle Shoals, Alabama, June 1976. Madison, Wis.: American Society of Agronomy, 1980. 910p.

Developed Countries	Third World	
19		McLaren, Arthur D., George H. Peterson, and J. Skuji, eds. Soil Biochemistry. New York: M. Dekker, 1967–90. 7 vols.
16		Meisinger, J. J., G. W. Randall, and M. L. Vitosh, eds. Nitrification Inhibitors—Potentials and Limitations. Papers presented at the Annual Meeting of the American Society of Agronomy, Chicago, Ill., Dec. 1978. Madison, Wis.: American Society of Agronomy and Soil Science Society of America, 1980. 129p.
	20	Miller, Raymond, and Roy L. Donahue. Soils: An Introduction to Soils and Plant Growth. 6th ed. Englewood Cliffs, N.J.: Prentice-Hall, 1990. 768p. (1st ed., 1958)
	17	Mohr, Edward C. J. Tropical Soils: A Comprehensive Study of Their Genesis. 3d rev. ed. The Hague: Mouton-Ichtiar Baru-Van Hoeve, 1973. 481p. (1st ed. 1954)
1	3	Mortvedt, J. J. et al., eds. Micronutrients in Agriculture. 2d ed. Madison, Wis.: Soil Science Society of America, 1991. 760p. (1st ed. 1972. "Proceedings of a Symposium, Muscle Shoals, Ala., Apr. 1971.")
2	1	Page, A. L. and R.H. Miller, eds. Methods of Soil Analysis. Part 2. 2d ed. Madison, Wis.: American Society of Agronomy, Soil Science Society of America, 1982–. (Agronomy Series no. 9) (1st ed.by C. A. Black, ed., 1965. 2 vols.)
8	10	Russell, Edward J. Russell's Soil Conditions and Plant Growth. 11th ed., edited by Alan Wild. Burnt Mill, U.K.: Longmans and New York: J. Wiley, 1988. 991p. (Rev. ed. of: Soil Conditions and Plant Growth, 10th ed., 1973. 1st ed., Longmans, 1912.)
6	7	Sanchez, Pedro A. Properties and Management of Soils in the Tropics = Suelos del Tropico: Caracteristicas y manejo. 1st ed. San Jose, Costa Rica: Instituto Interamericano de Cooperacion para la Agricultura, 1981. 634p. (Translation of: Properties and Management of Soils in the Tropics. New York: J. Wiley, 1976.)
9		Schlichting, Ernst and Udo Schwertmann. Pseudogley and Gley: Genesis and Use of Hydromorphic Soils. Weinheim, Germany: Verlag Chemie, 1973. 771p. (English, French or German.)
	11	Tisdale, Samuel L. Soil Fertility and Fertilizers. 4th ed. New York and London: Macmillan, 1985. 754p.
11		United States. Soil Conservation Service. Soil Classification: A Comprehensive System. 7th Approximation. Washington, D.C.: Soil Survey Staff, Soil Conservation Service, U.S. Dept. of Agriculture, 1960. 265p.
3	2	United States. Soil Conservation Service. Soil Survey Staff. Soil Taxonomy: A Basic System of Soil Classification for Making and Interpreting Soil Surveys. Washington, D.C.; Soil Conservation Service, U.S. Dept. of Agriculture, 1975. 754p. (USDA Agriculture Handbook no. 436)
	4	Westerman, R. L., ed. Soil Testing and Plant Analysis. 3d ed. Madison, Wis.: Soil Science Society of America, 1990. 784p.

Table 9.8. (Continued)

Developed Countries	Third World	
		(SSSA Special Publication no. 3) (Rev. ed. by Leo M. Walsh and James D. Beaton, eds, 1973. 491p.) (1st ed. 1967.)
12		Wilding, L. P., N. E. Smeck, and G. F. Hall, eds. Pedogenesis and Soil Taxonomy. Amsterdam and New York: Elsevier, 1983. 2 vols.
15		Yaalon, Dan H., ed. Paleopedology: Origin, Nature, and Dating of Paleosols. Jerusalem: International Society of Soil Science, 1971. 350p. (Symposium on the Age of Parent Materials and Soils, Amsterdam, 1970.)

same. Although the FAO soil classification system is of great importance especially in the developing world, the high ranking of the U.S. Soil Conservation Service, *Soil Taxonomy* (1975) as well as its *7th Approximation* (1960) in both lists shows that it is gaining in importance with wider application because of its greater detail. The only title specific to the tropics common to both lists is Sanchez's *Properties and Management of Soils in the Tropics,* (1981). On the Third World listing, four titles have tropical themes. Works on soil genesis and classification predominate on both lists, with broad overviews such as Russell's *Soil Conditions and Plant Growth* next in number and titles on soil fertility third. Fourteen of the top twenty titles on the Third World list are either in multiple or revised editions with eleven on the developed list.

The list of core monographs for soil science follows. These 930 titles are divided into first rank titles, which were highly cited in the literature and which the reviewers considered of great importance; second rank, which generally received fewer citations but which many reviewers still thought important; and, finally, third rank, titles which had few counts but received mixed responses distributed evenly between "important" to "of little value." Taken as a whole, this list represents the core soils monographs for research and instruction of the last forty years. Table 9.9 shows how the monographs are divided by rank in the two categories.

Table 9.9. Numbers pertaining to rankings and categories of monograph list

	First rank	Second rank	Third rank	Total
Developed Countries	96	196	315	607
Third World	112	302	253	667

Core List of Soils Monographs (927 titles)

Developed Countries		Third World
	A	
	Aandahl, A. R. et al., eds. Histosols, Their Characteristics, Classification, and Use; Proceedings of a Symposium, Miami Beach, Fla., Oct. 1972. Madison, Wis.: Soil Science Society of America, 1974. 136p. (SSSA Special Publication no. 6)	Second
Third	Abrol, Y. P., ed. Nitrogen in Higher Plants. Taunton, U.K.: Research Studies Press and New York: J. Wiley, 1990. 492p.	
Third	Abrol, Y. P. et al. Salt-Affected Soils and Their Management. Rome: Food and Agriculture Organization, 1988. 131p. (FAO Soils Bulletin no. 39)	First rank
First rank	Adams, Fred, ed. Soil Acidity and Liming. 2d ed. Madison, Wis.: American Society of Agronomy, 1984. 380p. (Agronomy no. 12) (1st ed. 1967)	First rank
Third	Adamson, Arthur W. Physical Chemistry of Surfaces. 5th ed. New York: J. Wiley, 1990. 777p. (1st ed. New York: Interscience Pub., 1960. 129p.) (Available in Spanish as Química Física. Barcelona: Reverté, 1979. 2 vols.)	
Third	Adkins, Clement J. Equilibrium Thermodynamics. 3d ed. Cambridge and New York: Cambridge University Press, 1983. 285p. (1st ed. London and New York: McGraw Hill, 1968. 283p.) (Available in Spanish as Termodinámica del Equilibrio. Barcelona: Reverté, 1977. 284p.)	
Third	Adriano, D. C. Trace Elements in the Terrestrial Environment. New York: Springer-Verlag, 1986. 533p.	Second
Third	Aduayi, E. A., and E. E. Ekong. General Agriculture and Soils. London: Cassell, 1981. 102p.	Third
	Ahn, Peter M. West African Soils. London: Oxford University Press, 1974. 332p.	Second
Second	Aiken, George R. et al., eds. Humic Substances in Soil, Sediment and Water: Geochemistry, Isolation and Characterization. New York: J. Wiley, 1985. 692p.	Third
	Akin, Wallace E. Global Patterns: Climate, Vegetation, and Soils. Norman, Okla.: University of Oklahoma Press, 1991. 370p.	Third
	Alexander, A., ed. Foliar Fertilization; Proceedings of the 1st International Symposium on Foliar Fertilization, Berlin, Mar. 1985. Dordrecht and Boston: Nijhoff, 1986. 488p.	Third
Second	Alexander, Lyle T., and John Cody. Genesis and Hardening of Laterite in Soils. Washington, D.C.: Soil Conservation Service, U.S. Dept. of Agriculture, 1962. 90p. (Technical Bulletin no. 1282)	Second

Developed Countries		Third World
First rank	Alexander, Martin. Introduction to Soil Microbiology. 2d ed. New York: J. Wiley, 1977. 467p. (1st ed., 1961.)	First rank
Third	Allen, Herbert E., and James R. Kramer, eds. Nutrients in Natural Waters. New York: J. Wiley, 1972. 457p.	Second
First rank	Allison, Franklin E. Soil Organic Matter and its Role in Crop Production. Amsterdam and New York: Elsevier, 1973. 639p.	First rank
Third	Alloway, B. J., ed. Heavy Metals in Soils. Glasgow: Blackie and New York: Halsted Press, 1990. 339p.	Second
Third	Altieri, Miguel A. et al. Agroecology: The Scientific Basis of Alternative Agriculture. Boulder, Colo.: Westview Press, 1987. 227p.	Third
	Altieri, Miguel A., and Susanna B. Hecht, eds. Agroecology and Small Farm Development. Boca Raton, Fla.: CRC Press, 1990. 262p.	Third
Third	American Society of Agronomy. Predicting Tillage Effects on Soil Physical Properties and Processes; Proceedings of a Symposium, Detroit, Mich., Nov.–Dec. 1980. Madison, Wis.: Amerian Society of Agronomy and Soil Science Society of America, 1982. 198p. (ASA Special Publication no. 44)	Second
Third	American Society of Civil Engineers, Irrigation and Drainage Division, Committee on Water Quality. Agricultural Salinity Assessment and Management. New York: American Society of Civil Engineers, 1990.	Second
	Anderson, J. M., and Ingram J. S. I., eds. Tropical Soil Biology and Fertility: A Handbook of Methods. 2nd. ed. Wallingford, U.K.: Commonwealth Agriculture Bureaux International, 1993. 221p.	Second
	Andriesse, J. P. Nature and Management of Tropical Peat Soils. Rome: Food and Agriculture Organization, 1988. 165p. (FAO Soils Bulletin no. 59)	Second
	Archer, John. Crop Nutrition and Fertiliser Use. Ipswich, U.K.: Farming Press, 1985. 258p.	Second
Second	Arendt, F., M. Hinsenveld and W. J. van den Brink, eds. Contaminated Soil '90; Proceedings of the 3d International Conference on Contaminated Soil, Karlsruhe, Germany, 1990. Dordrecht and Boston: Kluwer Academic Pub., 1990. 2 vols. 1454p.	
Second	Arkin, G. F., and H. M. Taylor, eds. Modifying the Root Environment to Reduce Crop Stress. St. Joseph, Mich.: American Society of Agricultural Engineers, 1981. 407p. (ASAE Monograph no. 4)	First rank
Third	Armson, K. A. Forest Soils: Properties and Processes. Toronto and Buffalo, N.Y.: University of Toronto Press, 1977. 390p.	Second

Developed Countries		Third World
	Asean Soil Conference, 3d, Kuala Lumpur, 1975. Theme: Soil Science for Agricultural Development. Kuala Lumpur: Government of Malaysia, 1977. 517p.	Third
Second	Assink, J. W., and W. J. van den Brink, eds. Contaminated Soil; Proceedings of the 1st International TNO Conference on Contaminated Soil, Utrecht, Netherlands, Nov. 1985. Dordrecht and Boston: M. Nijhoff, 1986. 923p.	
	Athaval, R. N., and V.B. Srivastava, eds. Approaches and Methodologies for Development of Groundwater Resources; Proceedings . . . Hyderabad, India: National Geophysical Research Institute, 1978. 405p. (1st Workshop)	Third
Third	Atkinson, D. et al. Mineral Nutrition of Fruit Trees; Proceedings of a symposium on Mineral Nutrition and Fruit Quality of Temperate Zone Fruit Trees, Canterbury, U.K., 1979. London and Boston: Butterworths, 1980. 435p.	Third
	Aubert, Huguette and M. Pinta. Trace Elements in Soils. Amsterdam and New York: Elsevier, 1977. 395p.	First rank
Third	Avery, B. W. Soil Classification for England and Wales (higher categories). Harpenden, U.K.: Soil Survey of England and Wales, 1980. 67p. (Technical Monograph Soil Survey no. 14)	
Second	Avery, B. W. Soils of the British Isles. Wallingford, U.K.: CAB International, 1990. 463p.	
	Ayanaba, A., and P. J. Dart, eds. Biological Nitrogen Fixation in Farming Systems of the Tropics; Proceedings of a symposium, International Institute of Tropical Agriculture, Ibadan, Nigeria, Oct. 1975. Chichester, U.K. and New York: J. Wiley, 1977. 377p.	Second
Second	Ayers, R. S., and D. W. Westcot. Water Quality for Agriculture. Rome: Food and Agriculture Organization, 1985. 174p. (FAO Irrigation and Drainage Paper no. 29, rev. 1)	First rank

B

Third	Bailey, S. W., ed. Hydrous Phyllosilicates: Exclusive of Micas. Washington, D.C.: Mineralogical Society of America, 1988. 725p.	
Third	Baker, Kenneth F., and William Snyder et al., eds. Ecology of Soil-Borne Plant Pathogens; Proceedings of the International Symposium on Factors Determining the Behavior of Plant Pathogens, Berkeley, 1963. Berkeley: University of California Press, 1965. 571p.	
Third	Baker, Robert A., ed. Contaminants and Sediments. Ann Arbor, Mich.: Ann Arbor Science, 1980. 2 vols.	

Developed Countries		Third World
Second	Bakker, H. de. Major Soils and Soil Regions in the Netherlands. The Hague and Boston: Junk, 1979. 211p. (1st published in Dutch in 1966)	
Third	Bal, L. Micromorphological Analysis of Soils. Wageningen, Netherlands: Centre for Agricultural Publishing and Documentation, 1973. 175p.	
	Balek, Jaroslav. Hydrology and Water Resources in Tropical Regions. New York: Elsevier Scientific Pub. Co., 1983. 271p.	Third
	Baligar, V. C., and R.R. Duncan, eds. Crops as Enhancers of Nutrient Use. San Diego, Calif.: Academic Press, 1990. 574p.	Third
	Bandung Symposium, 1969. Soils and Tropical Weathering; Proceedings . . . Paris: UNESCO, 1971. 149p.	Second
	Banin, A., and U. Kafkafi, eds. Agrochemicals in Soils. Oxford and New York: Pergamon Press, 1980. 448p.	Second
	Banta, Stephen J., ed. Wetland Soils: Characterization, Classification, and Utilization; Proceedings of a Workshop, 1984. Los Baños, Philippines: International Rice Research Institute, 1985. 559p.	Second
Second	Barber, Stanley A. Soil Nutrient Bioavailability: A Mechanistic Approach. New York: J. Wiley, 1984. 398p.	First rank
Second	Barnes, Kenneth K., ed. Compaction of Agricultural Soils. St. Joseph, Mich.: American Society of Agricultural Engineers, 1971. 471p. (ASAE Monograph no. 1)	First rank
Third	Barry, Roger G., and Richard J. Chorley. Atmosphere, Weather, and Climate. 5th ed. London and New York: Methuen, 1987. 460p. (1st ed. 1968.) (Available in Spanish as Atmósfera, Tiempo y Clima. 4th ed. Barcelona: Omega, n.d. 521p.)	Second
	Bartelli, L. J. Soil Surveys and Land Use Planning. Madison, Wis.: Soil Science Society of America and American Society of Agronomy, 1966. 196p. (Papers presented at the Annual meetings of the Soil Science Society of America and the American Society of Agronomy at Columbus, Ohio, Nov. 1965.)	Second
Second	Bartholomew, William V., and Francis E. Clark. Soil Nitrogen. Madison, Wis.: American Society of Agronomy, 1965. 615p. (Agronomy no. 10)	First rank
Third	Bates, Robert L., and Julia A. Jackson. Glossary of Geology. 3d ed. Alexandria, Va.: American Geological Institute, 1987. 788p.	Third
Third	Batten, James W., and J. Sullivan Gibson. Soils: Their Nature, Classes, Distribution, Uses, and Care. Rev. and enl. ed. University, Ala.: University of Alabama Press, 1977. 276p. (1st ed., 1970. 296p.)	Third

Developed Countries		Third World
	Baum, Eckhard, Peter Wolff, and Michael A. Zobisch, eds. The Extent of Soil Erosion: Regional Comparisons. Witzenhausen, Germany: German Institute for Tropical and Subtropical Agriculture, 1989. 201p.	Third
Third	Baum, Emanuel L., Earl O. Heady and J. Blackmore, eds. Methodological Procedures in the Economic Analysis of Fertilizer Use Data. Ames, Iowa: Iowa State College Press, 1956. 218p.	
Third	Beasley, R. P. Erosion and Sediment Pollution Control. 2d ed. Ames, Iowa: Iowa State University Press, 1984. 354p. (1st ed., 1972. 320p.)	Second
Third	Beatty, Marvin T., Lester Leslie, D. Swindale and Gary W. Petersen. Planning the Uses and Management of Land. Madison, Wis.: American Society of Agronomy, 1979. 1028p.	Second
Third	Beck, Theodor. Mikrobiologie des Bodens. Munich, Basel, Vienna: Bayerischer Landwirtschaftsverlag, 1968. 452p.	
Third	Beeson, Kenneth C., and Gennard Matrone. The Soil Factor in Nutrition: Animal and Human. New York: M. Dekker, 1976. 152p.	
Second	Bennett, Hugh H. Elements of Soil Conservation. 2d ed. New York: McGraw-Hill, 1955. 358p. (1st ed. 1939. Title: Soil Conservation.) (Available in Spanish as Elementos de Conservación del Suelo. Mexico: FCE, 1965. 429p.)	
Second	Bergersen, F. J., ed. Methods for Evaluating Biological Nitrogen Fixation. Chichester, U.K. and New York: J. Wiley, 1980. 702p.	First rank
Second	Bergey's Manual of Systematic Bacteriology. Edited by N. R. Krieg, and John G. Holt. Baltimore, Md.: Williams and Wilkins, 1984–. (1st ed., 1923 titled: Bergey's Manual of Determinative Bacteriology: A Key for the Identification of Organisms of the Class Schizomycetes. 442p. 8th ed. entered under: Bergey's Manual of Determinative Bacteriology.)	
Third	Berkeley, R. C. W. et al., eds. Microbial Adhesion to Surfaces. Chichester, U.K. and New York: Ellis Horwood and Halsted Press, 1980. 559p.	
	Beskos, D. E., T. Krauthammer and I. Vardoulakis, eds. Dynamic Soil-Structure Interaction; Proceedings of the International Symposium . . . Minneapolis, Minn., Sept. 1984. Rotterdam and Boston: Balkema, 1984. 176p.	Third
Third	Bethlenfalvay, G. J. and R.G. Linderman, eds. Mycorrhiae in Sustainable Agriculture; Proceedings of a symposium sponsored by Divisions S-3 and S-4 of the Soil Science Society of America, Division A-8 of the American Society of Agronomy, and Division C-2 of the Crop Science Society of America in Denver, Colo., Oct. 1991.	Third

186   Peter McDonald

Developed Countries		Third World
	Madison, Wisc.: American Society of Agronomy, Crop Science of America, and Soil Science Society of America, 1991. 124p.	
Second	Bezdicek, D. F. et al., eds. Organic Farming: Current Technology..; Proceedings of an International Symposium in Atlanta, Ga., Dec. 1981. Madison, Wis.: American Society of Agronomy, 1984. 192p. (ASA Special Publication no. 46)	First rank
	Birkeland, Peter W. Pedology, Weathering and Geomorphological Research. New York: Oxford University Press, 1974. 285p.	Second
First rank	Birkeland, Peter W. Soils and Geomorphology. New York: Oxford University Press, 1984. 372p. (Rev. ed. of Pedology, Weathering, and Geomorphological Research, 1974.)	Second
Third	Birot, Pierre. The Cycle of Erosion in Different Climates. Trans. by C. Ian Jackson and Keith M. Clayton. Berkeley: University of California Press, 1968. 144p.	Third
	Biswas, A. K., M. A. H. Samaho, M. H. Amer and M. A. Abu-Zeid, eds. Water Management for Arid Lands in Developing Countries. Oxford: Pergamon Press, 1980. 262p.	Second
Second	Bitton, Gabriel and Charles P. Gerba, eds. Groundwater Pollution Microbiology. New York: J. Wiley, 1984. 377p.	
First rank	Black, C. A. Soil-Plant Relationships. 2d ed. Malabar, Fla.: R.E. Krieger, 1984. 792p. (1st ed. 1957)	First rank
	Blaikie, Piers M. The Political Economy of Soil Erosion in Developing Countries. London: Longmans, 1985. 195p.	Second
	Blucher, Edgard, ed. Proceedings of the 3d Simposio Sobre o Cerrado, São Paulo, 1971. São Paulo, Brazil: da Universidade de São Paulo, 1971. 239p. (In Portuguese, with summaries in English.)	Third
Third	Boardman, J., I. D. L. Foster and J. A. Dearing, eds. Soil Erosion on Agricultural Land; Papers presented at a Workshop of the British Geomorphological Research Group, Annual Conference of the Institute of British Geographers, Coventry Polytechnic, Jan. 1989. Chichester, U.K. and New York: Wiley, 1990. 687p.	Third
Second	Boersma, L. L. et al., eds. Future Developments in Soil Science Research; A Collection of SSSA Golden Anniversary Contributions Presented at the Annual Meeting in New Orleans, LA, Nov.–Dec. 1986. Madison, Wis.: Soil Science Society of America, 1987. 537p.	
Second	Bohn, Hinrich L. Soil Chemistry. 2d ed. New York: J. Wiley, 1985. 341p. (1st ed., 1979.)	First rank
Third	Bollagg, Jean-Marc and G. Stotzky, eds. Soil Biochemistry vol. 6. New York: M. Dekker, 1990. 565p.	

Developed Countries		Third World
Second	Bolt, G. H., ed. Interactions at the Soil Colloid-Soil Solution Interface. NATO Advanced Study Institute on Interactions at the Soil Colloid-Soil Solution Interface, State University of Ghent, Aug. 1986. Dordrecht and Boston: Kluwer Academic, 1991. 603p. (NATO ASI Series E, Applied Sciences no. 190)	
First rank	Bolt, G. H., ed. Soil Chemistry. 1st and 2d rev. ed. Amsterdam and New York: Elsevier, 1976 and 1982.	First rank
Second	Bonneau, Maurice. Constituents and Properties of Soils. Trans. of Constituants et Proprietes du Sol. London and New York: Academic Press, 1982. 496p. (Pedologie no. 2) (1st ed. published Paris: Masson, 1979. 459p.)	Second
	Bonnet, Juan A. Edafologia de los Suelos Salinos y Sodicos. Rio Piedras, Puerto Rico: Universidad de Puerto Rico, 1960. 337p.	Third
	Booker Tropical Soil Manual: A Handbook for Soil Survey and Agricultural Land Evaluation in the Tropics and Subtropics. Edited by J.R. Landon. London: Booker Agriculture International Ltd. and New York: Longman, 1991, 1984. 450p. (Reprint)	First rank
Third	Bork, H. R., and W. Ricken. Soil Erosion, Holocene and Pleistocene Soil Development-Bodenerosion, Holozaene und Pleistozaene Bodenentwicklung. Cremlingen, Germany: Catena Verlag, 1983. 138p. (Catena Supplement no. 3) (In German with summaries in English and German.)	
	Bornemisza, Elemer and Alfredo Alvarado, eds. Soil Management in Tropical America; Proceedings of a Seminar, CIAT, Cali, Colombia, Feb. 1974. Raleigh, N.C.: North Carolina State University, 1975. 565p.	First rank
	Bosworth, Duane A., and Albert B. Foster. Approved Practices in Soil Conservation. 5th ed. Danville, Ill.: Interstate Printers and Pub., 1982. 470p. (1st ed. by A. B. Foster. Danville, Ill., 1955. 380p.)	Second
	Bouillon, A., ed. Etudes sur les Termites Africains. Un Colloque International, Universite Lovanium, Leopoldville, Congo, 1964. Leopoldville, Congo: Editions de l'Universite, 1964. 414p. (English, French and German.)	Third
Third	Boulaine, Jean. Histoire des Pédologues et de la Science des Sols. Paris: Institut National de la Recherche Agronomique, 1989. 285p.	Third
Second	Boullard, Bernard. Les Mycorrhizes . . . Paris: Masson et Cie, 1968. 135p.	
Second	Boullard, Bernard. Vie Intense et Cachée Sol. Essai de Pedobiologie Vegetale. Paris: Flammarion, 1967. 309p.	
Third	Bouwman, A. F. ed. Soils and the Greenhouse Effect; Pro-	Third

188   Peter McDonald

Developed Countries		Third World

ceedings of an international conference on behalf of the Netherlands' Ministry of Housing, Physical Planning and Environment. Chichester, U.K.: J. Wiley, 1990. 575p.

Bowen, G. D., and E. K. S. Nambiar, eds. Nutrition of Plantation Forests. London and Orlando, Fla.: Academic Press, 1984. 516p. — **Second**

Bowler, D. G. The Drainage of Wet Soils. Auckland, N.Z.: Hodder and Stoughton, 1980. 259p. — **Second**

Bowling, Dudley J. F. Uptake of Ions by Plant Roots. London and New York: Chapman and Hall, 1976. 212p. — **Second**

Second — Box, James E., and Luther C. Hammond, eds. Rhizosphere Dynamics. Boulder, Colo.: Westview Press, 1990. 322p. — **Second**

First rank — Brady, Nyle C. The Nature and Properties of Soils. 10th ed. New York: Macmillan, 1990. 639p. (1st ed. by Thomas Lyon, New York: Macmillan, 1922) — **First rank**

Third — Brengle, K. G. Principles and Practices of Dryland Farming. Boulder, Colo.: Colorado Associated University Press, 1982. 178p. — **Second**

Third — Bresler, Eshel, B. L. McNeal and D. L. Carter. Saline and Sodic Soils: Principles, Dynamics, Modeling. Berlin and New York: Springer-Verlag, 1982. 236p. — **Second**

First rank — Brewer, Roy. Fabric and Mineral Analysis of Soils. Huntington, N.Y.: R. E. Krieger, 1976. 482p. (Reprint of the 1964 ed. published by J. Wiley, 1964 with supplementary material.) — **First rank**

Third — Brewer, Roy and J. R. Sleeman. Soil Structure and Fabric. Melbourne: CSIRO, 1988. 172p. — **Second**

Third — Brian, M. V., ed. Production Ecology of Ants and Termites. Cambridge, U.K. and New York: Cambridge University Press, 1978. 409p.

First rank — Bridges, E. M. World Soils. 2d ed. Cambridge U.K. and New York: Cambridge University Press, 1978. 128p. (1st ed., 1970.) — **Second**

Second — Bridges, E. M., and Donald A. Davidson. Principles and Applications of Soil Geography. London and New York: Longmans, 1982. 297p. — **Third**

First rank — Brindley, G. W., and G. Brown, eds. Crystal Structures of Clay Minerals and Their X-ray Identification. London: Mineralogical Society, 1980. 495p. (Previously published as Mineralogical Society Monograph 2, under title: X-ray Identification and Crystal Structures of Clay Minerals.)

Third — Brink, A. B. A., and T. C. Partridge. Soil Survey for Engineering. Oxford, U.K.: Clarendon Press, 1982. 378p. — **Third**

Second — Brinkman, R. Ferrolysis: A Soil-Forming Process in Hydromorphic Conditions. Wageningen, Netherlands: Centre — **Second**

Developed Countries		Third World
	for Agricultural Publishing and Documentation, 1979. 106p.	
	Brinkman, R., and A. J. Smyth, eds. Land Evaluation for Rural Purposes: Summary of an Expert Consultation. Wageningen, Netherlands: International Institute for Land Reclamation and Improvement, 1973. 116p.	Third
Third	Bronger, A., and J. A. Catt, eds. Paleopedology: Nature and Application of Paleosols. Cremlingen-Destedt, Germany: Catena, 1989. 232p.	
	Browder, John O., ed. Fragile Lands of Latin America: Strategies for Sustainable Development; Proceedings of a Symposium on Fragile Lands of Latin America: The Search for Sustainable Uses held during the 14th Congress of the Latin American Studies Association, New Orleans, Mar. 1988. Boulder, Colo.: Westview Press, 1989. 301p.	Third
Third	Brown, Darrell et al. Reclamation and Vegetative Restoration of Problem Soils and Disturbed Lands. Park Ridge, N.J.: Noyes Data Corp., 1986. 560p.	Third
Second	Brown, Gordon E. et al. Spectroscopic Methods in Mineralogy and Geology, edited by Frank C. Hawthorne. Washington, D.C.: Mineralogical Society of America, 1988. 698p.	
Third	Brown, J. R., ed. Soil Testing: Sampling, Correlation, Calibration, and Interpretation. Madison, Wis.: Soil Science Society of America, 1987. 144p. (Papers presented during the annual meetings, Chicago, Dec. 1985)	First rank
	Brown, Leslie and J. Cocheme. Technical Report on a Study of the Agroclimatology of the Highlands of Eastern Africa. Rome: Food and Agriculture Organization, 1969. 330p.	Third
Second	Bruce, R. R., K. W. Flach and H. M. Taylor. Field Soil Water Regime; Proceedings of a symposium, during the annual meeting of the Soil Science Society of America and American Society of Agronomy, Aug. 1971. Madison, Wis.: Soil Science Society of America, 1973. 212p. (SSSA Special Publication no. 5)	
Third	Bruehl, George W. Soil-Borne Plant Pathogens. New York and London: Macmillan and Collier Macmillan, 1987. 368p.	Second
Third	Brunsden, D. Slopes, Form and Process. London: Institute of British Geographers, 1971. 178p.	
Second	Bullock, Peter and C. P. Murphy, eds. Soil Micromorphology; Proceedings of the 6th International Working-Meeting of the International Society of Soil Science, London, Aug. 1981. Berkhamsted, U.K.: A. B. Academic, 1983. 2 vols. 705p.	Third

Developed Countries		Third World

Developed Countries		Third World
Third	Bullock, Peter, N. Federoff, A. Jongerius, G. Stoops and T. Tursina. Handbook for Soil Thin Section Descriptions. Wolverhampton, U.K.: Waine Research Pub., 1985. 152p.	
Third	Bullock, Peter and Peter J. Gregory, eds. Soils in the Urban Environment. Oxford, U.K. and Boston: Blackwell Scientific Publications, 1991. 224p.	
First rank	Buol, S. W., F. D. Hole and Ralph McCracken. Soil Genesis and Classification. 3d ed. Ames, Iowa: Iowa State University Press, 1989. 446p.	Second
	Burges, Alan and F. Raw, eds. Soil Biology. London: Academic Press, 1967. 532p.	Second
	Buringh, P. Introduction to the Study of Soils in Tropical and Subtropical Regions. 4th ed. Wageningen, Netherlands: Centre for Agricultural Publishing and Documentation (PUDOC), 1981. 368p. (1st ed. 1968)	Second
Second	Burns, R. G., ed. Soil Enzymes. London and New York: Academic Press, 1978. 380p.	Second
	Burt, R. L. et al., eds. The Role of Centrosema, Desmodium, and Stylosanthes in Improving Tropical Pastures. Boulder, Colo.: Westview Press, 1983. 292p.	Third
Second	Butler, B. E. Soil Classification for Soil Survey. Oxford, U.K.: Clarendon Press and New York: Oxford University Press, 1980. 129p.	First rank
Third	Buurman, Peter, ed. Podzols. New York: Van Nostrand Reinhold, 1984. 450p.	

C

Developed Countries		Third World
Third	Cairney, T., ed. Reclaiming Contaminated Land. Glasgow, U.K.: Blackie, 1987. 260p.	
Third	Callot, G. et al. Mieux Comprendre les Interactions Solracine: Incidence sur la Nutrition Minerale. Paris: Institut National de la Recherche Agronomique, 1982. 325p.	
	Camargo, M. N., and F. H. Beinroth. Proceedings of the First International Soil Classification Workshop. Rio De Janeiro, Brazil: EMBRAPA, 1978. 376p.	Second
First rank	Campbell, Gaylon S. An Introduction to Environmental Biophysics. New York: Sringer-Verlag, 1977. 159p.	
Third	Campbell, Gaylon S. Soil Physics with Basic Transport Models for Soil Plant Systems. Amsterdam and New York: Elsevier, 1985. 150p.	
Third	Campbell, I. B., and G. G. C. Claridge. Antarctica: Soils Weathering Processes and Environment. Amsterdam and New York: Elsevier, 1987. 368p. (Developments in Soil Science no. 16)	
Second	Canada Dept. of Agriculture. The Canadian System of Soil	

Developed Countries		Third World
	Classification. 2d ed. Ottawa: Agriculture Canada, 1987. 164p. (Canada Dept. of Agric. Pub. no. 1646) (1st ed. 1978)	
	Cannell, Glen H., ed. Proceedings of an International Symposium on Rainfed Agriculture in Semi-Arid Regions, Riverside, Calif., Apr. 1977. Riverside, Calif.: University of California, 1977. 703p.	First rank
Third	Carroll, C. Ronald, Peter R. Vandermeer and Peter Rosset, eds. Agroecology. New York: McGraw-Hill, 1990. 641p.	
Second	Carson, Eugene W., ed. The Plant Root and its Environment; Proceedings of the Conference on the Plant Root and its Environment, Virginia Polytechnic Institute, 1971. Charlottesville, Va.: University Press of Virginia, 1974. 691p.	
Third	Carson, Michael A., and M. J. Kirkby. Hillslope Form and Process. Cambridge, U.K.: Cambridge University Press, 1972. 475p.	
Third	Carter, Vernon G., and Tom Dale. Topsoil and Civilization. Rev. ed. Norman, Okla.: University of Oklahoma Press, 1974. 292p. (1st ed. 1955.)	Second
Third	Carver, Robert E., ed. Procedures in Sedimentary Petrology. New York: J. Wiley, 1971. 653p.	Third
Third	Casenave, Alain and Christian Valentin. Les Etats de Surface de la Zone Sahelienne: Influence sur l'Infiltration. Paris: ORSTOM, 1989. 229p.	Third
Second	Cedergren, Harry R. Seepage, Drainage, and Flow Nets. 3d ed. New York: J. Wiley, 1989. 465p. (1st ed., 1967.)	Third
Third	Cerri, Carlos C., Diva Athiae and Daecio Sodrzeieski, eds. Proceedings of the Regional Colloquium on Soil Organic Matter Studies = Anais do Colaoquio Regional Sobre Mataeria OrgDanica do Solo, Piracicaba, Brazil, 1982. São Paulo, Brazil: Centro de Energia Nuclear na Agriculutra, Companhia de Promopcoao de Pesquisa Cientaifica e Tecnolaogica do Estado de São Paulo, 1982. 254p. (English, Portuguese, and Spanish.)	Second
First rank	Chapman, Homer D., ed. Diagnostic Criteria for Plants and Soils. Berkeley, Calif.: University of California, 1966. 793p.	First rank
Third	Chapman, Homer D., and Parker F. Pratt. Methods of Analysis for Soils, Plants, and Waters. Riverside: University of California, 1961. 309p. (Reprinted in 1982.) (Available in Spanish as Metodos de Analisis para Suelos, Plantas y Aguas. Mexico: Editorial Trillas, 1973. 195p.)	First rank
	Chapman, Valentine J. Salt Marshes and Salt Deserts of	Third

Developed Countries		Third World
	the World. London: L. Hill and New York: Interscience Pub., 1960. 392p.	
Third	Chauvel, Armand. Recherches sur la Transformation des Sols Ferrallitiques dans la Zone Tropicale a Saisons Contrastés: Evolution et Réorganisation des Sols Rouges de Moyenne Casamance (Sénégal). Paris: ORSTOM, 1977. 532p.	Third
Second	Chen, Y., and Y. Avnimelech, eds. The Role of Organic Matter in Modern Agriculture. Dordrecht and Boston: M. Nijhoff, 1986. 306p.	Second
Second	Cheng, H. H., ed. Pesticides in the Soil Environment: Process, Impacts and Modeling. Madison, Wis.: Soil Science Society of America, 1990. 530p.	Second
	Childers, Norman F. Nutrition of Fruit Crops: Tropical, Sub-Tropical, Temperate Tree and Small Fruits. New Brunswick, N.J.; Rutgers; The State University, 1966. 888p. (1st ed., 1954, has title: Mineral Nutrition.)	Second
Second	Childs, Ernest C. An Introduction to the Physical Basis of Soil Water Phenomena. London and New York: J. Wiley, 1969. 493p.	Second
	Chisholm, Anthony and Robert Dumsday, eds. Land Degradation: Problems and Policies; Proceedings of the Land Degradation and Public Policy workshop at Australian National University, Sept. 1985. Cambridge, U.K. and New York: Cambridge University Press, 1987. 404p.	Second
Third	Chorley, Richard J., ed. Spatial Analysis in Geomorphology. New York: Harper and Row, 1972. 393p.	Third
Third	Chorley, Richard J., and Barbara A. Kennedy. Physical Geography: A Systems Approach. London: Prentice-Hall, 1971. 370p.	
Second	Chudnovskii, A. F. Heat Transfer in the Soil = Fizika Teploobmena v Pochve. Jerusalem: Israel Program for Scientific Translations, 1962. 164p. (Trans. from Russian and edited by Israel Program for Scientific Translations.)	
Third	Clark, Edwin H., and Jennifer Eroding Soils: The Off-Farm Impacts. Washington, D.C.: Conservation Foundation, 1985. 252p.	Third
First rank	Clarke, George R. The Study of the Soil in the Field. 5th ed. Oxford, U.K.: Clarendon Press, 1971. 145p. (1st ed., 1936.)	First rank
Third	Clayton, C. R. I. Site Investigation. New York: Halsted Press, 1982. 424p.	
	Cochemé, J., and P. Franquin. Technical Report on a Study of the Semiarid Area South of the Sahara in West Africa = Rapport Technique sur une étude d'Agroclimatologie de l'Afrique seche au sud de Sahara en Afrique Occi-	Third

Developed   Third
Countries   World

dentale. Rome: Food and Agriculture Organization, 1967.
325p.

Third   Coineau, Yves. Introduction a l'Etude des Microarthro-
                  podes du Sol et de ses Annexes. Paris: Doin, 1974. 117p.

                  Coleman, David C., J. Malcolm Oades and Goro Uehara,   Second
                  eds. Dynamics of Soil Organic Matter in Tropical Ecosys-
                  tems. Papers presented at a workshop, Maui, Hawaii, Oct.
                  1988. Honolulu, Hawaii: Univeristy of Hawaii, 1989.
                  249p.

                  Le Colloque sur la Fertilité des Sols Tropicaux, Tan-   Third
                  anarive, Madagascar, Nov. 1967. Proceedings . . . Paris:
                  Institut de Recherches Agronomiques Tropicales et des
                  Culture Vivrieres, 1968. 2 vols.

Third   Colwell, J. D. Computations for Studies of Soil Fertility
                  and Fertilizer Requirements. Farnham Royal, U.K.: Com-
                  monwealth Agricultural Bureaux, 1978. 297p.

Second   Commonwealth Scientific and Industrial Research Organi-   Second
                  zation (Australia). Soils: An Australian Viewpoint. Mel-
                  bourne: CSIRO and London: Academic Press, 1983. 928p.

                  Conference on Chemistry and Fertility of Tropical Soils.   Third
                  Proceedings . . . , Kuala Lumpur, Malaysia, Nov. 1973.
                  Kuala Lumpur, Malaysia: Malaysian Society of Soil Sci-
                  ence, 1977. 300p.

                  Constantinesco, I. Soil Conservation for Developing Coun-   Second
                  tries. Rome: Food and Agriculture Organization, 1978.
                  92p. (FAO Soils Bulletin no. 30)

Second   Cook, Ray L., and Boyd G. Ellis. Soil Management: A   Second
                  World View of Conservation and Production. Rev. ed.
                  New York: J. Wiley, 1987. 413p. (1st ed. 1962)

Second   Cooke, G. W. Fertilizing for Maximum Yield. 3d ed. New
                  York: Macmillan, 1982. 465p. (1st ed. 1972.) (Available
                  in Spanish as Fertilización para Rendimientos Màximos.
                  Mexico: Cecsa, 1983. 383p.)

                  Cope, J. T., F. Young and J. L. Demeterio, eds. Proceed-   Third
                  ings of the IX International Forum on Soil Taxonomy and
                  Agrotechnology Transfer. Guam: Mangilao, 1985. 295p.

First rank   Cornell University Department of Agronomy, Soil Survey   First rank
                  Staff. Keys to Soil Taxonomy. 3d rev. ed. Ithaca, N.Y.
                  and Washington, D.C.: Cornell University and Agency for
                  International Development and the U.S. Dept. of Agricul-
                  ture, Soil Management Support Services, 1987. 280p.
                  (SMSS Technical Monograph no. 6) (1st ed. 1983). Also
                  available in French.

Third   Cottenie, A. Soil and Plant Testing as a Basis of Fertilizer First rank
                  Recommendations. Rome: Food and Agriculture Organiza-
                  tion, 1980. 120p. (FAO Soils Bulletin no. 38)

Developed   Third
Countries   World

Coughlan, K. J., and P. N. Truong, eds. Effects of Man-   Third
agement Practices on Soil Physical Properties. Proceedings
of a National Workshop, Toowoomba, Australia, Sept.
1987. Brisbane, Australia: Department of Primary Indus-
tries, Queensland Government, 1988. 268p.

Second   Craig, Robert F. Soil Mechanics. 5th ed. London: Chapman and Hall; New York: Van Nostrand Reinhold, 1992.
(1st ed., 74. 275p.)

Craswell, E. T., J. V. Remenyi and L. G. Nallana, eds.   Second
Soil Erosion Management. Proceedings of a Workshop,
PCARRD, Los Baños, Philippines, Dec. 1984. Canberra,
Australia: Australian Centre for International Agricultural
Research, 1985. 132p. (ACIAR Proceedings Series no. 6)

Craswell, E. T., and E. Pushparajah, eds. Management of   Third
Acid Soils in the Humid Tropics of Asia. Canberra and
Bangkok, Thailand: Australian Centre for International Agricultural Research and International Board for Soil Research and Management, 1989. 118p. (Research and country status papers presented at a workshop, Kuala Lumpur,
1989.)

Third   Crickmay, Colin H. The Work of the River: A Critical
Study of the Central Aspects of Geomorphogeny. London:
Macmillan, 1974. 271p.

Second   Curl, Elroy A., and Bryan Truelove. The Rhizosphere.   Second
Berlin and New York: Springer-Verlag, 1986. 288p.

Third   Curran, P., ed. Remote Sensing of Soils and Vegetation in
the USSR. London and New York: Taylor and Francis
Ltd., 1990. 203p.

Third   Curtis, Leonard F., F. M. Courtney and S. T. Trudgill.
Soils in the British Isles. London and New York: Longman, 1976. 364p.

D

Third   Dalton, G. E., ed. Study of Agricultural Systems; Proceedings of an International Symposium, Dept. of Agriculture
and Horticulture, University of Reading. London: Applied
Science Pub., 1975. 441p.

Dalzell, H. W. et al., eds. Soil Management: Compost   Second
Production and Use in Tropical and Subtropical Environments. Rome: Food and Agriculture Organization, 1987.
193p. (FAO Soils Bulletin no. 56)

Third   Davidescu, David and Velicica Davidescu. Evaluation of
Fertility by Plant and Soil Analysis = Testarea Stlarii de
Fertilitate prin Plantlla Psi Sol. Rev. and updated version.
Bucuresti, Romania and Turnbridge Wells, Kent: Editura

Developed Countries		Third World
	Academiei and Abacus Press, 1982. 560p. (Trans. from Romanian by Skaw Loren Kent.)	
	Davidson, Donald A. Soils and Land Use Planning. London and New York: Longmans, 1980. 129p.	Second
	Davies, Brian E., ed. Applied Soil Trace Elements. Chichester, U.K. and New York: J. Wiley, 1980. 482p.	Second
	Davies, D. Bryan, D. J. Eagle and J. B. Finney. Soil Management. 4th ed. Ipswich, U.K.: Farming Press, 1982. 287p. (1st ed., 1972. 254p.)	Second
Third	Davies, J. W. Mulching Effects on Plant Climate and Yield. Geneva: World Meterological Organization, 1975. 92p.	Third
Third	Davies, Nancy D. et al. A Guide to the Study of Soil Ecology. Scarborough, Canada: Prentice-Hall of Canada, 1973. 198p.	Third
Third	De Boodt, Marcel and D. Gabriels, eds. Assessment of Erosion; Proceedings of a workshop on Assessment of Erosion in USA and Europe, State University, Ghent, Belgium, Feb.–Mar. 1978. Chichester, U.K. and New York: J. Wiley, 1980. 563p.	
Third	De Boodt, Marcel, Michael H. B. Hayes and Adrien Herbillon, eds. Soil Colloids and their Associations in Soil Aggregates. NATO Advanced Research Workshop, Ghent, Belgium, 1984. New York: Plenum Press, 1990. 598p.	
	De Datta, S. K., and W.H. Patrick, eds. Nitrogen Economy of Flooded Rice Soils; Proceedings of a Symposium, Washington DC, 1983. Dordrecht and Boston: M. Nijhoff, 1986. 186p.	Second
Second	Delwiche, C. C., ed. Denitrification, Nitrification, and Atmospheric Nitrous Oxide. New York: J. Wiley, 1981. 286p.	
Second	Derbyshire, Edward D., K. J. Gregory and J. R. Hails. Geomophological Processes. Folkestone, U.K. and Boulder, Colo.: Dawson and Westview Presses, 1979. 312p.	
	Dewis, J., and F. Freitas. Physical and Chemical Methods of Soil and Water Analysis. Rome: Food and Agriculture Organization, 1970. 275p. (FAO Soils Bulletin no. 10)	Second
	Dhar, Nil R. World Food Crisis and Land Fertility Improvement. Calcutta, India: University of Calcutta, 1972. 247p.	Third
Second	Dickinson, C. H., and G. J. F Pugh. Biology of Plant Litter Decomposition. London and New York: Academic Press, 1974. 2 vols.	
Third	Die Mikromorphometrische Bodenanalyse. Stuttgart, Germany: Enke, 1967. 196p.	

Developed Countries		Third World
Third	Dindal, Daniel L., ed. Soil Biology as Related to Land Use Practices. Proceedings of the 7th International Soil Zoology Colloquium of the International Society of Soil Science, Syracuse, NY, 1979. Washington, D.C.; Office of Presticide and Toxic Substances, EPA, 1980. 880p.	
Third	Dindal, Daniel L., ed. Soil Biology Guide. New York: J. Wiley, 1990. 1349p.	Second
First rank	Dixon, Joe B., S. B. Weed and Richard C. Dinauer. Minerals in Soil Environments. 2d ed. Madison, Wis.: Soil Science Society of America, 1989. 1244p. (1st ed., 1977.)	First rank
Third	Dobereiner, Johanna and Fabio O. Pedrosa. Nitrogen-Fixing Bacteria in Nonleguminous Crop Plants. Madison, Wis.: Science Tech Pub. and Berlin and New York: Springer-Verlag, 1987. 155p.	
Third	Dobrovolskiy, G. V., and T. S. Urusevskaya. Geografiya Pochv (Soil Geography). Moscow: Moscow University Press, 1984. 416p.	
First rank	Doeksen, J., and J. van der Drift, eds. Soil Organisms; Proceedings of the Colloquium on Soil Fauna, Soil Microflora and their Relationships, Oosterbeek, Netherlands, 1962. Amsterdam: North-Holland, 1963. 453p.	Second
	Doll, E. C., and G. O. Mott, eds. Tropical Foragers in Livestock Production Systems; Proceedings of a symposium held during the annual meetings of the American Society of Agronomy, Crop Science Society of America, and Soil Science Society of America, Las Vegas, Nev., Nov. 1973. Madison, Wis.: American Society of Agronomy, 1975. 104p. (ASA Special Publication no. 24)	Third
	Dommergues, Yvon R., and H. G. Diem, eds. Microbiology of Tropical Soils and Plant Productivity. The Hague and Boston: Nijhoff, 1982. 328p.	Second
Second	Dommergues, Yvon R., and S. V. Krupa, eds. Interactions Between Non-pathogenic Soil Microorganisms and Plants. Amsterdam and New York: Elsevier, 1978. 475p.	
Second	Dommergues, Yvon R., and Francois Mangenot. Ecologie Microbienne du Sol. Paris: Masson, 1970. 796p.	Second
Second	Domsch, Klaus H., W. Gams, and T.-H. Anderson. Compendium of Soil Fungi. London and New York: Academic Press, 1980. 2 vols.	
	Donahue, Roy L. Soils: An Introduction to Soils and Plant Growth. 6th ed. Englewood Cliffs, N.J.: Prentice-Hall, 1990. 768p. (1st ed., 1958. 349p.) (Available in Spanish as Introducción al Estudio de los Suelos y Cecimiento de las Plantas. 4th ed. Naucalpan de Juárez: Prentice-Hall, 1988. 624p.)	First rank

Developed Countries		Third World

Developed Countries		Third World
Second	Donahue, Roy L., Roy H. Follett and Rodney W. Tulloch. Our Soils and Their Management: Increasing Production Through Environmental Soil and Water Conservation. 6th ed. Danville, Ill.: Interstate Printers and Pub., 1990. 594p. (1st ed., 1955.)	Second
	Doorenbos, J. and W. O. Pruitt. Guidelines for Predicting Crop Water Requirements. Rev. ed. Rome: Food and Agriculture Organization, 1977. 177p. (1st ed., 1975. 179p.) (FAO Irrigation and Drainage Paper no. 24)	Second
	Doorenbos, J., and A. H. Kassam. Yield Response to Water. Rome: Food and Agriculture Organization, 1979. 193p. (FAO Irrigation and Drainage Paper no. 33)	Second
	Dost, H., ed. Selected papers of the 3d International Symposium on Acid Sulphate Soils, Dakar, Senegal, Jan. 1986. Wageningen, Netherlands: International Institute for Land Reclamation and Improvement, 1988. 251p. (Three papers in French. Summaries in English and French.)	First rank
Second	Douglas, Lowell A., ed. Soil Micromorphology: A Basic and Applied Science; Proceedings of the 8th International Working Meeting of Soil Micromophology, San Antonio, Tex., July 1988. Amsterdam and New York: Elsevier, 1990. 716p.	Second
Third	Douglas, Lowell A., and Michael L. Thompson, eds. Soil Micromorphology and Soil Classification. Proceedings of a Symposium in Anaheim, California, Nov.–Dec. 1982. Madison, Wis.: Soil Science Society of America, 1985. 216p.	Third
Second	Dowdy, R. H. et al., eds. Chemistry in the Soil Environment. Proceedings of a Symposium, Fort Collins, Colo., Aug. 1979. Madison, Wis.: American Society of Agronomy, Soil Science Society of America, 1981. 259p.	Second
	Dregne, H. E. Soils of Arid Regions. Amsterdam and New York: Elsevier, 1976. 237p.	First rank
	Dregne, H. E., and W.O. Willis, eds. Dryland Agriculture. Madison, Wis.: American Society of Agronomy, 1983. 622p.	First rank
Third	Drescher, J., R. Horn and M. de Boodt, eds. Impact of Water and External Forces on Soil Structure. Selected papers of the 1st Workshop on Soilphysics and Soilmechanics, Hannover, Germany, 1986. Cremlingen-Destedt, Germany: Catena Verlag, 1988. 171p.	Third
Second	Drever, James I., ed. The Chemistry of Weathering; Proceedings of the NATO Advanced Research Workshop, Rodez, France, July 1984. Dordrecht and Boston: D. Reidel, 1985. 324p.	

Developed Countries		Third World
Third	Drew, James Van. Selected Papers in Soil Formation and Classification. Madison, Wis.: Soil Science Society of America, 1967. 428p.	
Third	Driessen, P. M., and Dudal R., eds. The Major Soils of the World. Wageningen, Netherlands and Leuren, Belgium: Centre for Agricultural Publishing and Documentation, 1991. 310p.	Third
	Drosdoff, Matthew, ed. Diversity of the Soils in the Tropics; Proceedings of a Symposium, Los Angeles, Calif., 1978. Madison, Wis.: American Society of Agronomy, Soil Science Society of America, 1978. 119p. (ASA Special Publication no. 34)	First rank
	Drosdoff, Matthew, ed. Soils of the Humid Tropics. Washington, D.C.: National Academy of Sciences, 1972. 219p. (Available in Spanish as Suelos de las Regiones Tropicales Humedas. Buenos Aires: Marymar, 1975. 272p.)	First rank
Third	Drury, Stephen A. A Guide to Remote Sensing: Interpreting Images of the Earth. Oxford, U.K. and New York: Oxford University Press, 1990. 199p.	
First rank	Duchaufour, Philippe. Ecological Atlas of Soils of the World = Atlas Ecologique des Sols du Monde. New York: Masson Pub., 1978. 178p. (Trans. from the French by G. R. Mehuys, C. R. De Kimpe, and Y. A. Martel.) (Available in Spanish as Atlas Ecológico de los Suelos del Mundo. Barcelona: Masson, n.d.)	Second
First rank	Duchaufour, Philippe. Pédogenese et Classification 2d ed., rev. and enl. Paris and New York: Masson, 1983. 491p. (Pedologie no. 1) (1st ed., 1977. 477p. English ed. titled: Pedology: Pedogenesis and Classification. London: Allen and Unwin, 1982. 448p.)	First rank
Second	Duchaufour, Philippe. Précis de Pédologie. 2d ed. Paris: Masson, 1965. 481p. (1st ed., 1960.)	
	Duchaufour, Philippe. Processus de Formation des Sols: Biochimie et Geochimie. Nancy: Centre Regional de Recherche et de Documentation Pedagogiques, 1972. 182p.	Third
Third	Duche, Jacques. La Biologie des Sols. Paris: Presses Universitaires de France, 1950. 128p.	Third
	Dudal, R., ed. Dark Clay Soils of Tropical and Subtropical Regions. Rome: Food and Agriculture Organization, 1965. 161p. (FAO/UNESCO Soil Map of the World Reference; FAO Agricultural Development Paper no. 83)	Second
First rank	Dudal, R. Definitions of Soil Units for the Soil Map of the World. Rome: Food and Agriculture Organization, 1968. 72p. (FAO World Soil Resources Report no. 33)	Second

Developed Countries		Third World

Third — Dullien, F. A. L. Porous Media: Fluid Transport and Pore Structure. 2d ed. San Diego, Calif.: Academic Press, 1992. 574p. (1st ed., New York: Academic Press, 1979. 396p.)

Third — Dunger, Wolfram. Tiere im Boden. Wittenberg Lutherstadt, Germany: A. Ziemsen, 1964. 265p.

Dutt, C. F. et al., eds. Rainfall Collection for Agriculture in Arid and Semi-Arid Regions; Proceedings, University of Arizona. Farnham Royal, U.K.: Commonwealth Agricultural Bureaux, 1981. 100p. — Third

Third — Dwyer, Francis P. J., and D. P. Mellor, eds. Chelating Agents and Metal Chelates. New York: Academic Press, 1964. 530p.

E

Eavis, B., and R. Struthers. Guidelines: Land Evaluation for Irrigated Agriculture. Rome: Food and Agriculture Organization, 1985. 243p. (FAO Soils Bulletin no. 55) — Second

Eden, Thomas. Elements of Tropical Soil Science. 2d ed. London: Macmillan; and New York: St. Martin's Press, 1964. 164p. (1st ed., London: Macmillan, 1956. 136p.) — Second

Second — Edwards, Clive A. et al., eds. Biological Interactions in Soil; Proceedings of a Workshop on Interactions between Soil-inhabiting Invertebrates and Microorganisms in Relation to Plant Growth, Ohio State University, Columbus, Ohio, Mar. 1987. Amsterdam and New York: Elsevier, 1988. (Reprinted from Agricultural Ecosystems and Environment, vol. 24, no. 1–3, 1988.) — Second

Third — Edwards, Clive A. et al., eds. Sustainable Agricultural Systems; Proceedings of the International Conference on Sustainable Agricultural Systems, Ohio State University, Columbus, Ohio, Sept. 1988. Ankeny, Iowa: Soil and Water Conservation Society, 1990. 696p.

Second — Edwards, Clive A., and G. K. Veeresh, eds. Soil Biology and Ecology in India; Proceedings of the 1st All India Symposium, University of Agricultural Sciences, Bangalore, Sept. 1976. Bangalore, India: University of Agricultural Sciences, 1978. 344p. — Third

Second — Ehrlich, Henry L. Geomicrobiology. 2d rev. ed. New York: M. Dekker, 1990. 646p. (1st ed., 1981. 393p.)

Second — Eisenbeis, Gerhard and Wilfried Wichard. Atlas on the Biology of Soil Arthropods. Berlin and New York: Springer-Verlag, 1987. 437p. (Trans. by Elizabeth A. Mole of Atlas

Developed Countries		Third World
	zur Biologie der Bodenarthropoden. Stuttgart and New York: G. Fischer, 1985. 434p.)	
	El-Swaify, S. A., E. W. Dangler and C. L. Armstrong. Soil Erosion by Water in the Tropics. Honolulu, Hawaii: College of Tropical Agriculture and Human Resources, University of Hawaii, 1982. 173p. (Univ. of Hawaii Research Extension Series no. 24)	Third
First rank	El-Swaify, S. A., W. C. Moldenhauer and A. Lo, eds. Soil Erosion and Conservation. International Conference on Soil Erosion and Conservation, Honolulu, Hawaii, Jan. 1983. Ankeny, Iowa: Soil Conservation Society of America, 1985. 793p.	Third
Second	Elliott, L. F., and F.J. Stevenson, eds. Soils for Management of Organic Wastes and Waste Waters; Proceedings of a Symposium at TVA National Fertilizer Development Center, Muscle Shoals, Ala., Mar. 1975. Madison, Wis.: Soil Science Society of America, American Society of Agronomy, and Crop Science Society of America, 1977. 650p.	
Third	Ellwood, D. C. et al., eds. Contemporary Microbial Ecology; Proceedings of the 2d International Symposium on Microbial Ecology, University of Warwick, U.K., Sept. 1980. London and New York: Academic Press, 1980. 438p.	
	Elprince, Adel M., ed. Chemistry of Soil Solutions. New York: Van Nostrand Reinhold, 1986. 411p.	Second
Second	Elrick, David E., and Daniel Hillel, eds. Scaling in Soil Physics, Principles and Applications; Proceedings of a symposium, Las Vegas, Nev., Oct. 1989. Madison, Wis.: Soil Science Society of America, 1990. 122p. (SSSA Special Publication no. 25)	Second
Third	Emerson, W. W., R. D. Bond and A. R. Dexter, eds. Modification of Soil Structure; Proceedings of the International Soil Science Society, Adelaide, Australia, Aug. 1976. London and New York: J. Wiley, 1978. 438p.	Second
Third	Engelhard, Arthur W., ed. Soilborne Plant Pathogens: Management of Diseases with Macro- and Microelements. St. Paul, Minn.: American Phytopathological Society Press, 1989. 217p.	Third
First rank	Engelstad, Orvis P., ed. Fertilizer Technology and Use. 3d ed. Madison, Wis.: Soil Science Society of America, 1985. 633p. (1st ed., ed. by M. H. McVickar et al., 1963)	First rank
Second	Epstein, Emanuel. Mineral Nutrition of Plants: Principles and Perspectives. New York: J. Wiley, 1972. 412p.	
Third	Erhart, Henri. La Genèse des Sols en Tant que Phénomene Geologique, Esquisse d'Une Théorie Géologique et Géo-	Third

Developed   Third
Countries   World

chimique, Biostasie et Rhexistasie. 2d ed., revue, corrigée
et augmentée. Paris: Masson et Cie, 1967. 177p. (1st ed.,
1956. 90p.)

F

Second	Fairbridge, Rhodes W., and Charles W. Finkl, eds. The Encyclopedia of Soil Science. Stroudsburg, Pa.: Dowden, Hutchinson and Ross, 1979. 646p. (Encyclopedia of Earth Sciences Series no. 12)	Second
Second	Fanning, Delvin S., and Mary C. B. Fanning. Soil: Morphology, Genesis, and Classification. New York: J. Wiley, 1989. 395p.	Second
Second	Farmer, V. C., ed. The Infrared Spectra of Minerals. London: Mineralogical Society, 1974. 539p. (1st ed. 1960.)	
	Fauck, Roger. Les Sols Rouges sur Sables et sur Gres d'Afrique Occidentale. Paris: ORSTOM, 1972. 257p.	Third
Second	Faulkner, Edward H. Plowman's Folly. Norman: University of Oklahoma Press, 1974. 156p. (10th printing of the 1943 ed. with new foreword.)	
Third	Fedoroff, N., and L.M. Bresson and M.A. Courty, eds. Soil Micromorphology = Micromorphologie des Sols: Actes de la VIIe Reunion Internationale . . . Plaisir. Paris: Association Francaise pour l'Etude du Sol, 1987. 686p. (7th International Working-Meeting on Soil Micromorphology, 1987, Paris)	Third
Third	Finck, Arnold. Fertilizers and Fertilization: An Introduction . . . Trans. from the German: Dunger und Dungung. Weinheim, FDR and Deerfield Beach, Fla.: Verlag Chemie, 1982. 438p.	Third
	Finck, Arnold. Tropische Böden: Einführung in die Bodenkundlichen Grundlagen Tropischer und Subtropischer Landwirtschaft. Hamburg: P. Parey, 1963. 188p.	Third
Third	Finkl, Charles W. Soil Classification. Stroudsburg, Pa.: Hutchinson Ross, 1982. 391p.	Second
Second	Fitter, A. H. et al., eds. Ecological Interactions in Soil: Plants, Microbes and Animals. Oxford and Boston: Blackwell Scientific Publications, 1985. 451p.	Second
First rank	FitzPatrick, Ewart A. Micromorphology of Soils. London: Chapman and New York: Hall, 1984. 433p. (Available in Spanish as Micromofología de Suelos. Mexico: Cecsa, 1990. 476p.)	Second
	FitzPatrick, Ewart A. Pedology: A Systematic Approach to Soil Science. Edinburgh, U.K.: Oliver and Boyd, 1971. 306p.	Second
Second	FitzPatrick, Ewart A. Soils: Their Formation, Classifica-	Second

Developed Countries		Third World
	tion, and Distribution. London and New York: Longmans, 1983. 353p. (Rev. ed. of: Pedology. 1972, 1971.)	
Third	Follett, Ronald F., ed. Soil Fertility and Organic Matter as Critical Components of Production Systems; Proceedings of a Symposium, Soil Science Society of America, Chicago, Dec. 1985. Madison, Wis.: American Society of Agronomy, 1987. 166p.	Second
Second	Follett, Ronald F., and Bobby A. Stewart, eds. Soil Erosion and Crop Productivity. Madison, Wis.: American Society of Agronomy, 1985. 533p.	Second
Third	Food and Agriculture Organization. Approaches to Soil Classification. Rome: Food and Agriculture Organization, 1968. 143p. (FAO. World Soil Resources Report no. 32)	
	Food and Agriculture Organization. Assessing Soil Degradation. Rome: Food and Agriculture Organization, 1977. 203p. (FAO Soils Bulletin no. 34)	Second
	Food and Agriculture Organization. Conservation in Arid and Semi-Arid Zones. Rome: Food and Agriculture Organization, 1976. 125p. (FAO Conservation Guide no. 3)	Second
Second	Food and Agriculture Organization. Drainage Materials. Rome: Food and Agriculture Organization, 1972. 126p. (FAO Irrigation and Drainage Paper no. 9) (4th printing 1976.)	Second
	Food and Agriculture Organization. Drainage of Heavy Soils. Rome: Food and Agriculture Organization, 1971. 114p. (FAO Irrigation and Drainage Paper no. 6) (5th printing 1976.)	Second
	Food and Agriculture Organization. Drainage of Salty Soils. Rome: Food and Agriculture Organization, 1973. 87p. (FAO Irrigation and Drainage Paper no. 16) (3d printing 1976.)	Second
	Food and Agriculture Organization. Drainage Testing. Rome: Food and Agriculture Organization, 1976. 185p. (FAO Irrigation and Drainage Paper no. 28) (2d printing 1984.)	Second
First rank	Food and Agriculture Organization. FAO-UNESCO Soil Map of the World. Revised Legend. Rome: Food and Agriculture Organization, 1988. 119p. (FAO World Soil Resources Report no. 52)	First rank
	Food and Agriculture Organization. A Framework for Land Evaluation. Rome: Food and Agriculture Organziation, 1976. (FAO Soils Bulletin no. 32)	First rank
	Food and Agriculture Organization. Irrigation, Drainage and Salinity: An International Source Book. London: Hutchinson, FAO, UNESCO, 1973. 510p.	First rank
	Food and Agriculture Organization. Land Evaluation Guide-	Second

Developed
Countries

Third
World

lines for Rainfed Agriculture. Report of an Expert Consultation, Rome, Dec. 1979. Rome: Food and Agriculture Organization, 1980. 118p. (FAO World Soil Resources Report no. 52) (Expert Consultation in Rome, Dec. 1979.)

First rank   Food and Agriculture Organization. Meeting of the Classification and Correlation of Soils from Volcanic Ash, Tokyo, 1964. Rome: Food and Agriculture Organization, 1965. 169p. (FAO World Soil Resources Bulletin no. 14)   First rank

Food and Agriculture Organization. Organic Materials and Soil Productivity in the Near East. Rome: Food and Agriculture Organization, 1982. (FAO Soils Bulletin no. 45)   Second

Food and Agriculture Organization. Organic Materials as Fertilizers. Rome: Food and Agriculture Organization, 1975. 167p. (FAO Soils Bulletin no. 27)   First rank

Food and Agriculture Organization. Organic Recycling in Africa. Rome: Food and Agriculture Organization, 1980. 304p. (FAO Soils Bulletin no. 43)   First rank

Food and Agriculture Organization. Organic Recycling in Asia. Rome: Food and Agriculture Organization, 1978. 312p. (FAO Soils Bulletin no. 36)   Second

Food and Agriculture Organization. A Practical Manual of Soil Microbiology Laboratory Methods. Rome: Food and Agriculture Organization, 1967. (FAO Soils Bulletin no. 7)   Second

Food and Agriculture Organization. Sandy Soils in the Near East and North Africa. Rome: Food and Agriculture Organization, 1973. 245p. (FAO Soils Bulletin no. 25)   Second

Food and Agriculture Organization. Shifting Cultivation and Soil Conservation in Africa. Rome: Food and Agriculture Organization, 1974. 248p. (FAO Soils Bulletin no. 24) (Papers presented at the FAO/SIDA/ARCN regional seminar, Ibadan, Nigeria, July 1973)   First rank

Food and Agriculture Organization. Soil Conservation and Management in Developing Countries. Rome: Food and Agriculture Organization, 1977. (FAO Soils Bulletin no. 33)   First rank

Food and Agriculture Organization. Soil Conservation for Developing Countries. Rome: Food and Agriculture Organization, 1976. (FAO Soils Bulletin no. 30)   First rank

Food and Agriculture Organization. Soil Erosion by Water: Some Measures for its Control on Cultivated Lands. Rome: Food and Agriculture Organization, 1965. 284p. (FAO Agricultural Development Paper no. 81)   Second

Food and Agriculture Organization. Soil Erosion by Wind and Measures for its Control on Agricultural Lands. Rome: Second

Developed Countries		Third World
	Food and Agriculture Organization, 1960. 87p. (FAO Agricultural Development Paper no. 71)	
	Food and Agriculture Organization. The Utilization of Secondary and Trace Elements in Agriculture; Proceedings of a Symposium, Geneva, 1987. Dordrecht and Boston: M. Nijhoff, 1987. 299p.	Third
Second	Food and Fertilizer Technology Center (Taiwan). The Fertility of Paddy Soils and Fertilizer Applications for Rice. Taipei, China: Food and Fertilizer Technology Center, 1976. 249p.	
	Food and Fertilizer Technology Center (Taiwan). Soil Taxonomy: Review and Use in the Asian and Pacific Region; Proceedings at the International Workshop, Taichung, Taiwan, 1985. Taipei, Taiwan: Food and Fertilizer Technology Center for the Asian and Pacific Region, 1985. 158p.	Third
	Forbes, Terence R., David G. Rossiter and A. van Wambeke. Guidelines for Evaluating the Adequacy of Soil Resource Inventories. Washington, D.C.: Soil Conservation Service, U.S. Dept. of Agriculture, 1982. 50p.	Third
Second	Foster, G. R., ed. Soil Erosion: Prediction and Control. Proceedings of a Symposium at Purdue University, West Lafayette, Ind., May 1976. Ankeny, Iowa: Soil Conservation Society of America, 1977. 393p. (SCSA Special Publication no. 21)	
Second	Foster, R. C., A. D. Rovira and T. W. Cook. Ultrastructure of the Root-Soil Interface. St. Paul, Minn.: American Phytophathological Society, 1983. 157p.	
First rank	Foth, Henry D. Fundamentals of Soil Science. 8th ed. New York: J. Wiley, 1990. 360p. (1st ed. 1943.) (Available in Spanish as Fundamento de la Ciencia del Suelo. Trans. by Antonio M. Amprosio. 6th ed. Mexico: Cecsa, 1985. 448p.)	First rank
Second	Foth, Henry D., and Boyd G. Ellis. Soil Fertility. New York: J. Wiley, 1988. 212p.	First rank
Second	Foth, Henry D., and John W. Schafer. Soil Geography and Land use. New York: J. Wiley, 1980. 484p.	Second
	Francis, Charles A., ed. Multiple Cropping Systems. New York and London: Macmillan, 1986. 383p.	Third
Second	Franz, Herbert. Bodenzoologie als Grundlage der Bodenpflege: Mit Besonderer Berucksichtigung der Bodenfauna in den Ostalpen und im Donaubecken. Berlin: Akademie-Verlag, 1950. 316p. (Soil Zoology as the Foundation of Soil Management with Special Emphasis on Soil Fauna in the Eastern Alps and the Danube Valley)	
Third	Fraysse, Georges, ed. Remote Sensing Application in Agriculture and Hydrology. Proceedings of An Advanced	

Developed Countries		Third World

	Seminar, Joint Research Centre of the Commission of the European Communites in the framework of the Ispra Courses, Ispra, Italy, Nov.–Dec. 1977. Rotterdam, Netherlands: A. A. Balkema, 1980. 502p.	
	Freney, J. R., and J. R. Simpson, eds. Gaseous Loss of Nitrogen from Plant-Soil Systems. The Hague and Boston: Nijhoff, 1983. 317p.	Second
	Frenkel, Haim and Avraham Meiri, eds. Soil Salinity: Two Decades of Research in Irrigated Agriculture. New York: Van Nostrand Reinhold, 1985. 459p.	Second
Second	Frevert, R. K., ed. Soil and Water Conservation Engineering. 3d ed. New York: J. Wiley, 1981. 525p. (1st ed., 1955)	Third
First rank	Fridland, Vladimir M. Pattern of the Soil Cover. Trans. from Russian by N. Kaner. Jerusalem: Israel Program for Scientific Translations, 1976. 291p.	Third
Second	Fridland, Vladimir M., ed. Soil Combinations and Their Genesis = Pochvennye Combinatsii iikh Genezis. New Delhi: Amerind Pub. Co., 1976. 272p.	Third
Third	Frimmel, F. H., and R. F. Christman, eds. Humic Substances and Their Role in the Environment. Report of the Dahlem Workshop on Humic Substances and Their Role in the Environment, Berlin, 1987. Chichester, U.K. and New York: J. Wiley, 1988. 271p.	Second
Third	Frissel, M. J., ed. Cycling of Mineral Nutrients in Agricultural Ecosystems. Proceedings of the 1st International Environmental Symposium of the Royal Netherlands Development Society, Amsterdam, 1976. Amsterdam and New York: Elsevier, 1978.	Second
Third	Fukuoka, Masanobu. The One-Straw Revolution: An Introduction to Natural Farming. Emmaus, Penn: Rodale Press, 1978. 181p. (Trans. from the Japanese by C. Pearce, T. Kurosawa, and L. Korn.)	Third

G

Third	Gachon, Louis, ed. Phosphore et Potassium dans les Relations Sol-Plante: Conséquences sur la Fertilisation. Paris: Institut National de la Recherche Agronomique, 1988. 566p.	
First rank	Garrels, Robert M., and Charles L. Christ. Solutions, Minerals, and Equilibria. New York: Harper and Row, 1965. 450p. (Based on Mineral Equilibria at Low Temperatures and Pressure by R. M. Garrels, 1960)	
Second	Garrett, Stephen D. Soil Fungi and Soil Fertility: An Introduction to Soil Mycology. 2d ed. Oxford and New York: Pergamon Press, 1981. 150p. (1st ed., 1963. 165p.)	

Developed Countries		Third World
	Garver, Cynthia L., and D. R. Mohan Raj, eds. Management of Vertisols for Improved Agricultural Production; Proceedings of an IBSRAM inaugural workshop (International Board for Soil Research and Management), ICRISAT Center, India, Feb. 1985. Patancheru, Andhra Pradesh, India: International Crops Research Institute for the Semi-Arid Tropics, 1989. 278p.	Third
First rank	Geiger, Rudolf. Klima der Bodennahen Luftschicht = The Climate Near the Ground. Cambridge, Mass.: Harvard University Press, 1965. 611p.	
Second	Gerasimov, I. P., and M. A. Glazovskaya. Fundamentals of Soil Science and Soil Geography. Trans. of Osnovy Pochvovedeniya i Geografiya Pochv. Jerusalem: Israel Program for Scientific Translations, 1965. 382p.	Second
Second	Gerasimov, I. P., Serge V. Zonn and V. M. Fridland. Genesis and Geography of Soils. Trans. of Genezis i Geografiya Pochv . . . New Delhi, India: Indian National Scientific Documentation Centre, 1975. 313p. (1st ed.: Genezis i Geografisisa Pochv . . . , edited by I.P. Gerasimov. Moskow: Nauka, 1964. 164p.)	Third
Third	Germon, J. C., ed. Management Systems to Reduce Impact of Nitrates. London and New York: Elsevier, 1989. 274p. (Based on a meeting, Brussels, Sept. 1987.)	
Third	Gerrard, John, ed. Alluvial Soils. New York: Van Nostrand Reinhold, 1987. 305p.	Second
Second	Gerrard, John. Soils and Landforms: An Integration of Geomorphology and Pedology. London and Boston: Allen and Unwin, 1981. 219p.	Second
Third	Gest, Howard. The World of Microbes. Madison, Wis.: Science Tech Pub. and Menlo Park, Calif.: Benjamin Cummings, 1987. 249p.	
Second	Geus, Jan G. de. Fertilizer Guide for the Tropics and Subtropics. 2d ed. Zurich: Centre d'Etude de l'Azote, 1973. 774p. (1st ed.: Fertilizer Guide for Tropical and Subtropical Farming. 1967.)	Second
	Ghildyal, B. P., and R. P. Tripathi. Soil Physics. New York: J. Wiley, 1987. 656p.	Second
Third	Gibson, Alan H., and William E. Newton, eds. Current Perspectives in Nitrogen Fixation; Proceedings of the 4th International Symposium on Nitrogen Fixation, Canberra, Australia, Dec. 1980. Amsterdam and New York: Elsevier/ North-Holland Biomedical Press, 1981. 534p.	Second
Second	Gieseking, John E., ed. Soil Components. New York: Springer-Verlag, 1975. 2 vols.	Second
	Gil, N. Watershed Development: With Special Reference to Soil and Water Conservation. Rome: Food and Agricul-	Second

Developed   Third
Countries   World

ture Organization, 1979. 267p. (FAO Soils Bulletin no.
44)

Third   Gill, William R., and Glen E. Vanden Berg. Soil Dy-
             namics in Tillage and Traction. Washington, D.C.: Ag-
             ricultural Research Service, U.S. Dept. of Agricul-
             ture, 1967. 511p. (USDA Agricultural Handbook no.
             316)

             Glantz, Michael H., ed. Desertification: Environmental   Third
             Degradation In and Around the Arid Lands. Boulder,
             Colo.: Westview Press, 1977. 346p.

Third   Glazovskaya, M. A. Soils of the World = Pochvy Mira.   Third
             New Delhi, India: Amerind Pub. Co., 1983.

Third   Gliessman, Stephen R., ed. Agroecology: Researching the Third
             ecological Basis for Sustainable Agriculture. New York:
             Springer-Verlag, 1990. 380p.

Second   Glinka, Konstantin D. Treatise on Soil Science (Pochvove- Second
             denie). 4th rev. ed. Jerusalem: Israel Program for Scien-
             tific Translations, 1963. 674p. (Trans. from Russian by A.
             Gourevitch.)

Third   Glinski, Jan and Jerzy Lipiec. Soil Physical Conditions and Second
             Plant Roots. Boca Raton, Fla.: CRC Press, 1990. 250p.

Third   Glinski, Jan and Witold Stepniewski. Soil Aeration and Its Second
             Role for Plants. Boca Raton, Fla.: CRC Press, 1985. 229p.

Third   Goldschmidt, Victor M. Geochemistry. Edited by Alex
             Muir. Oxford, U.K.: Clarendon Press, 1954. 730p.

             Goosen, D. Aerial Photo Interpretation in Soil Survey.   Second
             Rome: Food and Agriculture Organization, 1967. 55p.
             (FAO Soils Bulletin no. 6)

Second   Goring, Cleve A. I., and J. W. Hamaker. Organic Chemi-   Second
             cals in the Soil Environment. New York: M. Dekker,
             1972. In 2 vols. 968p.

             Goudie, Andrew. Duricrusts in Tropical and Subtropical   Third
             Landscapes. Oxford, U.K.: Clarendon Press, 1973. 174p.

Second   Graff, Otto and John E. Satchell, eds. Progress in Soil
             Biology. Proceedings of a Colloquium on Dynamics of
             Soil Communities, Brunswick and Volkenrode, Germany,
             1966 = Beitrage zur Bodenbiologie = Travaux Recents
             de la Biologie de Sol. Braunschweig: Vieweg and Amster-
             dam: North Holland Pub. Co., 1967. 656p.

Second   Graham, Edward H. Natural Principles of Land Use. New
             York: Greenwood Press, 1969. 274p. (1st ed. 1944.)

             Graham, Peter H., and Susan C. Harris, eds. Biological   Second
             Nitrogen Fixation Technology for Tropical Agriculture.
             Proceedings of a Workshop, CIAT, Colombia, Mar. 1981.
             Cali, Colombia: Centro Internacional de Agricultura Tropi-
             cal, 1982. 726p.

Developed Countries		Third World
Second	Gray, T. R. G., and D. Parkinson, eds. The Ecology of Soil Bacteria. Proceedings of an International Symposium. Liverpool, U.K.: Liverpool University Press, 1968. 681p.	Second
Second	Greenland, D. J. The Chemistry of Soil Processes. Chichester, U.K. and New York: J. Wiley, 1981. 714p.	First rank
Second	Greenland, D. J., and M. H. B. Hayes. The Chemistry of Soil Constituents. Chichester, U.K. and New York: J. Wiley, 1978. 469p.	First rank
	Greenland, D. J., ed. Characterization of Soils in Relation to Their Classification and Management for Crop Production: Examples from Some Areas of the Humid Tropics. Oxford, U.K.: Clarendon Press, 1981. 446p.	First rank
	Greenland, D. J., and R. Lal, eds. Soil Conservation and Management in the Humid Tropics; Proceedings of a Conference Sponsored by the International Institute for Tropical Agriculture and the Agricultural Research Council, Ibadan, Nigeria, 1975. Chichester, U.K. and New York: J. Wiley, 1977. 283p.	First rank
Third	Gregg, Sidney J., and K. S. W. Sing. Adsorption, Surface Area, and Porosity. 2d ed. London and New York: Academic Press, 1982. 303p. (1st ed. 1967.)	Third
Second	Griffin, David M. Ecology of Soil Fungi. London: Chapman and Hall, 1972. 193p.	Second
Third	Griffiths, John C. Scientific Method in Analysis of Sediments. New York: McGraw-Hill, 1967. 508p.	Second
	Griffiths, John F., ed. Climates of Africa. Amsterdam and New York: Elsevier, 1972. 604p.	Third
First rank	Grim, Ralph E. Clay Mineralogy. 2d ed. New York: McGraw-Hill, 1968. 596p. (1st ed. 1953)	
Third	Gudehus, G., F. Darve and I. Vardoulakis. Constitutive Relations for Soils; Results of the International Workshop on Constitutive Relations for Soils, Grenoble, Sept. 1982. Rotterdam and Boston: A. A. Balkema, 1984. 497p.	
Third	Guenzi, W. D. et al., eds. Pesticides in Soil and Water. Madison, Wis.: Soil Science Society of America, 1974. 562p.	Second
	Gupta, U.S., ed. Physiological Aspects of Dryland Farming. New Delhi: Oxford and IBH Pub. Co., 1975. 391p.	Second

H

Third	Haan, C. T., H.P. Johnson and D.L. Brakensiek, eds. Hydrologic Modeling in Small Watersheds. St. Jospeh, Mich.: American Society of Agricultural Engineers, 1982. 533p. (ASAE Monograph Series no. 5)	
Third	Hadas, A. et al., eds. Physical Aspects of Soil Water and Salts in Ecosystems; Proceedings of a Symposium on Soil-	Second

Developed Countries		Third World
	Water Physics and Technology, Rehovot, Israel, 1971. Berlin and New York: Springer-Verlag, 1973. 460p.	
Second	Hagan, Robert M., Howard R. Haise and Talcott W. Edminster, eds. Irrigation of Agricultural Lands. Madison, Wis.: American Society of Agronomy, 1967. 1180p. (Agronomy Monograph Series no. 11)	Second
Second	Hallsworth, E. G., and D.V. Crawford., eds. Experimental Pedology; Proceedings of the 11th Easter School in Agricultural Science, University of Nottingham, 1964. London: Butterworths, 1965. 413p.	Second
	Hamdi, Y. A. Application of Nitrogen-Fixing Systems in Soil Management. Rome: Food and Agriculture Organization, 1982. 188p. (FAO Soils Bulletin no. 49)	Second
	Handbook on Reference Methods for Soil Analysis. 3d ed. Athens, Ga.: Council on Soil Testing and Plant Analysis, 1992. 202p. (1974 ed.: Handbook on Reference Methods for Soil Testing.)	Second
Second	Hanks, Ronald J., and G. L. Ashcroft. Applied Soil Physics: Soil, Water and Temperature Applications. Berlin and New York: Springer-Verlag, 1980. 159p.	Second
	Hansen, Jens A., and Kaj Henriksen, eds. Nitrogen in Organic Wastes Applied to Soils. London and San Diego, Calif.: Academic Press, 1989. 381p.	Second
	Hardy, Frederick. Suelos Tropicales: Pedologia Tropical con Enfasis en America. Districto Federale, Mexico: Herrero Hermanos, 1970. 334p.	Second
Second	Hargrove, W. L., ed. Cropping Strategies for Efficient use of Water and Nitrogen. Proceedings of a Symposium, Atlanta, Ga., Nov.–Dec. 1987. Madison, Wis.: American Society of Agronomy, Crop Science Society of America, and Soil Science Society of America, 1988. 218p. (ASA Special Publication no. 51)	Second
Second	Harley, J. L., and R. Scott Russell, ed. The Soil-Root Interface; Proceedings of an International Symposium, Oxford, England, Mar. 1978. London and New York: Academic Press, 1979. 448p.	Second
Second	Harley, J. L. The Biology of Mycorrhiza. 2d ed. London: L. Hill, 1969. 334p. (1st ed., and New York: Interscience Pub., 1959. 233p.)	
	Harmsen, K. Behaviour of Heavy Metals in Soils. Wageningen, Netherlands: Centre for Agricultural Publishing and Documentation (PUDOC), 1977. 171p.	Third
Third	Hattori, Tsutomu. Microbial Life in the Soil: An Introduction. New York: M. Dekker, 1973. 427p.	Third
First rank	Hauck, Roland D., ed. Nitrogen in Crop Production; Proceedings of a Symposium, Sheffield, Ala., May 1982. Mad-	First rank

Developed Countries		Third World
	ison, Wis.: American Society of Agronomy, 1984. 804p.	
	Hauser, G. F. Guide to the Calibration of Soil Tests for Fertilizer Recommendations. Rome: Food and Agriculture Organization, 1973. 71p. (FAO Soils Bulletin no. 18)	Second
	Hayes, Mary J., and June H. Cooley, eds. Tropical Soil Biology: Current Status of Concepts. Athens, Ga.: International Association for Ecology, 1987. 58p.	Second
Third	Hayes, Michael H. B. et al., eds. Humic Substances II: In Search of Structure. Papers from the Second Conference of the International Humic Substances Society, University of Birmingham, July, 1984. Chichester, U.K. and New York: J. Wiley, 1989. 764p.	Third
Second	Haynes, R. J. Mineral Nitrogen in the Plant-Soil System. Orlando, Fla.: Academic Press, 1986. 483p.	Second
Third	Head, K. H. Manual of Soil Laboratory Testing. 2d ed. New York: Halsted Press, 1992. 1 vol. (1st ed., New York: J. Wiley, 1980–82. 3 vols.)	
Third	Heinonen, Reijo. Soil Management and Crop Water Supply. 4th ed. Uppsala, Sweden: Swedish University, 1985. 105p. (3d ed., 1979. 106p.)	Second
	Henin, Stephane. Cours de Physique du Sol. Paris: OR-STOM, 1976–77. 2 vols. (Initiations, Documentations Techniques no. 28–29)	Second
	Henin, Stephane, R. Gras and G. Monnier. Le Profil Cultural, l'etat Physique de sol et ses Consequences Agronomiques. 2d ed. Paris: Masson et Cie, 1969. 332p. (1st ed., 1960)	Third
Third	Hignett, Travis P., ed. Fertilizer Manual. Dordrecht and Boston: Nijhoff/W. Junk Pub., 1985. 363p.	Second
First rank	Hillel, Daniel, ed. Applications of Soil Physics. New York: Academic Press, 1980. 385p.	First rank
First rank	Hillel, Daniel. Fundamentals of Soil Physics. New York: Academic Press, 1980. 413p.	First rank
Second	Hillel, Daniel. Soil and Water: Physical Principles and Processes. New York: Academic Press, 1971. 288p.	First rank
Second	Hillel, Daniel and David E. Elrick, eds. Scaling in Soil Physics: Principles and Applications. Proceedings of a symposium, Las Vegas Nev., 1989. Madison, Wis.: Soil Science Society of America, 1990. 122p. (SSSA Special Publication no. 25)	Second
Third	Hodgson, J. M., ed. Soil Survey Field Handbook: Describing and Sampling Soil Profiles. Reprinted with minor amendments. Harpenden, U.K.: Soil Survey of England and Wales, 1976. 99p. (1st ed., 1974. 99p.)	
Second	Hodgson, John M. Soil Sampling and Soil Description.	Second

Developed Countries		Third World
	Oxford, U.K.: Clarendon Press, 1978. 241p. (Available in Spanish as Muestreo y Descripción de Suelos. Barcelona: Reverté, 1987. 240p.)	
Third	Holding, A. J. et al., eds. Soil Organisms and Decomposition in Tundra; Proceedings of the Microbiology, Decomposition and Invertebrate Working Groups Meeting, University of Alaska, Fairbanks, Aug. 1973. Stockholm, Sweden: Tundra Biome Steering Committee, 1974. 398p.	
Second	Hole, Francis D., and James B. Campbell. Soil Landscape Analysis. Totowa, N.J.: Rowman and Allanheld, 1985. 196p.	
First rank	Hook, Donald D., and R. M. M. Crawford, eds. Plant Life in Anaerobic Environments. Ann Arbor, Mich.: Ann Arbor Science Pub., 1978. 564p.	
	Hopfen, Hans J. Farm Implements of Arid and Tropical Regions. Rev. ed. Rome: Food and Agriculture Organization, 1969. 159p. (FAO Agricultural Development Paper no. 91) (Available in Spanish as Aperos de Labranza para las Regiones Áridas y Tropicales. 3d ed., rev. Rome: Food and Agriculture Organization, 1982. 154p.)	Third
Third	Hossner, L. R., ed. Reclamation of Surface-Mined Lands. Boca Raton, Fla.: CRC Press, 1988. Vols. I, 219 p. and II, 249p.	Third
First rank	Huang, P. M., and M. Schnitzer, eds. Interactions of Soil Minerals with Natural Organics and Microbes; Proceedings of a Symoposium, Washington, D.C., Aug. 1983. Madison, Wis.: Soil Science Society of America, 1986. 606p.	Second
Third	Hucker, T. W. G., and G. Catroux, eds. Phosphorus in Sewage Sludge and Animal Waste Slurries; Proceedings of the EEC Seminar, June 1980. Boston: Kluwer Pub. Co., 1981. 443p.	
	Hudson, Norman W. Soil and Water Conservation in Semi-Arid Areas. Rome: Food and Agriculture Organization, 1987. 185p. (FAO Soils Bulletin no. 57)	Second
First rank	Hudson, Norman W. Soil Conservation. 2d ed. Ithaca, N.Y.: Cornell University Press, 1981. 324p. (Available in Spanish as Conservación del Suelo. Barcelona: Reverté, 1982. 352p.)	First rank
Second	Hunt, Charles B. Geology of Soils: Their Evolution, Classification, and Uses. San Francisco, Calif.: W. H. Freeman, 1972. 344p.	Third
Third	Hyams, Edward. Soil and Civilization. London and New York: Thames and Hudson, 1952. 312p.	Second

Developed
Countries

I

Third

Ikawa, H., and G. Y. Tsuji, eds. Soil Classification and Agrotechnology Transfer for Agriculture Development. Proceedings of the 3d International Forum on Soil Taxonomy and Agrotechnology Transfer, Yaoundae, Cameroon, 1983. Honolulu, Hawaii: HITAHR, College of Tropical Agriculture and Human Resources, University of Hawaii, 1987. 254p. (Benchmark Soils Project Technical Report no. 10)

Third

Iler, Ralph K. The Colloid Chemistry of Silica and Silicates. Ithaca, N.Y.: Cornell University Press, 1955. 324p.

Third

Institut Geografii. (USSR.) Micromorphological Methods in the Investigation of Genesis of Soils = Mikromorfologicheskie metody v issledovanii genezisa pochvk. Edited by S.V. Zonn. New Delhi: Indian National Scientific Documentation Centre, 1977. 198p.

Second

Institute of Soil Science, Academia Sinica, ed. Soils of China, edited and translated by Sun Ou Li Chingkwei. English ed. Beijing, China: Science Press, 1990. 873p. (Trans. of 2d Chinese ed.; 1st ed., 1978, titled: Chung-kuo Tu Jang.)

Third

Institute of Soil Science, Academia Sinica. Symposium on Paddy Soil. Beijing, China: Science Press and Berlin: Springer Verlag, 1981. 864p.

Third

International Atomic Energy Agency. Agrochemical Residue—Biota Interactions in Soil and Aquatic Ecosystems; Proceedings . . . Vienna, Austria, 1978. Vienna, Austria: International Atomic Energy Agency, 1980. 305p.

Third

International Atomic Energy Agency. Isotope and Radiation Techniques in Soil Physics and Irrigation Studies; Proceedings of an International Symposium on . . . , Aix-en-Provence, Apr. 1983. Vienna, Austria: International Atomic Energy Agency, 1983. 597p.

Third

International Atomic Energy Agency. Proceedings of the Symposium on Soil Organic Matter Studies, Braunschweig, Germany, 1976. Vienna, Austria: International Atomic Energy Agency, 1977. 2 vols. (English, French, Russian or Spanish with abstracts in English.)

Second

International Board for Soil Research and Management. First Training Workshop on Acid Tropical Soils Management and Land Development Practices; Papers of a Workshop, Yurimaguas Experiment Station, Peru, 1988. Bangkok, Thailand: IBSRAM, 1988. 242p. (IBSRAM Technical Notes no. 2)

Third

International Board for Soil Research and Management. Management of Acid Tropical Soils for Sustainable Agri-

Second

Developed Countries		Third World
	culture; Proceedings of an IBSRAM workshop, Yurimaguas, Peru and Brasilia, 1985. Bangkok, Thailand: IBSRAM, 1987. 299p.	
	International Clay Conference, 5th, Mexico City, 1975. Proceedings . . . , edited by S.W. Bailey. Wilmette, Ill.: Applied Pub., 1976. 691p. (English or French.)	
Third	International Clay Conference, 7th, Bologna and Pivia, Italy, 1981. Proceedings . . . , edited by H. Van Olphen and F. Veniale. Amsterdam and New York: Elsevier, 1982. 827p. (Developments in Sedimentology no. 35)	Third
Third	International Clay Conference, 8th, Denver, Colo., 1985. Proceedings . . . , edited by Leonard G. Schultz, H. Van Olphen and Frederick A. Mumpton. Bloomington, Ind.: Clay Minerals Society, 1987. 456p.	Third
	International Conference on Geomechanics in Tropical Soils; Proceedings, Singapore, 1988. Rotterdam and Brookfield, Vt.: A. A. Balkema, 1988–1990. 2 vols. (Sponsored by Nanyang Technological Institute, International Society for Soil Mechanics and the South East Asian Geotechnical Society)	Third
Third	International Congress of Soil Science, 5th, 1954, Leopoldsville, Belgian Congo. Transactions. Leopoldsville, Congo: International Society of Soil Science, 1954.	Third
First rank	International Congress of Soil Science, 6th, Paris, 1956. Rapports. Paris: C. Bernard, 1956. 5 vols.	First rank
First rank	International Congress of Soil Science, 7th, Madison, 1960. Transactions. Amsterdam: Elsevier, 1961. 4 vols. (English, French or German with summaries in the three languages.)	First rank
First rank	International Congress of Soil Science, 8th, Bucharest, 1964. Transactions . . . = Comptes Rendus. Berichte. Bucharest: Publishing House of the Academy of the Socialist Republic of Romania, 1965. 5 vols. (In English, French, or German.)	First rank
First rank	International Congress of Soil Science, 9th, Adelaide, Australia, 1968. Proceedings . . . New York: Elsevier, 1968. 4 vols. (English, French, or German, summaries in all three languages.)	First rank
First rank	International Congress of Soil Science, 10th, Moscow, 1974. Proceedings . . . Prague: Ustav Vedeckotechnick ych Informaci, 1974. 555p. (Papers in English or German with summaries in Czech and Russian.)	First rank
First rank	International Congress of Soil Science, 11th, University of Alberta, Edmonton, June 1978. Proceedings . . . Alberta, Canada: University of Alberta, 1978. 3 vols.	First rank
Third	International Congress of Soil Science, 12th, New Delhi,	First rank

Developed Countries		Third World
	India, 1982. Desertification and Soils Policy: Symposium Papers 3 . . . New Delhi, India: Indian Agricultural Research Institute, 1982. 166p. (Abstracts in English, French and German)	
Third	International Congress of Soil Science, 13th, Hamburg, Germany, 1986. XIII Kongress Mezhdunarodnogo Obshchestva Pochvovedov . . . : Informatsionnyi Material, edited by V. A. Kovda et al. Pushchino: Nauka, 1987. 48p.	Second
	International Congress on Phosphorus Compounds, 2d; Proceedings . . . 2d International Congress, Boston, Mass., Apr. 1980. Paris: Institut Mondial du Phosphate, 1980. 826p.	Third
	International Fertilizer Development Center. Fertilizer Efficiency Research and Technology Transfer Workshop for Africa South of the Sahara. Proceedings . . . Douala, Cameroon, Jan. 1985. Muscle Shoals, Ala: International Fertilizer Development Center, 1985. 474p. (International Fertilizer Development Center Special Publication no. SP-5)	Third
	International Fertilizer Development Center. Sulfur in the Tropics. Muscle Shoals, Ala.: International Fertilizer Development Center and the Sulphur Institute, 1979. 69p. (IFDC Technical Bulletin no. 12.)	Third
Third	International Interactive Workshop on Soil Resources: Their Inventory, Analysis and Interpretation for Use in the 1990's. Proceedings . . . , Minneapolis, Minn., Mar. 1988. St. Paul, Minn.: University of Minnesota, 1988. 259p.	
	International Network on Soil Fertility and Fertilizer Evaluation for Rice. Efficiency of Nitrogen Fertilizers for Rice. Proceedings of the Meeting of the International Network on Soil Fertility and Fertilizer Evaluation for Rice, Griffith, N.S.W., Australia, Apr. 1985. Los Baños, Philippines: International Rice Research Institute, 1987. 260p.	Second
Third	International Peat Congress, 6th, 1980, Duluth, Minn.. Proceedings . . . Duluth, Minn.: International Peat Society, 1981. 735p.	Second
	International Peat Congress, 7th, Dublin, Ireland, 1983. Proceedings . . . Dublin, Ireland: International Peat Society, 1984. 4 vols.	Third
	International Potash Institute. Nutrient Balances and the Need for Fertilizers in Semi-arid and Arid Regions; Proceedings of the 17th Colloquium of the International Potash Institute, Rabat and Marrakech, Morocco, 1983. Worblaufen-Berne, Switzerland: International Potash Institute, 1983. 394p.	Second
Third	International Potash Institute. Potassium in Soil; Proceedings of the 9th Colloquium of the International Potash In-	Second

Developed Countries		Third World

stitute, Landshut, Federal Republic of Germany, 1972. Berne, Switzerland, 1972. 220p.

International Potash Institute. Role of Fertilization in the Intensification of Agricultural Production; Proceedings of the 9th Congress of the International Potash Institute, Antibes, 1970 = Role de la Fertilisation dans l'Intensificaion de la Production Agricole = Die Rolle der Düngung in der Intensivierung der Landwirtschaft. Berne, Swizterland: International Potash Institute, 1973. 489p.   — Third

International Potash Institute. 14th Congress of the International Potash Institute, Sevilla, Spain, 1979. Soils in Mediterranean Type Climates and Their Yield Potential; Proceedings . . . Berne, Switzerland: International Potash Institute, 1979. 451p.   — Third

International Rice Research Institute. Field Problems of Tropical Rice. Rev. ed. Los Baños, Philippines: International Rice Research Institute, 1983. 172p. (1st ed., 1970 by K.E. Mueller.)   — Second

International Rice Research Institute. Organic Matter and Rice. Los Baños, Philippines: International Rice Research Institute, 1984. 631p. (Papers from the International Conference on Organic Matter and Rice, Los Baños, Philippines, Sept.–Oct. 1982.)   — Second

First rank   International Rice Research Institute. Priorities for Alleviating Soil-Related Constraints to Food Production in the Tropics; Papers from an international symposium at IRRI, June 1979. Los Baños, Philippines: International Rice Research Institute, 1980. 468p.   First rank

Second   International Rice Research Institute. Soils and Rice; Proceedings of a Symposium, IRRI, the Philippines. Los Baños, Philippines: International Rice Research Institute, 1978. 825p.   Second

International Seminar on Fertilizer Use Efficiency. Proceedings . . . Lahore, Pakistan, Nov. 1985. Rawalpindi: Fauji Fertilizer Co., 1986. 227p.   — Third

Third   International Seminar on Lateritisation Processes, Trivandrum, India, 1979. Proceedings . . . Rotterdam, Netherlands: A. A. Balkema, 1981. 450p.   Third

Third   International Seminar on Lateritisation Processes, 2d, Sao Paulo, Brazil, 1982. Proceedings . . . , edited by A. J. Melfi, and A. Carvalho. São Paulo, Brazil: Instituto Astron Domico e Geogaisico, University of São Paulo, 1983. 590p.

Second   International Seminar on Soil Environment and Fertility Management in Intensive Agriculture, Tokyo, Japan, 1977. Proceedings . . . Tokyo, Japan: Society of the Science of   Second

Developed Countries		Third World
	Soil and Manure, National Institute of Agricultural Sciences, 1977. 821p.	
Second	International Society of Soil Science and the British Society of Soil Science. Biological Processes and Soil Fertility; Proceedings of a Symposium . . . The Hague and Boston: M. Nijhoff and W. Junk, 1984.	Third
	International Society of Soil Science, Aberdeen, 1966. Soil Chemistry and Fertility; Proceedings of a Symposium . . . Amsterdam: International Society of Soil Science, 1967. 415p.	Third
Third	International Society of Soil Science. Proceedings of the Symposium on Water and Solute Movement in Heavy Clay Soils, Wageningen, Netherlands, 1984, edited by J. Bouma, and P. A. C. Raats. Wageningen, Netherlands: International Institute for Land Reclamation and Improvement, 1984. 363p.	Third
Third	International Society of Soil Science. Working Group on Soil Information Systems. Soil Information Systems. Proceedings of the Meeting of the ISSS Group on Soil Information Systems, Wageningen, Netherlands, Sept. 1975. Wageningen: Center for Agricultural Publishing and Documentation (PUDOC), 1975. 87p.	
Third	International Society of Soil Science. Working Group on Soil Information Systems. Uses of Soil Information Systems; Proceedings of the Australian Meeting of the ISSS Working Group on Soil Information Systems, Canberra, Australia, Mar. 1976, edited by Alan W. Moore and Stein W. Bie. Wageningen: Center for Agricultural Publishing and Documentation (PUDOC), 1977. 109p.	
	International Soil Classification Workshop, 2d, Malaysia and Thailand, 1978. Proceedings . . . Bangkok, Thailand: Soil Survey Division, Land Development Dept., 1979. 429p.	Third
	International Soil Classification Workshop, 5th, Sudan, 1982. Taxonomy and Management of Vertisols and Aridisols; Proceedings . . . Khartoum, Sudan: Soil Survey Administration, 1985. 1 vol.	Second
Third	International Soil Zoology Colloquium, 4th, Dijon, 1970. Organismes du Sol et Production Primaire. Comptes Rendus de IVe Colloque du Comité de Zoologie de l'Association Internationale de Science du Sol. Paris: Institut National de la Recherche Agronomique, 1971. 590p.	
Third	International Symposium on Acid Sulphate Soils. Acid Sulphate Soils; Proceedings . . . , Wageningen, Netherlands, Aug. 1972. Wageningen: International Institute for Land Reclamation and Improvement, 1973. 2 vols. (Inter-	

Developed Countries		Third World
	national Institute for Land Reclamation and Improvement no. 24)	
	International Symposium on Distribution, Characteristics and Utilization of Problem Soils. Proceedings of a Symposium on Tropical Agriculture Research, Tsukuba, Japan, Oct. 1981. Yatabe, Japan: Tropical Agriculture Research Center, Ministry of Agriculture, Forestry and Fisheries, 1982. 417p. (Tropical Agriculture Research Series no. 15)	Third
Third	International Symposium on Soil Classification, 3d. Proceedings of the symposium: Classification des Sols, Ghent, Belgium, May–June 1962. Ghent: Société Belge de Pédologie, 1965. 201p.	Third
	International Symposium on Soil Fertility Evaluation, Delhi, 1971. Proceedings . . . , organized by the Indian Society of Soil Science in cooperation with the International Society of Soil Science, Indian Council of Agricultural Research, the Indian Society of Agronomy, and the Fertilizer Association of India. New Delhi: Indian Society of Soil Science, 1971. 112p.	Second
	International Workshop on Soil Taxonomy and Characterization of Tropical Soils, National Chung Hsing University, Taichung, Taiwan, 1985. Soil Taxonomy: Review and Use in the Asian and Pacific Regions. Taipei and Taiwan, Republic of China: Food and Fertilizer Technology Center, 1985. 158p.	Third
	International Workshop on Soils. Research to Resolve Selected Problems of Soils in the Tropics; Proceedings, Townsville, Australia, 1983. Townsville, Australia: Australian Centre for International Agricultural Research, 1983. 187p.	Third
Third	Irrigation Association. Irrigation: Formerly Sprinkler Irrigation. 5th ed. Silver Spring, Md.: Irrigation Association, 1983. 686p. (2d ed., Sprinkler Irrigation, edited by Guy O. Woodward. Washington, D.C.: Sprinkler Irrigation Assocation, 1959. 377p.)	Third
Third	Iskandar, I. K., ed. Modeling Wastewater Renovation and Land Treatment. New York: J. Wiley, 1981. 802p.	
Second	Israelsen, Orson W., and Glen E. Stringham. Irrigation Principles and Practices. 4th ed. New York: J. Wiley, 1980. 450p. (1st ed., 1965. 447p.) (Available in Spanish as Principios y Aplicaciones del Riego. 2d ed. Barcelona: Reverté, 1965. 420p.)	Second
Third	Ivanova, Y. N., ed. Genesis and Classification of Semidesert Soils = Genezis i Klassifikatssisa Polupustynnykh	Third

Developed Countries		Third World

Pochv. Jerusalem: Israel Program for Scientific Translations, 1970. 261p. (Trans. from Russian by A. Gourevitch. Edited by Y. Harver.)

Third — Ivanova, Y. N., and N. A. Nogina, eds. Investigations in the Field of Soil Genesis = Issledovaniya v Oblasti Genezisa Pochv. New Delhi, India: Indian National Scientific Documentation Centre, 1977. 418p. (Trans. from the Russian ed. published by Izdatel'stvo Akademii Nauk SSSR, Moscow, 1963.)

J

Third — Jacks, Graham V. Multilingual Vocabulary of Soil Science = Vocabulaire Multilingue de la Science du Sol. 2d ed., rev. Rome: Food and Agriculture Organization, 1960. 428p. (1st ed., 1954. 439p.) — Second

Jacks, Graham V., ed. Soil Chemistry and Fertility. Conference on Soil Chemistry and Fertility, Aberdeen, Scotland, Sept. 1966. Amsterdam: International Society of Soil Science, 1967. 415p. (In English and French.) — Second

Jackson, I. J. Climate, Water and Agriculture in the Tropics. 2d ed. Harlow, U.K.: Longman; and New York: J. Wiley, 1989. 377p. (1st ed., 1977. 248p.) — Third

First rank — Jackson, Marion L. Soil Chemical Analysis: Advanced Course. A Manual of Methods Useful for Instruction and Research in Soil Chemistry, Physical Chemistry of Soils, Soil Fertility, and Soil Genesis. 2d ed. Madison, Wis.: Dept. of Science, University of Wisconsin, 1974. 895p. (1st ed. 1956) — First rank

Jacob, Arthur and H. V. Uexkull. Fertilizer Use: Nutrition and Manuring of Tropical Crops. Trans. by C. L. Whittles. 2d ed. Hannover, Germany: Verlagsgesellschaft für Ackerbau, 1960. 617p. (1st ed. entitled, The Use of Fertilizers in Tropical and Sub-Tropical Agriculture, by A. Jacob and V. Coyle. London: E. Benn, 1931. 272p.) — Second

Third — Jacobs, L. W., ed. Selenium in Agriculture and the Environment. Proceedings of a Symposium sponsored by the American Society of Agronomy and Soil Science Society of America, New Orleans, Dec. 1986. Madison, Wis.: American Society of Agronomy and Soil Science Society of America, 1989. 233p. (SSSA Special Publication no. 23)

Third — Jamagne, Marcel et al. Bases et Techniques d'Une Cartographie des Sols. Paris: Institut National de la Recherche Agronomique, 1967. 142p. (Annales Agronomiques no. 18) — Third

Developed   Third
Countries   World

	James, David W. Manual de Investigaciones Sobre Fertil-idad de Suelos: U Programa Nacional para Investigaciaon y Extensiaon. La Paz, Bolivia: Ministerio de Asuntos Campesinos y Agropecurios, Instituto Boliviano de Tecnologaia Agropecuaria, Consortium for International Development, 1980. 128p.	Third
Third	James, David W., R. J. Hanks and J. J. Jurinak. Modern Irrigated Soils. New York: J. Wiley, 1982. 235p.	Second
Second	Jenkinson, D. S., and K. A. Smith, eds. Nitrogen Efficiency in Agricultural Soils. London and New York: Elsevier, 1988. 450p. (Seminar on Nitrogen Efficiency in Agricultural Soils and the Efficient Use of Fertilizer Nitrogen, Edinburgh, Sept. 1987.)	Second
First rank	Jenny, Hans. The Soil Resource: Origin and Behavior. Corrected 2d printing. New York: Springer-Verlag, 1983. 377p. (1st published in 1980.) (Ecological Studies no. 37)	First rank
	Jones, M. J., and A. Wild. Soils of the West African Savanna. Harpenden, U.K.: Commonwealth Bureau of Soils, Commonwealth Agricultural Bureau, 1975. 246p. (Technical Communication no. 55)	Third
Second	Jongerius, A., and G. K. Rutherford, eds. Glossary of Soil Micromorphology: English, French, German, Spanish and Russian. Wageningen, Netherlands: Centre for Agricultural Publishing and Documentation, 1979. 138p.	Second
	Jordan, Carl F. Nutrient Cycling in Tropical Forest Ecosystems: Principles and Their Application in Management and Conservation. Chichester, U.K. and New York: J. Wiley, 1985. 190p.	Second
	Joseph, K. T., ed. Proceedings of a Conference on Classification and Management of Tropical Soils, Kuala Lumpur, Malaysia, 1977. Kuala Lumpur, Malaysia: Malaysian Society of Soil Science, 1980. 667p.	Third
Third	Jungerius, Peter D. Soils and Geomorphology. Cremlingen, Germany: Catena Verlag, 1985. 174p.	Second
First rank	Jury, William A., Wilford R. Gardner and Walter H. Gardner. Soil Physics. 5th ed. New York: J. Wiley, 1991. 328p. (1st ed. by L. D. Baver. New York: J. Wiley & Sons, Inc.; London: Chapman and Hall, Ltd., 1940. 370p.)	First rank

K

| Third | Kabata-Pendias, Alina, and Henryk Pendias. Trace Elements in Soils and Plants. 2d ed. Boca Raton, Fla.: CRC Press, 1991. 365p. (1st ed., 1984. 315p.) | Second |
| Third | Kahn, Shahamat U. Pesticides in the Soil Environment. Amsterdam and New York: Elsevier, 1980. 240p. | Second |

Developed Countries		Third World
	Kalpage, F. S. Tropical Soils: Classification, Fertility and Management. Delhi, India: Macmillan Co. of India, 1974. 282p.	Third
	Kang, B. T., and J. van der Heide, eds. Nitrogen Management in Farming Systems in Humid and Subhumid Tropics; Proceedings of a Symposium . . . , International Institute of Tropical Agriculture, Ibadan, Nigeria, Oct. 1984. Haren, Netherlands and Ibadan, Nigeria: Institute for Soil Fertility and International Institute of Tropical Agriculture, 1985. 362p.	Second
	Kanwar, Jaswant S., ed. Soil Fertility: Theory and Practice. New Delhi, India: Indian Council of Agricultural Research, 1976. 533p.	Second
	Kanwar, Jaswant S., and M. S. Mudahar. Fertilizer Sulfur and Food Production. Dordrecht and Boston: M. Nijhoff/ W. Junk, 1986. 247p.	Second
Third	Karavayeva, N. A. Zabolachivanie i Evoliutsiia Pochv (Bogging and Evolution of Soils). Moscow: Nauka Pub., 1982. 295p.	
	Katyal, J. C., and N. S. Randhawa. Micronutrients. Rome: Food and Agriculture Organization, 1983. 92p. (FAO Fertilizer and Plant Nutrition Bulletin no. 7)	Second
	Kawaguchi, Keizaburo and K. Kyuma. Paddy Soils in Tropical Asia: Their Material Nature and Fertility. Honolulu, Hawaii: University of Hawaii Press, 1977. 258p.	Second
	Keleti, Cornelius, ed. Nitric Acid and Fertilizer Nitrates. New York: M. Dekker, 1985. 378p.	Third
Second	Keller, W. D. The Principles of Chemical Weathering: An Integrated Review . . . Rev. ed. Columbia, Mo.: Lucas Brothers, 1962. 111p. (1st ed. 1957.)	
	Kelley, Hubert W. Keeping the Land Alive: Soil Erosion—Its Causes and Cures. Rome: Food and Agriculture Organization, 1983. 78p. (FAO Soils Bulletin no. 50)	Second
	Kelley, Walter P. Alkali Soils: Their Formation, Properties, and Reclamation. New York: Van Nostrand Reinhold, 1951. 176p.	Second
	Kellogg, Charles E. Agricultural Development: Soil, Food, People, Work. Madison, Wis.: Soil Science Society of America, 1975. 233p.	First rank
Second	Kevan, D. Keith. Soil Animals. New York: Philosophical Library, 1962. 237p.	Third
First rank	Kevan, D. Keith, ed. Soil Zoology; Proceedings of the 2d Easter School in Agricultural Science, University of Nottinham, 1955. London: Butterworths, 1955. 512p.	First rank
	Kezdi, Arpad. Soil Physics = Bodenphysik. Amsterdam	Third

Developed Countries		Third World
	and New York: Elsevier, 1974. 294p. (Trans. by I. Lazanyi.)	
First rank	Khasawneh, F. E., E. C. Sample and E. J. Kamprath, ed. The Role of Phosphorus in Agriculture; Proceedings of a Symposium, National Fertlizer Development Center, Tennessee Valley Authority, Muscle Shoals, Alabama, June 1976. Madison, Wis.: American Society of Agronomy, 1980. 910p.	First rank
Third	Kilmer, Victor J. Handbook of Soils and Climate in Agriculture. Boca Raton, Fla.: CRC Press, 1982. 445p.	Second
First rank	Kilmer, Victor J., S. E. Younts and N. C. Brady. The Role of Potassium in Agriculture. Madison, Wis.: American Society of Agronomy, 1968. 509p.	First rank
Third	Kirby, M. J., and R. P. C. Morgan, eds. Soil Erosion. New York: J. Wiley, 1980. 312p.	Second
First rank	Kirkham, Don and W. L. Powers. Advanced Soil Physics. Malabar, Fla.: R. E. Krieger Pub. Co., 1984. 534p.	First rank
Third	Kittrick, J. A., ed. Mineral Classification of Soils; Proceedings of A Symposium, Soil Science Society of America, Atlanta, Ga., Nov. 1981 and Anaheim, Calif., Nov. 1982. Madison, Wis.: Soil Science Society of America, American Society of Agronomy, 1985. 178p.	Second
Third	Kittrick, J. A., ed. Soil Mineral Weathering. New York: Van Nostrand Reinhold, 1986. 271p.	
Third	Kittrick, J. A., D. S. Fanning and L. R. Hossner, eds. Acid Sulfate Weathering. Proceedings of A Symposium, Fort Collins, Colo.. Aug. 1979. Madison, Wis.: Soil Science Society of America, 1982. 234p. (SSSA Special Publication no. 10)	
First rank	Klute, Arnold, ed. Methods of Soil Analysis. Part 1. Physical and Mineralogical Methods. 2d ed. Madison, Wis.: American Society of Agronomy and Soil Science Society of America, 1986. 1188p. (Agronomy Series no. 9) (Part 2 see A. L. Page.)	Second
	Knorr, Dietrich, ed. Sustainable Food Systems. Westport, Conn.: AVI Pub. Co., 1983. 416p.	Third
	Knuti, Leo L., Milton Korpi and J. C. Hide. Profitable Soil Management. Englewood Cliffs, N.J.: Prentice-Hall, 1962. 376p.	Third
Third	Kofoed, A. Dam, J. H. Williams and P. L'Hermite, eds. Efficient Land Use of Sludge and Manure; Proceedings of a Round-Table Seminar, Bryup-Askov, Denmark, June 1985. London and New York: Elsevier, 1986. 245p.	
Second	Kononova, M. M. Soil Organic Matter: Its Nature, Its Role in Soil Formation and in Soil Fertility. 2d ed. Oxford	First rank

Developed Countries		Third World
	and New York: Pergamon Press, 1966. 544p. (1st ed. 1961) (Available in Spanish as Materia Orgánica del Suelo: Su Naturaleza, Propiedades y Métodos de Investigación. Barcelona: Oikos-Tau, n.d.)	
Second	Kononova, M. M., ed. Microorganisms and Organic Matter of Soils. Jerusalem: Israel Program for Scientific Translations, 1970. 306p. (Trans. from Russian by N. Kaner and S. Nemchonok. Edited by D. Greenberg.)	Second
Third	Koolen, Adrianus J. Agriculture Soil Mechanics. Berlin and New York: Springer-Verlag, 1983. 241p.	Third
Second	Koorevar, P. et al. Elements of Soil Physics. Amsterdam and New York: Elsevier, 1983. 228p.	Second
	Kovda, Viktor A., and E.V. Lobova, eds. Geography and Classification of Soils in Asia = Geografiya i Klassifikatsiya Pochv Azii. Jerusalem, Israel: Israel Program for Scientific Translations, 1968. 267p.	Third
	Kowal, Jan M., and A. Kassam. Agricultural Ecology of Savanna: A Study of West Africa. Oxford and New York: Clarendon Press and Oxford University Press, 1978. 403p.	Third
Third	Kozlowski, Theodore T., ed. Water Deficits and Plant Growth. New York: Academic Press, 1968–1983. 7 vols.	
Third	Kozlowski, Theodore T., and C. E. Ahlgren. Fire and Ecosystems. New York: Academic Press, 1974. 542p.	Third
	Kral, David M., ed. Soil Erosion and Conservation in the Tropics. Proceedings of a Symposium sponsored by the American Society of Agronomy and the Soil Science Society of America, Fort Collins, Col., Aug. 1979. Sherri Hawkins, assistant ed. Madison, Wis.: American Society of Agronomy and Soil Science Society of America, 1982. 149p. (ASA Special Publication no. 43)	Third
Third	Kramer, James R., and Herbert E. Allen, eds. Metal Speciation: Theory, Analysis and Application. Chelsea, Mich.: Lewis Pub., 1988. 357p.	
Second	Kramer, Paul J. Water Relations of Plants. 3d rev. ed. New York: Academic Press, 1983. 489p. (1st ed. 1949: 2d ed. 1969 titled: Plant and Soil Water Relationships. 347p.)	Second
Second	Krasilnikov, N. A. Soil Microorganisms and Higher Plants. Trans. of Microorganizmy Pochvy i Vysshie Rasteni by Y. Halperin. Moscow: Academy of Sciences of the USSR, 1958. 474p.	
	Krause, Rudiger and Franz Lorenz. Soil Tillage in the Tropics and Subtropics. Rev. Eng. ed. Rev. and translated by Willem B. Hoogmoed. Eschborn, Germany: GTZ, 1984. 320p. (1st ed., 1979. 310p.)	Second
Second	Krauskopf, Konrad B. Introduction to Geochemistry. 2d ed. New York: McGraw-Hill, 1979. 617p. (1st ed. 1967.)	

Developed Countries		Third World
Third	Krupennikov, I. A. History of Soil Science. Moscow: Nauka Pub., 1981. 325p.	
Third	Kruse, E. G., C. R. Burdick and Y. A. Yousef, eds. Environmentally Sound Water and Soil Management. New York: American Society of Civil Engineers, 1982. 536p.	Third
First rank	Kubiena, Walter L. The Soils of Europe: Illustrated Diagnosis and Systematics with Keys and Descriptions for Easy Identification of the Most Important Soil Formations of Europe with Consideration of the Most Frequent Synonyms. Translation of Bestimmungsbuch und Systematik der Boden Europas. Madrid: Consejo Superior de Investigaciones Cientificas, 1953. 317p.	
Third	Kubiena, Walter L. Grundzuge der Geopedologie und der Formenwandel der Boden. Wien, Austria: Verlagsunion Agrar, Osterreichischer Agraverlag, 1986. 128p.	Third
First rank	Kubiena, Walter L. Micromorphological Features of Soil Geography. New Brunswick, N.J.: Rutgers University Press, 1970. 254p.	Second
	Kubyshev, V. A. Mechanical and Technological Bases for the Protection of Soil from Erosion = Mekhaniko-tekhnologicheskie Osnovy Zashchity Pochv ot Erozii. Moscow: VIM, 1983. 179p.	Third
Second	Kuhnelt, Wilhelm. Soil Biology, with Special Reference to the Animal Kingdom = Bodenbiologie. 2d ed., rewritten and enl. London: Faber and Faber, 1976. 483p. (Based on the original tranlsation by Norman Walker. 1961. 397p.)	Second
Third	Kumada, Kyloichi. Chemistry of Soil Organic Matter. Tokyo and New York: Japan Scientific Societies Press and Elsevier, 1987. 241p.	Third
Third	Kuntze, H., G. Roeschmann and G. Schwerdtfeger. Bodenkunde. 4th ed. Stuttgart, Germany: Ulmer, 1988. 400p. (2d ed., 1981. 407p.)	
Third	Kunze, George W., J. S. White and R. H. Rust. Mineralogy in Soil Science and Engineering. Madison, Wis.: Soil Science Society of America, 1968. 106p. (SSSA Special Publication no. 3)	Second
Third	Kutilek, Miroslav and Don R. Nielsen. Soil Hydrology. Cremlingen-Destedt, Germany: Catena Verlag, 1990. 250p.	Third
Second	Kutilek, Miroslav and J. Sutor, eds. Water in Heavy Soils; Proceedings of the Symposium . . . , Bratislava, Sept. 1976. Prague: Czechoslovak Scientific Technical Society, 1976. 3 vols.	
Third	Kuznetsov, Sergei I., Mikhail V. Ivanov and Natal'ya N. Lyalikova. Introduction to Geological Microbiology. New York: McGraw-Hill, 1963. 252p. (Translator: Paul T.	

Developed   Third
Countries   World

Broneer. Editor of English ed.: Carl H. Oppenheimer.
Trans. of Vvedenie v Geologischeskuiu Mikrobiologiiu.)

L

Third   Laatsch, Willi. Dynamik der Mitteleuropaischen Mineralboden. 4th ed. Dresden: T. Steinkopff, 1957. 280p. (Previous eds. published under title: Dynamik der Deutschen
Acker- und Waldboden. 2d ed., 1944. 289p.)

Lal, R. et al., eds. Land Clearing and Development in the   Second
Tropics; Proceedings . . . International Symposium, Intern-
tional Institute for Tropical Agriculture, Ibadan, Nigeria,
Novemeber 1982. Rotterdam and Boston: A. A. Balkema,
1986. 450p.

Lal, R. Soil Erosion in the Tropics: Principles and Man-   Second
agement. New York: McGraw-Hill, 1990. 580p.

Lal, R. Soil Erosion Problems on an Alfisol in Western   Second
Nigeria and Their Control. Ibadan, Nigeria: Communications and Information Office, IITA, 1976. 160p.

Lal, R., and S. Lal, eds. Pesticides and Nitrogen Cycle.   Third
Boca Raton, Fla.: CRC Press, 1988. 3 vols.

Second   Lal, R., and B. A. Stewart. Soil Degradation. New York   Second
and Berlin: Springer-Verlag, 1989. 346p. (Advances in
Soil Science no. 11)

Second   Lal, R., ed. Soil Erosion Research Methods. Ankeny,   Second
Iowa and Wageningen, Netherlands: Soil and Water Conservation Society and International Soil Science Society,
1988. 244p.

Lal, R., and D. J. Greenland, eds. Soil Physical Properties   Second
and Crop Production in the Tropics. Chichester and New
York: J. Wiley, 1979. 551p.

Lal, R., and E. W. Russell, eds. Tropical Agricultural Hy-   Third
drology: Watershed Management and Land Use; Proceedings of a Conference, International Institute of Tropical
Agriculture, Ibadan, Nigeria, Nov. 1979. Chichester, U.K.,
and New York: J. Wiley, 1981. 482p.

Lal, R. et al., eds. Tropical Land Clearing for Sustainable   Second
Agriculture. Proceedings of an IBSRAM Inaugural Workshop, August–Sept. 1985, Jakarta and Bukittingii, Indonesia. Bangkok, Thailand: International Board for Soil
Research and Management, 1987. 226p.

Larson, W. E. et al., eds. Mechanics and Related Pro-   Third
cesses in Structured Agricultural Soils; Proceedings of the
NATO Advanced Research Workshop, St. Paul, Minnesota, Sept. 1988. Dordrecht, FDR and Boston: Kluwer Academic Pub., 1989. 273p. (NATO ASI Series no. 172)

Third   Larson, William E. et al., eds. Soil and Water Resources:

Developed Countries		Third World
	Research Priorities for the Nation. Madison, Wis.: Soil Science Society of America, 1981. 229p.	
	Latham, Marc, ed. Soil Management under Humid Conditions in Asia. Proceedings of the 1st Regional Seminar, Oct. 1986. Bangkok, Thailand: IBSRAM, 1987. 466p.	Third
	Latham, Marc et al., eds. Management of Vertisols under Semi-Arid Conditions. Proceedings . . . Nairobi, Kenya, Dec. 1986. Bangkok: IBSRAM, 1987. 344p. (IBSRAM Proceedings no. 6)	Third
Third	Lebrun, Philippe et al., eds. New Trends in Soil Biology; Proceedings of the 8th International Colloquium of Soil Zoology, Louvain-la-Neuve, Belgium, 1982 = Tendances Nouvelles en Biologie de Sol; Comptes Rendus . . . = Neue Richtlinien der Bodenzoologie; Verhandlungen . . . Louvain-la-Neuve: Universite Catholique de Louvain, 1983. 709p.	Third
Third	Lee, J. A., S. McNeill and I. H. Rorison, eds. Nitrogen as an Ecological Factor. The 22d Symposium of the British Ecological Society, Oxford, 1981. Oxford and Boston: Blackwell Scientific Publications, 1983. 470p.	
Third	Lee, Kenneth E. Earthworms: Their Ecology and Relationships with Soils and Land Use. Sydney, Australia and Orlando, Fla.: Academic Press, 1985. 411p.	
Second	Lee, Kenneth E., and T. G. Wood. Termites and Soils. London and New York: Academic Press, 1971. 251p.	Second
	Lee, Richard. Forest Hydrology. New York: Columbia University Press, 1980. 349p.	Third
Third	Leeper, Geoffrey W. Managing the Heavy Metals on the Land. New York: M. Dekker, 1978. 121p.	
Third	Lemee, G. Precis de Biogeographie. Paris: Masson et·Cie, 1967. 358p.	
	Leroux, Marcel. Le Climat de l'Afrique Tropicale: The Climate of Tropical Africa. Paris: Editions Champion, 1983. 633p.	Third
	Letey, John, ed. Soil and Plant Interactions with Salinity: Kearney Foundation Five-Year Report 1980–1985. Berkeley: Agricultural Experiment Station, University of California, Division of Agriculture and Natural Resources, 1986. 138p. (Special Publication, Agricultural Experiment Station, University of California, Division of Agriculture and Natural Resources no. 3315)	Second
	Levy, Rachel, ed. Chemistry of Irrigated Soils. New York and Toronto: Van Nostrand Reinhold, 1984. 418p.	Second
Second	Lillesand, Thomas M., and Ralph W. Kiefer. Remote Sensing and Image Interpretation. 2d ed. New York: J. Wiley, 1987. 721p. (1st ed., 1979. 612p.)	

Developed Countries		Third World
First rank	Lindsay, Willard L. Chemical Equilibria in Soils. New York: J. Wiley, 1979. 449p.	First rank
Third	Lipiec, Jerzy. Soil Physical Conditions and Plant Roots. Boca Raton, Fla.: CRC Press, 1990. 250p.	Second
Third	Liverovskii, Juri A. Problemy Genezisa i Geografii Pochv. Moscow: Nauka, 1987. 247p.	
Second	Lockeretz, William, ed. Environmentally Sound Agriculture; Selected papers from the 4th International Conference of the International Federation of Organic Agriculture Movements, Cambridge, Mass., Aug. 1982. New York: Praeger, 1983. 426p.	Third
	Loehr, Raymond C., ed. Food, Fertilizer and Agricultural Residues; Proceedings of the 9th Cornell Agricultural Waste Management Conference, 1977. Ann Arbor, Mich.: Ann Arbor Science Pub., 1977. 727p.	Second
	Logan, Terry J. et al., eds. Effects of Conservation Tillage on Groundwater Quality: Nitrates and Pesticides. Chelsea, Mich.: Lewis Pub., 1987. 292p.	Second
Third	Lohm, U., and T. Persson, eds. Soil Organisms as Components of Ecosystems; Proceedings of the 6th International Soil Zoology Colloquium of the International Society of Soil Science, Uppsala, Sweden, June 1976. Stockholm: Swedish Natural Science Research Council (NFR), 1977. 614p.	
Second	Loneragan, J. F., A. D. Robson and R. D. Graham, eds. Copper in Soils and Plants; Proceedings of the Golden Jubilee International Symposium, Murdoch University, Perth, Western Australia, May 1981. Sydney, Australia and New York: Academic Press, 1981. 380p.	Second
Third	Lopes-Real, J. M., and R.D. Hodges, eds. The Role of Microorganisms in Sustainable Agriculture; Proceedings of the 2d International Conference on Biological Agriculture, Wye College, University of London, 1984. Berkhamsted, U.K.: A. B. Academic, 1986. 246p.	Second
Second	Loughnan, F. C. Chemical Weathering of the Silicate Minerals. New York: Elsevier, 1969. 154p.	
Third	Loveday, J., ed. Methods for Analysis of Irrigated Soils. Slough, U.K.: Commonwealth Agricultural Bureaux, 1974. 208p. (Technical Communication of the Commonwealth Bureau of Soils no. 54)	
Third	Lowrance, R. Richard et al., eds. Nutrient Cycling in Agricultural Ecosystems; Proceedings of the International Symposium, University of Georgia, 1980. Athens, Ga.: University of Georgia, 1983. 602p.	Second
Third	Lowrison, George C. Fertilizer Technology. Chichester, U.K.: Horwood Press and New York: Halsted Press, 1989. 542p.	Third

Developed Countries		Third World
Second	Lynch, James M. Soil Biotechnology: Microbiological Factors in Crop Productivity. Oxford and Boston: Blackwell Scientific, 1983. 191p.	Second
Second	Lynch, James M., and Nigel J. Poole. Microbial Ecology: A Conceptual Approach. New York: J. Wiley, 1979. 266p.	

M

	Malik, Kauser A., S. H. Mujtaba Naqvi and M. I. H. Aleem, eds. Nitrogen and the Environment; Proceedings of International Symposium/Workshop, Lahore, Pakistan, Jan. 1984. Faisalabad, Pakistan: Nuclear Institute for Agriculture and Biology, 1985. 437p.	Third
	Maliwal, G. L. Salt Tolerance of Crops and Plant Metabolism in Saline Substrate: An Annotated Bibliography, 1940–1980. Dehradun, India: International Book Distributers, 1982. 197p.	Third
	Management of Acid Tropical Soils for Sustainable Agriculture; Proceedings of an IBSRAM Inaugural Workshop, Apr.–May, 1985, Yurimaguas, Peru and Brasilia, Brazil. Bangkok, Thailand: IBSRAM, 1987. 299p. (IBSRAM Proceedings no. 2)	Third
First rank	Marbut, Curtis F. Soils: Their Genesis and Classification: A memorial volume of lectures given in the Graduate School of the United States Department of Agriculture in 1928. Madison, Wis.: Soil Science Society of America, 1951. 134p.	First rank
Third	Marini-Bettolo, G. B., ed. Towards a Second Green Revolution: From Chemical to New Biological Technologies in Agriculture in the Tropics. Amsterdam and New York: Elsevier, 1987. 527p.	Third
Second	Marshall, Charles E. The Physical Chemistry and Mineralogy of Soils. New York: J. Wiley, 1964. 2 vols.	First rank
Second	Marshall, Theo J., and J. W. Holmes. Soil Physics. 2d ed. Cambridge, U.K. and New York: Cambridge University Press, 1988. 374p. (1st ed., 1979. 345p.)	Second
Third	Mason, Brian H. Principles of Geochemistry. 4th ed. New York: J. Wiley, 1982. 344p. (1st ed. 1952)	
Third	McFarlane, M. J. Laterite and Landscape. London and New York: Academic Press, 1976. 151p.	
Third	McGarity, J. W., E.H. Hoult and H.B. So, eds. The Properties and Utilization of Cracking Clay Soils; Proceedings of a Symposium, University of New England, 1981. Armidale, Australia: University of New England, 1984. 386p.	
Third	McKyes, Edward. Agricultural Engineering Soil Mechanics. Amsterdam and New York: Elsevier, 1989. 292p.	

Developed Countries		Third World
	McKyes, Edward. Soil Cutting and Tillage. Amsterdam and New York: Elsevier, 1985. 217p.	Third
First rank	McLaren, Arthur D., George H. Peterson, J. Skuji, and E. A. Paul, eds. Soil Biochemistry. New York: M. Dekker, 1967–75. 4 vols.	First rank
Second	McRae, Stuart G. Practical Pedology: Studying Soils in the Field. Chichester, U.K.: Horwood and New York: Halsted Press, 1988. 253p.	Second
Second	McRae, Stuart G., and C. P. Burnham. Land Evaluation. Oxford, U.K.: Clarendon Press, 1981. 239p.	Third
	McVickar, Malcom H. et al., eds. Agricultural Anhydrous Ammonia; Proceedings of the Anhydrous Ammonia Symposium, St. Louis, 1965. Memphis, Tenn.: Agricultural Ammonia Institute, 1966. 314p.	Second
	Medrana, Generosa and Reyna Quisumbing, eds. Rainfed Agriculture in Perspective; Proceedings of the International Workshop on the Development of Rainfed Agriculture, Iloilo City, Philippine, July 1983. Los Baños, Philippines: Philippine Council for Agriculture and Resources Research and Development of the National Science and Technology Authority, 1984. 140p.	Third
	Meigs, Peveril. Geography of Coastal Deserts. Paris: UNESCO, 1966. 140p.	Third
First rank	Meisinger, J. J., G. W. Randall and M. L. Vitosh, eds. Nitrification Inhibitors—Potentials and Limitations. Papers presented at the annual meeting of the American Society of Agronomy, Chicago, Ill., Dec. 1978. Madison, Wis.: American Society of Agronomy and Soil Science Society of America, 1980. 129p.	First rank
Third	Melchior, Daniel C., and R. L. Bassett, eds. Chemical Modeling of Aqueous Systems II; Proceedings of a Symposium . . . Division of Geochemistry at the 196th National Meetings of the American Chemical Society, Los Angeles, Sept. 1988. Washington, D.C.: American Chemical Society, 1990. 566p.	
	Mendez, J., and M. I. Lojo. Humus. Salamanca, Spain: Universidad de Salamanca, 1972. 187p. (In Spanish.) (Acta Salmanticensia. Ciencias, 41)	Third
Third	Metting, F. Blaine, ed. Soil Microbial Ecology: Applications in Agricultural and Environmental Management. New York: Marcel Dekker, Inc., 1993. 646p.	Third
Third	Migrations Organo-Minerales dans les Sols Temperes, Nancy, France, Septembre, 1979. Proceedings . . . Paris: Editions due Centre National de la Recherche Scientifique, 1981. 500p. (Colloques Internationaux du Centre National de la Recherche Scientifique no. 303)	

Developed Countries		Third World
	Mikola, Peitsa, ed. Tropical Mycorrhiza Research. Oxford, U.K.: Clarendon Press and Oxford University Press, 1980. 270p.	Third
Third	Miller, Raymond W., and Roy L. Donahue. Soils: An Introduction to Soils and Plant Growth. 6th. Englewood Cliffs, N.J.: Prentice Hall, 1990. 768p. (1st ed. by R. L. Donahue, 1958. 349p.)	
First rank	Millot, Georges. Geology of Clays: Weathering, Sedimentology, Geochemistry. New York: Springer-Verlag, 1970. 429p. (Trans. of Geologie des Argiles: Alterations, Sedimentologie, Geochimie. Paris: Masson, 1964. 499p.)	
	Milthorpe, F., ed. Plant-Water Relations in Arid and Semiarid Conditions; Proceedings of the symposium in Madrid, 1959. Paris: UNESCO, 1962. 352p.	Second
Third	Mitchell, Colin W. Terrain Evaluation: An Introductory Handbook . . . Harlow, Eng.: Longmans; and New York: Wiley, 1991. 441p. (1st ed., London: Longmans, 1973. 221p.)	
	Mitchell, James K. Fundamentals of Soil Behavior. New York: J. Wiley, 1976. 422p.	Second
	Mitsui, Shingo. Inorganic Nutrition, Fertilisation and Soil Amelioration for Lowland Rice. 4th ed. Tokyo: Yokendo Ltd., 1960. 107p. (1st ed., 1955. 107p.)	Second
	Mohr, Edward C. J. Tropical Soils: A Comprehensive Study of Their Genesis. 3d rev. ed. The Hague: Mouton-Ichtiar Baru-Van Hoeve, 1973. 481p. (1st ed. 1954)	First rank
	Mokwunye, A. Uzo and Paul L. G. Vlek, eds. Management of Nitrogen and Phosphorus Fertilizers in Sub-Saharan Africa. Proceedings of a Symposium, Lome, Togo, Mar. 1985. Dordrecht, Boston and Lancaster: M. Nijhoff, 1986. 362p.	Third
Second	Money, D. C. Climate, Soils and Vegetation. 3d ed. Slough, U.K.: University Tutorial Press, 1978, 1985 printing. 286p. (1st ed., 1965. 272p.)	Second
Second	Monteith, John L., ed. Vegetation and the Atmosphere. London and New York: Academic Press, 1975–76. 2 vols.	
First rank	Monteith, John L., and M. H. Unsworth. Principles of Environmental Physics. 2d ed. London and New York: Edward Arnold, 1990. 291p. (1st ed. 1973.)	
Third	Moore, Alan W., and Stein W. Bie, eds. Uses of Soil Information Systems. Proceedings of the Australian Meeting of the ISSS Working Group on Soil Information Systems, Canberra, Australia, Mar. 1976. Wageningen, Netherlands: Centre for Agricultural Publishing and Documentation (PUDOC), 1977. 109p.	Third

Developed Countries		Third World
Second	Morgan, R. P. C. Soil Erosion and Its Control. New York: Van Nostrand Reinhold, 1986. 311p. (1st ed., Longman, 1979. "Soil Erosion.")	Second
Second	Morgan, R. P. C., ed. Soil Conservation; Proceedings of the International Conference on Soil Conservation. Chichester, U.K. and New York: J. Wiley, 1981. 576p.	
	Morrison, R. J., and D. M. Leslie, eds. Proceedings of South Pacific Regional Forum on Soil Taxonomy, Suva, Fiji, Nov. 1981. Suva, Fiji: University of the South Pacific, 1982. 445p.	Third
First rank	Mortvedt, J. J., editorial chairman. Micronutrients in Agriculture. 2d ed. Madison, Wis.: Soil Science Society of America, 1991. 670p. (1st ed. 1972. "Proceedings of a Symposium, Muscle Shoals, Ala., Apr. 1971.")	First rank
	Moss, R. P., ed. The Soil Resources of Tropical Africa; A Symposium of the African Studies Association of the United Kingdom, University College, London, Sept. 1965. London: Cambridge University Press, 1968. 226p.	Second
Second	Mückenhausen, Eduard. Die Bodenkunde und Ihre Geologischen, Geomorphologischen, Mineralogischen und Petrologischen Grundlagen. Frankfurt am Main, Germany: DLG-Verlag, 1985. 579p.	
Second	Mückenhausen, Eduard. Die Wichtigsten Boden der Bundesrepublik Deutschland. 2 rev. ed. Frankfurt, Germany: Verlag Commentator, 1959. 146p. (1st ed., 1957. 146p.)	
First rank	Mückenhausen, Eduard. Entstehung, Eigenschaften und Systematik der Boden der Bundessrepublik Deutschland. 2d ed. Frankfurt, Germany, 1977. 204p. (1st ed., 1962. 148p.)	
Third	Muirhead, W. A., and E. Humphreys, eds. Root Zone Limitations to Crop Production on Clay Soils. Symposium of the Australian Society of Soil Science, Riverina Branch, Australia, Sept. 1984. Melbourne: CSIRO, 1985. 372p.	
Third	Muller, Georg. Bodenbiologie. Jena, Germany: G. Fischer, 1965. 889p.	
Second	Munson, Robert D., ed. Potassium in Agriculture; Proceedings of an International Symposium, Atlanta, Ga., July 1985. Madison, Wis.: American Society of Agronomy, 1985. 1223p.	First rank
Third	Murphy, C. P. Thin Section Preparation of Soils and Sediments. Berkhamsted, Eng.: A. B. Academic Pub., 1986. 149p.	

N

	Nahon, Daniel. Introduction to the Petrology of Soils and Chemical Weathering. New York: Wiley, 1991. 313p.	Third

Developed Countries		Third World
	Nair, P. K. R. Soil Productivity Aspects of Agroforestry. Nairobi, Kenya: International Council for Research in Agroforestry, 1984. 92p.	Third
	Narayana, N., and Chandrakant Chhotalal Shah. Physical Properties of Soils. Bombay, India: Manaktalas, 1966. 227p.	Third
Second	National Conservation Tillage Conference. Conservation Tillage; Proceedings of a National Conference, Des Moines, Iowa. Ankeny, Iowa: Soil Conservation Society of America, 1973. 241p.	
Third	National Fertilizer Development Center (U.S.). The Role of Phosphorus in Agriculture; Proceedings of a Symposium, June 1976 at the National Fertilizer Development Center, Tennessee Valley Authority, Muscle Shoals, Alabama. Madison, Wis.: American Society of Agronomy, 1980. 910p.	Second
	National Research Council (U.S.). Committee on Tropical Soils. Soils of the Humid Tropics. Washington, D.C.: National Academy of Sciences, 1972. 219p.	Second
Third	National Research Council (U.S.) Panel on Nitrates. Nitrates: An Environmental Assessment. Washington, D.C.: National Academy of Sciences, 1978. 723p.	
Second	National Research Council (U.S.). Committee on the Role of Alternative Farming Methods in Modern Production Agriculture. Alternative Agriculture. Washington, D.C.: National Academy Press, 1989. 448p.	First rank
	National Seminar on Waterlogging and Salinity, Lahore, Pakistan, 1986. Proceedings and recommendations . . . Lahore, Pakistan: Soil Salinity Research Institute, 1986. 259p.	Third
Second	Nelson, D. W. et al., eds. Chemical Mobility and Reactivity in Soil Systems; Proceedings of a Symposium, Atlanta Ga., 1981. Madison, Wis.: Soil Science Society of America, American Society of Agronomy, 1983. 262p.	
	Nerpin, Sergei V., and A. F. Chudnovskii. Physics of the Soil = Fizika Pochvy. Jerusalem: Israel Program for Scientific Translations, 1970. 2 vols. 466p.	Third
Second	Newton, William E., and Nyman C.J., eds. Proceedings of the 1st International Symposium on Nitrogen Fixation, Pullman, WA, 1974. Pullman, Wash.: Washington State University Press, 1976. 2 vols.	Third
Second	Newton, William E., J. R. Postgate and C. Rodriguez-Barrueco, eds. Recent Developments in Nitrogen Fixation; Proceedings of the 2d International Symposium on Nitrogen Fixation, Salamanca, Spain, 1976. London and New York: Academic Press, 1977. 622p.	Second

Developed Countries		Third World
Second	Nicholas, D. J. D., and Adrian R. Egan, eds. Trace Elements in Soil-Plant-Animal Systems; Proceedings of a Symposium, Waite Agricultural Research Institute, Glen Osmond, Australia, Nov. 1974. New York and London: Academic Press, 1975. 417p.	First rank
Third	Nichols, J. D., P. L. Brown and W. J. Grant. Erosion and Productivity of Soils Containing Rock Fragments; Proceedings of a Symposium sponsored by Divisions S-5 and S-6 of the Soil Science Society of America, Anaheim, Calif., Nov.–Dec. 1982. Madison, Wis.: Soil Science Society of America, 1984. (SSSA Special Publication no. 13)	
Third	Nielsen, Donald R., and J. Bouma, eds. Soil Spatial Variability; Proceedings of a Workshop of the ISSS and SSSA, Las Vegas, Nov.–Dec. 1984. Wageningen, Netherlands: Centre for Agricultural Publishing and Documentation (PUDOC), 1985. 243p.	
Second	Nielsen, Donald R., and J. G. MacDonald, eds. Nitrogen in the Environment. New York: Academic Press, 1978. 2 vols.	Second
	NifTAL Project. Dynamics of Soil Organic Matter in Tropical Ecosystems. Honolulu: Dept. of Agronomy and Soil Science, University of Hawaii, 1989. 249p. (Papers presented at a workshop, Maui, Hawaii, Oct. 1988, University of Hawaii.)	Second
	Nobe, Kenneth C., and Rajan K. Sampath, eds. Irrigation Water Management in Developing Countries: Current Issues and Approaches. Boulder, Colo.: Westview Press, 1986. 500p.	Third
Third	Northcote, Keith H. A Factual Key for the Recognition of Australian Soils. 4th ed. Glenside, Australia: Rellim Technical Publications, 1979. 124p.	
Third	Novak, Bohumir, ed. Studies About Humus; Transactions of the International Symposium Humus et Planta VIII, Prague, 1983 = Koklady o Gumuse; Sbornik dokladov Mezhdunarodnogo Simpoziuma Humus et Planta VIII, Praga, 1983. Prague: Research Institute of Crop Production, 1983. 2 vols.	
Third	Nriagu, Jerome O., ed. Environmental Biogeochemistry; Proceedings of the 2d International Symposium on Environmental Biogeochemistry. Ann Arbor, Mich.: Ann Arbor Science, 1976. 2 vols. 797p.	Third
Third	Nriagu, Jerome O., ed. Zinc in the Environment. New York: J. Wiley, 1980. 2 vols.	
Third	Nye, Peter H., and P.B. Tinker. Solute Movement in the Soil-Root System. Oxford, U.K.: Blackwell Scientific Publications, 1977. 342p.	

Developed Countries		Third World
First rank	Nye, Peter H., and D. J. Greenland. The Soil Under Shifting Cultivation. Farnham Royal, U.K.: Commonwealth Agricultural Bureaux, 1960. 156p.	First rank

O

	Ochse, J. J. Tropical and Subtropical Agriculture. New York: Macmillan, 1961. 2 vols. 1446p. (Available in Spanish as Cultivo Mejoramiento de Plantas Tropicales y Subtropicales. Mexico: Limusa-Noriega, n.d. 703p.)	Second
Third	Oelhaf, Robert C. Organic Agriculture: Economic and Ecological Comparisons of Organic and Conventional Farming. Montclair, N.J.: Allanheld, Osmun, 1978. 271p.	Third
	Oke, T. R. Boundary Layer Climates. 2d. London and New York: Methuen, 1987. 435p. (1st ed., 1978)	Third
	Oldeman, L. R., and M. Frere. Technical Report on a Study of the Agroclimatology of the Humid Tropics of Southeast Asia. Rome: Food and Agriculture Organization, 1982. 229p. (Cover title: A Study of the Agroclimatology of the Humid Tropics of Southeast Asia: Technical Report.)	Third
Third	Olson, Gerald W. Field Guide to Soils and the Environment: Applications of Soil Surveys. New York: Chapman and Hall, 1984. 219p.	Second
Third	Olson, Gerald W. Soils and the Environment: A Guide to Soils Surveys and Their Applications. New York: Chapman and Hall, 1981. 178p.	
Third	Orlov, Dmitri S. Humus Acids of Soils. New Delhi, India: Amerind Pub. Co., 1985. 378p.	Second
Second	Oschwald, W. R. ed. Crop Residue Management Systems; Proceedings of a Symposium, sponsored by the American Society of Agronomy, the Crop Science Society of America, and the Soil Science Society of America. Madison, Wis.: American Society of Agronomy, 1978. 248p. (ASA Special Publication no. 31)	Second
	Oswal, M. C. Textbook of Soil Physics. Ghaziabad, India and New York: Vikas and Advent Books, 1983. 214p.	Third

P

| First rank | Page, A. L., and R.H. Miller, eds. Methods of Soil Analysis. Part 2. Chemical and Microbiological Properties. 2d ed. Madison, Wis.: American Society of Agronomy, Soil Science Society of America, 1982. (Agronomy Series no. 9) (1st ed. edited by C.A. Black, 1965. 2 vols.) (Part 1, see A. Klute.) | First rank |
| Second | Page, A. L., T.J. Logan and J.A. Ryan, eds. Land Application of Sludge: Food Chain Implications. Chelsea, | |

Developed Countries		Third World
	Mich.: Lewis Pub., 1987. 168p. (Workshop sponsored by the EPA; University of California, Riverside; and Ohio State University in Las Vegas, Nov. 1985)	
Third	Palmer, Robert G., and Frederick R. Troeh. Introductory Soil Science: Laboratory Manual. 2d ed. Ames, Iowa: Iowa State University Press, 1977. 136p. (1st ed., 1966. 95p.)	
	Panel on Volcanic Ash Soils in Latin America, Turrialba, Costa Rica, 1969. Volcanic Ash Soils in Latin America. Turrialba, Costa Rica: Training and Research Center of the IAAIS, 1969. 350p.	Second
Third	Papadakis, Juan. Soils of the World. Amsterdam and New York: Elsevier, 1969. 208p. (1st ed. 1964.)	Second
First rank	Papendick, R. I., P. A. Sanchez and G. B. Triplett, eds. Multiple Cropping. Madison, Wis.: American Society of Agronomy, 1976. 378p. (ASA Special Publication no. 27)	First rank
Second	Parker, C. A. et al., eds. Ecology and Management of Soil-Borne Plant Pathogens; Proceedings of the 4th International Congress of Plant Pathology, University of Melbourne, Australia. St. Paul, Minn.: American Phytopathological Society, 1985. 358p.	Second
Second	Parkinson, D., and J. S. Waid, eds. The Ecology of Soil Fungi. Proceedings of the International Symposium on Ecology of Soil Fungi, 1958. Liverpool, U.K.: Liverpool University Press, 1960. 324p.	Second
Second	Parr, J. F., W. R. Gardner and L. F. Elliott, eds. Water Potential Relations in Soil Microbiology; Proceedings of a Symposium, Chicago, Ill., Dec. 1978. Madison, Wis.: Soil Science Society of America, 1981. 151p. (SSSA Special Publication no. 9)	
	Parry, M. L., T.R. Carter and N.T. Konijin, eds. The Impact of Climatic Variations on Agriculture. Dordrecht and Boston: Kluwer Academic Pub., 1988. 2 vols.	Third
	Pathak, Prabhakar, S. A. El-Swaify, and Sardar Singh, eds. Alfisols in the Semi-Arid Tropics; Proceedings of the Consultants' Workshop on the State of the Art and Management Alternatives for Optimizing the Productivity of SAT Alfisols and Related Soils, ICRISAT Center, India, Dec. 1983. Patancheru, Andhra Pradesh, India: ICRISAT, 1987. 188p.	Third
Second	Paton, T. R. The Formation of Soil Material. London and Boston: Allen and Unwin, 1978. 143p.	
First rank	Paul, Eldor A., and F. E. Clark. Soil Microbiology and Biochemistry. San Diego, Calif.: Academic Press, 1989. 273p.	First rank
	Pearson, Robert W. Soil Acidity and Liming in the Humid	First rank

Developed Countries		Third World
	Tropics. Ithaca, N.Y.: Cornell University, 1975. 66p. (Cornell International Agriculture Bulletin no. 30)	
Third	Pereira, H. C. Land Use and Water Resources in Temperate and Tropical Climates. Cambridge, U.K.: Cambridge University Press, 1973. 246p.	Third
	Perloff, William H., and William Baron. Soil Mechanics: Principles and Applications. New York: J. Wiley, 1976. 745p.	Third
	Perrier, Alain et al., eds. Les Besoins en Eau des Cultures = Crop Water Requirements; Proceedings of an International Conference, UNESCO, Versailles, France. Paris: Institut National de La Recherche Agronomique, 1985. 927p.	Third
Third	Persons, Benjamin S. Laterite: Genesis, Location, Use. New York: Plenum Press, 1970. 103p.	Third
Third	Petrusewicz, K., ed. Secondary Productivity of Terrestrial Ecosystems: Principles and Methods; Proceedings of the Working Meeting, Jab Yonna, 1966. Warsaw: Paanstwowe Wydawn. Naukowe, 1967. 2 vols. 879p.	
Second	Phillips, Ronald E., and Shirley H. Phillips, eds. No-Tillage Agriculture, Principles and Practices. New York: Van Nostrand Reinhold, 1984. 306p. (Available in Spanish as Agricultura sin Laboreo. Principios y Aplicaciones. Barcelona: Bellaterra, 1986. 316p.)	Second
Second	Phillipson, John, ed. Methods of Study in Quantitative Soil Ecology: Population, Production and Energy Flow. Oxford: Blackwell Scientific Publications, 1971. 297p.	
Third	Phillipson, John, ed. Methods of Study in Soil Ecology; Proceedings of a Symposium organized by UNESCO and the International Biological Programme. Paris: UNESCO, 1970. 303p.	
Third	Pitty, Alistair F. The Nature of Geomorphology. London and New York: Methuen, 1982. 161p.	
	Plaisance, Georges and A. Cailleux. Dictionary of Soils = Dictionnaire des Sols, translated from French by Margaret D. Saidi. New Delhi, India: Amerind Pub. Co., 1981. 1109p. (Entries are in French, definitions in English.)	Second
Third	Ponomareva, Vera V. Theory of Podzolization. Trans. by A. Gourevitch. Jerusalem: Israel Program for Scientific Translations, 1969. (Available from National Scientific and Technical Information Service, Springfield, Va.)	
First rank	Postgate, John R. Nitrogen Fixation. 2d ed. London: Edward Arnold, 1987. 73p. (1st ed., 1978. 67p.) (Available in Spanish as Fijación del Nitrógeno. Barcelona: Omega. 84p.	
	Le Potassium dans les Cultures et les Sols Tropicaux (Potassium in Tropical Crops and Soils). Compte Rendu 10ᵉ	Third

Developed Countries		Third World
	Colloque de l'Institut International de la Potasse, Abidjan, Republique de Cote d'Ivoire, Dec. 1973. Berne, Switzerland: Institut International de la Potasse, 1974. 603p.	
First rank	Povoledo, D., and H. L. Golterman, eds. Humic Substances: Their Structure and Function in the Biosphere; Proceedings of an International Meeting, Nieuwersluis, Netherlands, May 1972. Wageningen: Centre for Agricultural Publishing and Documentation, 1973. 368p.	Second
First rank	Power, J. F., ed. The Role of Legumes in Conservation Tillage Systems; Proceedings of a National Conference, University of Georgia, Athens, Apr. 1987. Ankeny, Iowa: Soil Conservation Society of America, 1987. 153p.	
Second	Pramer, David and E. L. Schmidt. Experimental Soil Microbiology. Minneapolis, Minn.: Burgess Pub. Co., 1964. 107p.	
Third	Pratt, P. F. et al. Nitrate in Effluents from Irrigated Lands; Final Report to the National Science Foundation from University of California. Springfield, Va.: National Technical Information Services, 1979. 822p.	
Third	Prevot, Andre R. Humus, Biogenese, Biochemie, Biologie. Saint-Mande, France: Editions de la Tourelle, 1970. 344p.	
	Primavesi, Ana. Manejo Ecologico del Suelo: La Agricultura en Regiones Tropicales. Trans. from English by M. Mauricio Prelooker. 5th ed. Buenos Aires, Argentina: Libreria "El Ateneo" Editorial, 1984. 499p. (1st ed., Sao Paulo, Brazil, 1980.)	Third
	Primavesi, Ana, ed. Progressos em Biodinamica e Productividade do Solo: Exertos de trabalhos; Realizado na Universidade Federal de Santa Maria, Instituto de Solos e Culturas, Santa Maria, Rs., Brasil, Julho 1968. Santa Maria, Brazil: Pallotti, 1968. 553p. (Text in Portugese, Spanish, English or French. Summaries in Portugese and English.)	Third
Third	Pritchett, William L. Properties and Management of Forest Soils. 2d ed. New York: J. Wiley, 1987. 494p. (1st ed., 1979. 500p.) (Available in Spanish as Suelos Forestales: Propriedades, Conservación y Mejoramiento. Trans by José Hurtado Vega. Mexico: Limusa-Noriaga, 1986. 634p.)	
	Probert, M. E. et al. The Properties and Management of Vertisols. Wallingford, U.K. and Bangkok, Thailand: CAB International and IBSRAM, 1987. 36p.	Third
	Pushparajah, E., and S. H. A. Hamid, eds. International Conference on Phosphorus and Potassium in the Tropics; Proceedings . . . , Kuala Lampur, Malaysia, 1981. Kuala Lampur: Malaysian Society of Soil Science, 1982. 590p.	Second

Developed Countries		Third World

Developed
Countries

Third
World

Q

Second   Quispel, A., ed. The Biology of Nitrogen Fixation. Amsterdam and New York: Elsevier, 1974. 769p.

R

Radcliffe, D. Guidelines: Land Evaluation for Rainfed Agriculture. Rome: Food and Agriculture Organization, 1984. 249p. (FAO Soils Bulletin no. 52)   Second

Rakhmanov, Viktor V. Role of Forests in Water Conservation. Jerusalem: Israel Program for Scientific Translations, 1966. 192p. (Trans. from Russian and edited by A. Gourevitch and L. M. Hughes.)   Third

Rambler, Mitchell B., Lynn Margulis and Rene Fester eds. Global Ecology: Toward a Science of the Biosphere. Boston: Academic Press, 1989. 204p.   Third

Second   Rapoport, E. H., ed. Progresos en Biologia del Suelo: Actas; Proceedings of the 1st Coloquio Latinamericano de Biologia del Suelo, Bahia Blanca, Argentina, 1965. Montevideo: Centro de Cooperacion Cientificia de la Unesco para America Latina, 1966. 715p.

Rauschkolb, Roy S. Land Degradation. Rome: Food and Agriculture Organization, 1971. 105p. (FAO Soils Bulletin no. 13)   Second

Regional Seminar on Methods of Amelioration of Saline and Waterlogged Soils, Baghdad, 1970. Salinity seminar, Baghdad, Iraq, Dec. 1970. Rome: Food and Agriculture Organization, 1971. 254p. (FAO Irrigation and Drainage Paper no. 7)   Second

Remote Sensing and Tropical Land Management; Proceedings of a Workshop of the Commonwealth Geographical Bureau, McGill University, Montreal, Aug. 1983. New York: J. Wiley, 1986. 365p.   Second

Rendig, Victor V., and Howard M. Taylor. Principles of Soil-Plant Interrelationships. New York: McGraw-Hill, 1989. 275p.   Second

Third   Reuss, J. O., and D. W. Johnson. Acid Deposition and the Acidification of Soils and Water. New York: Springer-Verlag, 1986. 119p.   Third

Third   Rich, Charles I., and G. W. Kunze, eds. Soil Clay Mineralogy; A Symposium. Chapel Hill, N.C.: University of North Carolina Press, 1964. 330p.   Third

Third   Richards, Bryant N. Introduction to the Soil Ecosystem. London and New York: Longman, 1974. 266p.   Second

Third   Richards, Bryant N. The Microbiology of Terrestrial Eco-   Third

Developed Countries		Third World
	sytems. Essex, U.K.: Longman and New York: J. Wiley, 1987. 399p.	
First rank	Richards, R. A., ed. Diagnosis and Improvement of Saline and Alkali Soils. Washington, D.C.: Regional Salinity Laboratory, U.S. Govt. Print. Office, 1954. 160p. (USDA Agriculture Handbook no. 60) (1st issued 1947.)	First rank
Third	Rieger, Samuel. The Genesis and Classification of Cold Soils. New York: Academic Press, 1983. 230p.	
	Riehl, Herbert. Tropical Meteorology. New York: McGraw-Hill, 1954. 392p.	Third
	Robertson, G. P., R. Herrera and T. Rosswall, eds. Nitrogen Cycling in Ecosystems of Latin America and the Caribbean. The Hague and Hingham, Mass.: Nijhoff/W. Junk Pub. and Distributors for the U.S. and Canada, Kluwer Boston, 1982. 430p. (Reprinted from Plant and Soil, Vol. 67, 1982.)	Third
Third	Robson, A. D., ed. Soil Acidity and Plant Growth. Sydney, Australia and San Diego, Calif.: Academic Press, 1989. 306p.	
Third	Rode, Aleksei A. Podzol-Forming Process = Podzoloobrazovatel'nyi Protsess. Jerusalem: Israel Program for Scientific Translations, 1970. 387p. (Trans. from Russian by A. Gourevitch.)	
Third	Rolston, Dennis E., and Francis E. Broadbent. Field Measurement of Denitrification. Ada, Okla.: R. S. Kerr Environmental Research Laboratory, 1977. 75p. (Environmental Protection Technology Series no. EPA-600/2-77-233)	
Third	Roots and the Soil Environment; Meeting at University of St. Andrews, Sept. 1989. Wellesbourne, U.K.: Association of Applied Biologists, 1989. 448p.	Second
Third	Rorison, I. H., ed. Ecological Aspects of the Mineral Nutrition of Plants; Proceedings of a Symposium of The British Ecological Society, Sheffield, U.K. Oxford and Edinburgh: Blackwell Scientific Publications, 1969. 484p.	Second
Second	Rose, C. W. Agricultural Physics. Oxford, U.K.: Pergamon Press, 1966. 226p.	
Second	Rosenberg, Norman J., Blaine L. Blad and Shashi B. Verma. Microclimate: The Biological Environment. 2d ed. New York: J. Wiley, 1983. 495p. (1st ed., 1974. 315p.)	
	Rosswall, T., ed. Nitrogen Cycling in West African Ecosystems; Proceedings of a Workshop, International Institute for Tropical Agriculture, Ibadan, Nigeria, Dec. 1978. Stockholm, Sweden: Royal Swedish Academy of Sciences, 1980. 450p. (Papers in English and French.)	Third
	Rudd, Robert L. Pesticides and the Living Landscape.	Third

Developed Countries		Third World
	Madison, Wis.: University of Wisconsin Press, 1964. 320p.	
	Ruhe, Robert V. Erosion Surfaces of Central African Interior High Plateaus. Brussels: Institut National Pour l'Etude Agronomique du Congo Belge, 1954.	Third
Second	Ruhe, Robert V. Geomorphology: Geomorphic Processes and Surficial Geology. Boston: Houghton Mifflin, 1975. 246p.	Third
First rank	Russell, Edward J. Russell's Soil Conditions and Plant Growth. 11th ed., edited by Alan Wild. Burnt Mill, U.K.: Longmans and New York: J. Wiley, 1988. 991p. (Rev. ed. of Soil Conditions and Plant Growth. 10th ed. 1973. 1st ed. Longmans, 1912)	First rank
	Russell, J. S., and E. L. Greacen, eds. Soil Factors in Crop Production in a Semi-Arid Environment. St. Lucia, Australia: University of Queensland Press, 1977. 327p.	Second
Third	Russell, J. S., and R. F. Isbell, eds. Australian Soils: The Human Impact. St. Lucia, Australia: University of Queensland Press, 1986. 522p.	
Second	Russell, Robert S. Plant Root Systems: Their Functions and Interaction with the Soil. London and New York: McGraw-Hill, 1977. 298p.	
	Ruthenberg, Hans. Farming Systems in the Tropics. 3d ed. Oxford and New York: Clarendon Press, 1980. 424p. (1st ed., 1971. 313p.)	Second

S

Developed Countries		Third World
	Saltzman, Sarina and Bruno Yaron, eds. Pesticides in Soil. New York: Van Nostrand Reinhold, 1986. 379p.	Third
First rank	Sanchez, Pedro A. Properties and Management of Soils in the Tropics. New York: J. Wiley, 1976. 618p. (Also available in Spanish as Suelos del Tropico: Caracteristicas y Manejo. San Jose, Costa Rica: Instituto Interamericano de Cooperacion para la Agricultura, 1981. 634p.)	First rank
	Sanchez, Pedro A., and Luis E. Tergas, eds. Pasture Production in Acid Soils of the Tropics; Proceedings of a Seminar, CIAT, Cali, Colombia, Apr. 1978. Cali, Colombia: Centro Internacional de Agricultura Tropical, 1979. 488p.	First rank
First rank	Sanders, F. E., Barbara Mosse and P. B. Tinker, eds. Endomycorrhizas; Proceedings of a Symposium, University of Leeds, July 1974. London and New York: Academic Press, 1975. 626p.	
	Sarmiento, Guillermo. The Ecology of Neotropical Savannas. Cambridge, Mass.: Harvard University Press,	Third

Developed Countries		Third World
	1984. 235p. (Trans. by Otto Solbrig of Estructura y Funcionamiento de Sabanas Neotropicales.)	
Third	Sauchelli, Vincent. Trace Elements in Agriculture. New York: Van Nostrand Reinhold, 1969. 248p.	Second
Third	Sawhney, B. L., and K. Brown, eds. Reactions and Movement of Organic Chemicals in Soils; Proceedings of a Symposium, Atlanta, Ga., 1987. Madison, Wis.: Soil Science Society of America, American Society of Agronomy, 1989. 474p.	
	Schaller, Frank W., and George W. Bailey, eds. Agricultural Management and Water Quality. Ames, Iowa: Iowa State University Press, 1983. 472p. (Proceedings at the national conference, Ames, May 1981)	Third
	Scharpenseel, H. W., M. Schomaker and A. Ayoub, eds. Soils on a Warmer Earth: Effects of Expected Climate Change on Soil Processes, With Emphasis on the Tropics and Sub-Tropics; Proceedings of an International Workshop, Nairobi, Kenya, 1990. New York: Elsevier, 1990. 274p.	Third
First rank	Scheffer, Fritz and Paul Schachtschabel. Lehrbuch der Bodenkunde. 13th ed. Stuttgart, Germany: F. Enke Verlag, 1991. 490p. (Earlier ed. published as vol. 1 of author's Lehrbuch der Agrikulturchemie und Bodenkunde. 3d., Stuttgart: F. Enke, 1952.)	
Third	Schippers, B., and W. Gams, eds. Soil-Borne Plant Pathogens; Proceedings of the 4th International Symposium on Factors Determining the Behavior of Plant Pathogens in Soil, Munich, 1978. London and New York: Academic Press, 1979. 686p.	Third
First rank	Schlichting, Ernst and Udo Schwertmann. Pseudogley and Gley: Genesis and Use of Hydromorphic Soils. Weinheim, Germany: Verlag Chemie, 1973. 771p. (English, French or German.)	First rank
	Schlippe, Pierre de. Shifting Cultivation in Africa: The Zande System of Agriculture. London: Routledge and Paul, 1956. 304p.	Third
	Schmidt, E. L. A Practical Manual of Soil Microbiology Laboratory Methods. Rome: Food and Agriculture Organization, 1967. 69p. (FAO Soils Bulletin no. 7)	Second
First rank	Schnitzer, Martin and S. U. Khan. Humic Substances in the Environment. New York: M. Dekker, 1972. 327p.	
First rank	Schnitzer, Martin and S. U. Khan. Soil Organic Matter. Amsterdam and New York: Elsevier, 1978. 319p.	First rank
Third	Segalen, P. Le Fer dans les Sols. Paris: ORSTOM, 1964. 150p. (Initiations, Documentations, Techniques no. 4)	Third
Third	Sekhon, G. S., ed. Potassium in Soils and Crops; Proceed-	Second

Developed Countries		Third World

ings of a Symposium, Potash Research Institute of India, New Delhi, Nov. 1978. New Delhi, India: Potash Research Institute of India, 1978. 432p.

Third — Sellers, William D. Physical Climatology. Chicago, Ill.: University of Chicago Press, 1965. 272p.

Second — Shainberg, I., and J. Shalhevet, eds. Soil Salinity Under Irrigation: Processes and Management. Berlin and New York: Springer-Verlag, 1984. 349p. — First rank

Shalhevet, J., and J. Kamburov. Irrigation and Salinity: A World-Wide Survey. New Delhi: International Commission on Irrigation and Drainge, 1976. 106p. — Second

Second — Sheals, J. G., ed. The Soil Ecosystem: Systematic Aspects of the Environment, Organisms and Communities. A symposium . . . London: Systematics Association, 1969. 247p. — Third

Sheng, T.C. Soil Conservation for Small Farmers in the Humid Tropics. Rome: Food and Agricultural Organization, 1989. 104p. (FAO Soils Bulletin no. 60) — Third

Shulgin, A. M. Soil Climate and Its Control = Klimat Pochvy i ego Regulirovanie. New Delhi, India: National Scientific Documentation Centre, 1978. 406p. — Third

Sillanpaa, Mikko. Micronutrient Assessment at the Country Level: An International Study. Rome: Food and Agriculture Organization and the Finnish International Development Agency, 1990. 208 p. (FAO Soils Bulletin no. 63) — Third

Sillanpaa, Mikko. A Study on the Response of Wheat to Fertilizers. Rome: Food and Agriculture Organization, 1971. 131p. (FAO Soils Bulletin no. 12) — Third

Silva, James A. ed. Soil-Based Agrotechnology Transfer. Manoa, Hawaii: Benchmark Soils Project, Dept. of Agronomy and Soil Science, University of Hawaii, 1985. 269p. — Third

Simpson, Ken. Fertilizers and Manures. London and New York: Longmans, 1986. 254p. — Second

Third — Simpson, Ken. Soil. London and New York: Longmans, 1983. 238p. — Second

Third — Singer, Michael J., and Donald N. Munns. Soils: An Introduction. New York: Macmillan, 1987. 492p. — Second

Third — Slack, Archie V. Chemistry and Technology of Fertilizers. New York: Interscience Pub., 1967. 142p. (Revised reprint from: Encyclopedia of Chemical Technology, Vol. 9, 2d ed. 150p.)

Second — Slatyer, R. O. Plant-Water Relationships. London and New York: Academic Press, 1967. 366p. — Second

Third — Smeck, Neil E., and Edward J. Ciolkosz, eds. Fragipans: Their Occurrence, Classification, and Genesis. Proceedings of a Symposium sponsored by Divisions S-5 and S-9 of the Soil Science Society of America, Atlanta, Ga., Dec.

Developed Countries		Third World
	1987. Madison, Wis.: Soil Science Society of America, 1989. 153p. (SSSA Speical Publication no. 24)	
	Smedema, Lambert K., and David W. Rycroft. Land Drainage: Planning and Design of Agricultural Drainage Systems. Ithaca, N.Y.: Cornell University Press, 1983. 376p.	Third
Third	Smith, David G., ed. The Cambridge Encyclopedia of Earth Sciences. New York: Crown, 1981. 496p.	Third
Second	Smith, Guy D. The Guy Smith Interviews: The Rationale for Concepts in Soil Taxonomy. Washington, D.C.: U.S. Dept. of Agriculture and Ithaca, N.Y.: Cornell University, 1986. 259p.	Third
Third	Smith, Guy H. Conservation of Natural Resources. 4th ed. New York: J. Wiley, 1971. 685p. (1st ed. 1950.)	
Third	Smith, Keith A., ed. Soil Analysis: Modern Instrumental Techniques. 2d ed. New York: M. Dekker, 1991. 659p. (1st ed., 1983. 562p.)	Third
Third	Smith, Keith A., and Chris E. Mullins, ed. Soil Analysis: Physical Methods. New York: M. Dekker, 1991. 620p.	Third
	Smyth, A. J. The Preparation of Soil Survey Reports. Rome: Food and Agriculture Organization, 1970. 52p. (FAO Soils Bulletin no. 9)	Second
	Smyth, A. J. Selection of Soil for Cocoa. Rome: Food and Agriculture Organization, 1966. 76p. (FAO Soils Bulletin no. 5)	Third
Third	Soil Resource Inventories and Development Planning; Proceedings of Workshops organized by the Soil Resource Inventory Study Group at Cornell University, Apr. 1977 and Dec. 1978. Washington, D. C.: Soil Management Support Services, Soil Conservation Service, USDA, 1981. 407p. (Technical Monograph/Soil Management Support Services no. 1) (Previously published as Agronomy Mimeo 77–23 and 79–23 of the Department of Agronomy, Cornell University, 1979. 332p.)	
First rank	Soil Science Society of America. Glossary of Soil Science Terms. Madison, Wis.: Soil Science Society of America, 1987. 44p.	First rank
Second	Soil Science Society of America. Soil Taxonomy: Achievementsz and Challenges. Madison, Wis.: Soil Science Society of America, 1984. 76p.	First rank
	Soil, Crop and Water Management Systems for Rainfed Agriculture in the Sudano-Sahelian Zone. Proceedings of International Workshop, ICRISAT, Sahelian Center, Niamey, Niger, Jan. 1987. Patancheru, India: International Crops Research Institute for the Semi-Arid Tropics, 1989. 385p.	Second
	Sombroek, W. G. Amazon Soils: A Reconnaissance of the	Third

Developed Countries		Third World
	Soils of the Brazilian Amazon Region. Wageningen: Centre for Agricultural Publication and Documentation (PUDOC), 1966. 292p. (Agricultural Research Report no. 672)	
Third	Soon, Y. K., ed. Soil Nutrient Availability: Chemistry and Concepts. New York: Van Nostrand Reinhold, 1985. 353p.	Third
	Sopher, Charles D., and Jack Vernon Baird. Soils and Soil Management. 2d ed. Reston, Va.: Reston Pub. Co., 1982. 312p. (1st ed., 1978)	Second
Third	Spanner, Douglas C. Introduction to Thermodynamics. London and New York: Academic Press, 1964. 278p.	
Second	Sparks, Donald L., ed. Soil Physical Chemistry. Boca Raton, Fla.: CRC. Press, 1986. 308p.	
Second	Sparks, Donald L. Kinetics of Soil Chemical Processes. San Diego, Calif.: Academic Press, 1989. 210p.	
First rank	Sposito, Garrison. The Surface Chemistry of Soils. Oxford and New York: Clarendon Press and Oxford University Press, 1984. 234p.	First rank
Third	Sprague, Milton A., and Glover B. Triplett, eds. No-Tillage and Surface-Tillage Agriculture. New York: J. Wiley, 1986. 467p.	Third
Second	Sprent, Janet I., and Peter Sprent. Nitrogen Fixing Organisms: Pure and Applied Aspects. 2d ed. London and New York: Chapman and Hall, 1990. 256p. (Rev. ed. of Biology of Nitrogen-Fixing Organisms, 1979.)	Second
Third	Stace, H. C. T. et al. A Handbook of Australian Soils. Glenside, Australia: Rellim Technical Publications for the Commonwealth Scientific and Industrial Research Organisation and the International Society of Soil Science, 1968. 435p.	
Third	Stallings, James H. Soil Conservation. Englewood Cliffs, N.J.: Prentice-Hall, 1957. 575p. (Available in Spanish as Suelo: Su uso y Mejoramiento. Mexico: Cecsa, 1969. 484p.)	
	Stamp, L. Dudley and W. T. W. Morgan. Africa: A Study in Tropical Development. 3d ed. New York: J. Wiley, 1972. 520p. (1st ed., 1953. 568p.)	Third
	Steele, J. Gordon. Soil Survey Interpretation and Its Use. Rome: Food and Agriculture Organization, 1967. 68p. (FAO Soils Bulletin no. 8)	Second
Third	Steila, Donald and Thomas E. Pond. The Geography of Soils: Formation, Distribution, and Management. 2d ed. Savage, Md.: Rowman and Littlefield, 1989. 239p. (1st ed., Englewood Cliffs, N.J.: Prentice-Hall, 1976. 222p.)	
First rank	Stevenson, Frank J. Humus Chemistry: Genesis, Composition, Reactions. New York: J. Wiley, 1982. 443p.	Second

Developed Countries		Third World
First rank	Stevenson, Frank J. Cycles of the Soil: Carbon, Nitrogen, Phosphorus, Sulfur Micronutrients. New York: J. Wiley, 1986. 380p.	Second
Second	Stevenson, Frank J., ed. Nitrogen in Agricultural Soils. Madison, Wis.: American Society of Agronomy, Crop Science of America, Soil Science Society of America, 1982. 940p. (Agronomy no. 22)	First rank
Second	Stewart, B. A. et al. Control of Water Pollution from Cropland. Washington, D.C.: U.S. Dept. of Agriculture and Environmental Protection Agency, 1975–76. 2 vols. (ASAE Publication no. PROC-275)	
First rank	Stewart, G. A. Land Evaluation; Proceedings of a CSIRO Symposium, organized in Cooperation with UNESCO, Aug. 1968. Melbourne, Australia: Macmillan of Australia, 1968. 392p.	First rank
First rank	Stewart, W. D. P., ed. Nitrogen Fixation of Free-Living Micro-Organisms; Papers presented at an IBP Synthesis Meeting on Nitrogen Fixation, Edinburgh, Sept. 1973. Cambridge, U.K. and New York: Cambridge University Press, 1975. 471p.	Second
Third	Stiegeler, Stella E., ed. A Dictionary of Earth Sciences. Totowa, N.J.: Rowman and Allanheld, 1983. 301p.	Second
Third	Stoops, Georges and Hari Eswaran, eds. Soil Micromorphology. New York: Van Nostrand Reinhold, 1986. 345p.	Third
Second	Strahler, Arthur N. Elements of Physical Geography. 3d ed. New York: J. Wiley, 1984. 538p. (1st ed. 1978.) (Available in Spanish as Geographia Fisica. 3d ed. Barcelona: Omega, n.d. 636p.)	
Third	Strzemski, Micha. Ideas Underlying Soil Systematics = Mysli Przewodnie Systematyki Gleb. Warsaw, Poland: Foreign Scientific Publications Dept. of the Polish National Center for Scientific Technical and Economic Information, 1975. 540p. (Trans. from Polish by J. Bachrach.)	
First rank	Stucki, Joseph W., B. A. Goodman and U. Schwertmann, eds. Iron in Soils and Clay Minerals. NATO Advanced Study Institute on Iron in Soils and Clay Minerals, Bad Windsheim, Germany, 1985. Dordrecht and Boston: D. Reidel, 1988. 893p. (NATO ASI Series vol. 217)	
First rank	Stumm, Werner and James J. Morgan. Aquatic Chemistry: An Introduction Emphasizing Chemical Equilibria in Natural Waters. 2d ed. New York: J. Wiley, 1981. 780p. (1st ed. 1970.)	
Third	Subba Rao, Nanjappa S. Biofertilizers in Agriculture. 2d ed. New Delhi, India: Oxford University Press and IBH Pub. Co., 1988. 208p. (1st ed., 1982. 186p.)	Second

Developed Countries		Third World
	Submicroscopy of Soils and Weathered Rocks; Proceedings of the 1st Workshop of the International Working-Group on Submicroscopy of Undisturbed Soil Materials, Wageningen, Netherlands, 1980. Wageningen, Netherlands: Centre for Agricultural Publishing and Documentation (PUDOC), 1981. 320p.	Third
Second	Sutton, Oliver G. Micrometerology: A Study of Physical Processes in the Lowest Layers of the Earth's Atmosphere. New York: McGraw-Hill, 1953. 333p.	
Second	Swift, M. J., and O. W. Heal. Decomposition in Terrestrial Ecosystems. Oxford, U.K.: Blackwell Scientific Publications, 1979. 372p.	
	Swindale, Leslie D., ed. Soil-Resource Data for Agricultural Development; Proceedings of the Papers presented at a Seminar, Hyderabad, India, Jan. 1976. Honolulu, Hawaii: University of Hawaii, 1978. 306p.	Second
Third	Sybesma, Christiaan. Biophysics: An Introduction. Dordrecht and Boston: Kluwer Academic Pub., 1989. 320p. (Rev. ed. of: An Introduction to Biophysics, 1977.)	
Third	Sylvia, D. M., L. L. Hung and J. H. Graham, eds. Mycorrhizae in the Next Decade: Practical Applications and Research Priorities; Proceedings of the 7th North American Conference on Mycorrhizae, Gainesville, Fla., May 1987. Gainesville, Fla.: University of Florida, 1987. 364p.	
	Symposium on Paddy Soil, Nanjing, China, Oct. 1980. Proceedings . . . Nanjing, China: Organizing Committee Academia Sinica, 1980. 100p.	Third
	Symposium on Tropical Ecology, New Delhi, 1971. Papers from a symposium . . . : With an Emphasis on Organic Productivity. Compiled by Priscilla M. Golley and Frank B. Golley. Athens, Ga.: University of Georgia, 1972. 418p.	Third
	Sys, C. Caracterisation Morphologique et Physico-chimique de Profils Types de l'Afrique Centrale. Brussels: l'Institut National pour l'Etude Agronomique du Congo, 1972. 497p.	Third
	Szegi, J. Cellulose Decomposition and Soil Fertility. Budapest, Hungary: Akademiai Kiado, 1988. 186p.	Third
Third	Szegi, J., ed. Proceedings of the 9th International Symposium on Soil Biology and Conservation of the Biosphere, University of Forestry and Timber Industry, Aug. 1985. Budapest, Hungary: Akademiai Kiado, 1987. 2 vols.	Third
Third	Szegi, J., ed. Soil Biology and Conservation of the Biosphere; Proceedings of the 8th meeting of the Soil Biology Section of the Hungarian Society for Soil Science, God-	Third

Developed Countries		Third World

ollo, Aug. 1981. Budapest, Hungary: Akademiai Kiado, 1984. 2 vols.

T

Developed Countries		Third World
Second	Tabatabai, M. A., ed. Sulfur in Agriculture. Madison, Wis.: American Society of Agronomy, 1986. 668p.	First rank
	Takahashi, H., ed. Nitrogen Fixation and Nitrogen Cycle. Tokyo, Japan: University of Tokyo Press, 1975. 161p.	Third
Third	Talsma, T., and J. R. Philip, eds. Salinity and Water Use; Proceedings of the 2d National Symposium on Hydrology, Canberra, Australia, 1971. London: Macmillan, 1971. 296p.	
	Tamhane, R. V., D. P. Motiramani and Y. P. Bali. Soils: Their Chemistry and Fertility in Tropical Asia. New Delhi, India: Prentice-Hall, 1964. 475p. (Adapted from Soils: An Introduction to Soils and Plant Growth by R.L. Donahue.)	Third
Third	Tan, Kim H., ed. Andosols. New York: Van Nostrand Reinhold, 1984. 418p.	Third
Third	Tan, Kim H. Principles of Soil Chemistry. New York: M. Dekker, 1982. 267p.	
Second	Tate, Robert L. Soil Organic Matter: Biological and Ecological Effects. New York: J. Wiley, 1987. 291p.	Second
Third	Tate, Robert L., and Donald D. Klein, eds. Soil Reclamation Processes: Microbiological Analyses and Applications. New York: M. Dekker, 1985. 349p.	
First rank	Taylor, H. M., Wayne R. Jordan and Thomas R. Sinclair. Limitations to Efficient Water Use in Crop Production. Madison, Wis.: American Society of Agronomy, 1983. 538p.	
Third	Taylor, Sterling A. Physical Edaphology: The Physics of Irrigated and Nonirrigated Soils. Rev. and ed. by Gaylen L. Ashcroft. San Francisco, Calif.: W. H. Freeman, 1972. 533p.	Second
Third	Tedrow, John C. F. Soils of the Polar Landscapes. New Brunswick, N.J.: Rutgers University Press, 1977. 638p.	
	Tessens, Eddy and S. Jusop. Quantitative Relationships between Mineralogy and Properties of Tropical Soils. Serdang, Malaysia: Universiti Pertanian, 1983. 190p.	Third
	Theng, B. K. G. The Chemistry of Clay-Organic Reactions. New York: J. Wiley, 1974. 343p.	Second
Second	Theng, B. K. G. Formation and Properties of Clay-Polymer Complexes. Amsterdam and New York: Elsevier, 1979. 362p.	
First rank	Theng, B. K. G., ed. Soils with Variable Charge. Lower	First rank

Developed Countries		Third World
	Hutt, N.Z.: New Zealand Society of Soil Science, 1980. 448p.	
	Thomas, Michael F. Tropical Geomorphology: A Study of Weathering and Landform Development in Warm Climates. London: Macmillan and J. Wiley, 1974. 332p.	Second
	Thomas, Michael F., and G. W. Whittington, eds. Environment and Land Use in Africa. London: Methuen, 1969. 554p.	Second
	Thomasson, A. J., ed. Soils and Field Drainage. Harpenden: Soil Survey of England and Wales, 1975. 80p. (Technical Monograph no. 7)	Third
Second	Thompson, Louis M., and Frederick M. Troeh. Soils and Soil Fertility. 4th ed. New York: McGraw-Hill, 1978. 516p. (1st ed. 1952) (Available in Spanish as Suelos y su Fertilidad. 4th ed. Barcelona: Reverté, 1980. 660p.)	
Third	Thorne, D. Wynne and Marlowe D. Thorne. Soil, Water and Crop Production. West Port, Conn.: AVI Pub. Co., Inc., 1979. 353p.	Second
	Tian-ren, Yu, ed. Physical Chemistry of Paddy Soils. Beijing, Berlin: Science Press and New York: Springer, 1985. 217p. (Originally written in Chinese.)	Second
	Tinker, P. B., ed. Soils and Agriculture. New York: J. Wiley, 1981. 151p.	First rank
Second	Tinsley, J., and J. F. Darbyshire, eds. Biological Processes and Soil Fertility. Meeting of Commission III and IV of the International Society of Soil Science jointly with the British Society of Soil Science. The Hague and Boston: M. Nijhoff/ W. Junk, 1984. 403p.	
First rank	Tisdale, Samuel L. Soil Fertility and Fertilizers. 4th ed. New York and London: Macmillan, 1985. 754p. (1st ed., 1956. 430p.)	First rank
Third	Torrey, J. G., and D. T. Clarkson, eds. The Development and Function of Roots; 3d Cabot Symposium, Harvard University, Apr. 1974. London and New York: Academic Press, 1975. 618p.	
Second	Tout, E. A., and G. S. Argosino, eds. Soil Physics and Rice. Proceedings of the Workshop on Physical Aspects of Soil Management in Rice-Based Cropping Systems, IRRI, Dec. 1984. Los Baños, Philippines: International Rice Research Institute, 1985. 430p.	
	Trewartha, Glenn T. The Earth's Problem Climates. 2d ed. Madison, Wis.: University of Wisconsin Press, 1981. 371p. (1st ed., 1961. 334p.)	Third
Third	Troeh, Frederick R., J. Arthur Hobbs and Roy L. Donahue. Soil and Water Conservation. 2d ed. Englewood	Third

Developed Countries		Third World
	Cliffs, N.J.: Prentice-Hall, 1991. 1991p. (Rev. ed. of: Soil and Water Conservation for Productivity and Environmental Protection, 1980)	
Third	Trudgill, Stephen T. Soil and Vegetation Systems. 2d ed. Oxford: Clarendon Press; and New York: Oxford University Press, 1988. 211p. (1st ed., 1977. 180p.)	Third
Third	Turner, Neil and John B. Passioura, eds. Plant Growth, Drought and Salinity. Melbourne, Australia: CSIRO, 1986. 210p.	

U

Third	Uehara, Goro, ed. A Multidisciplinary Approach to Agrotechnology Transfer. Proceedings of a Workshop on Agrotechnology Transfer by the Benchmark Soils Project and the Hawaii Institute of Tropical Agriculture and Human Resources, Honolulu, Aug. 1980. Honolulu, Hawaii: HITAHR, College of Tropical Agriculture and Human Resources, University of Hawaii, 1984. 164p. (Benchmark Soils Project Technical Report no. 7)	Third
	Uehara, Goro and G. Gillman. The Mineralogy, Chemistry, and Physics of Tropical Soils with Variable Charge Clays. Boulder, Colo.: Westview Press, 1981. 170p.	First rank
Second	Ulrich, B., and M. E. Sumner, eds. Soil Acidity. Berlin and New York: Springer-Verlag, 1991.	Second
	UNESCO. Natural Resources of Humid Tropical Asia. Paris: UNESCO, 1974. 456p. (French title: Resources Naturelles de l'Asie Tropicale Humide.)	Third
	Unger, Paul W. et al., eds. Predicting Tillage Effects on Soil Physical Properties and Processes; Proceedings of a Symposium sponsored by Divisions A-3, S-6, and S-1 of the American Society of Agronomy and the Soil Science Society of America. Madison, Wis.: American Society of Agronomy and Soil Science Society of America, 1982. 198p. (ASA Special Publication no. 44)	Second
	Unger, Paul W. Tillage Systems for Soil and Water Conservation. Rome: Food and Agriculture Organization, 1984. 278p. (FAO Soils Bulletin no. 54)	Second
Second	United Kingdom. Ministry of Agriculture, Fisheries and Food. Soil Physical Conditions and Crop Production; Proceedings of a Conference organized by the Soil Scientists of the Agricultural Development and Advisory Service, Jan. 1972. London: HMSO, 1975. 505p. (M.A.F.F. Technical Bulletin no. 29)	
Third	United States. Dept. of Agriculture. Soil Survey Laboratory Methods and Procedures for Collecting Soil Samples.	

Developed Countries		Third World
	Rev. Washington, D.C.: U.S. Dept. of Agriculture, 1972. 63p.	
First rank	United States. Dept. of Agriculture. Soil Survey Manual. Washington, D.C.: U.S. Govt. Print. Off., 1951. 503p. (USDA, Dept. of Agriculture Handbook no. 18) (Rev. and enlargement of USDA Miscellaneous publication 274, the Soil Survey Manual, issued Sept. 1937.)	First rank
Third	United States. Dept. of Agriculture. Soil: The Yearbook of Agriculture 1957. Washington, D.C.: U.S. Govt. Print. Off, 1957. 784p.	
Third	United States. Soil Conservation Service. Drainage of Agricultural Land: A Practical Handbook for Planning, Design, Construction and Maintenance of Agricultural Drainage Systems. Port Washington, N.Y.: Water Information Center, 1975. 430p.	First rank
Third	United States. Soil Conservation Service. Engineering Field Manual for Conservation Practices. 2d print. Washington, D.C.: Soil Conservation Service, U.S. Dept. of Agriculture, 1975. 1024p. (1st ed. 1969)	
First rank	United States. Soil Conservation Service. Soil Classification: A Comprehensive System. 7th Approximation. Washington, D.C.: Soil Conservation Service, U.S. Dept. of Agriculture, 1960. 265p.	First rank
Second	United States. Soil Conservation Service. Soil Survey Staff. Soil Survey Laboratory Methods Manual. Washington, D.C.: Soil Conservation Service, 1992. 400 p. (U.S. Soil Conservation Service. Soil Survey Investigations Report No. 42.)	
First rank	United States. Soil Conservation Service. Soil Survey Staff. Soil Taxonomy: A Basic System of Soil Classification for Making and Interpreting Soil Surveys. Washington, D.C.: Soil Conservation Service, U.S. Dept. of Agriculture, 1975. 754p. (USDA Agricultural Handbook no. 436)	First rank
Third	U.S.S.R. Ministry of Agriculture. Classification and Diagnostics of Soils of the USSR = Klassifikastssisisa i Diagnotika Pochv SSSR (Trans. from the Russian). New Delhi, India: Amerind Pub. Co., 1986. 288p.	
Third	The Utilization of Secondary and Trace Elements in Agriculture; Proceedings of a symposium, Geneva, Jan. 1987. Dordrecht and Boston: M. Nijhoff, 1987. 299p.	

V

| | van der Heide, J. ed. Nutrient Management for Food Crop Production in Tropical Farming Systems; Proceeding of a | Third |

Developed Countries		Third World
	symposium, Malang, Indonesia, 1987. Haren, Netherlands: 1989. 394p.	
	van der Meer, H. G. et al., eds. Animal Manure on Grassland and Fodder Crops, Fertilizer or Waste? Proceedings of an International Symposium of the European Grassland Federation, Wageningen, Netherlands, Aug.–Sept. 1987. Dordrecht, Netherlands and Boston: M. Nijhoff, 1987. 388p.	Third
First rank	Van Olphen, H. An Introduction to Clay Colloid Chemistry for Clay Technologists, Geologists, and Soil Scientists. New York: Interscience Pub., 1963. 301p.	Second
First rank	Van Olphen, H., and J. J. Fripiat, eds. Data Handbook for Clay Materials and Other Non-metallic Minerals: Providing Those Involved in Clay Research and Industrial Application with Sets of Authoritative Data Describing the Physical and Chemical Properties and Mineralogical Composition of the Available Reference Materials. Oxford, U.K. and New York: Pergamon Press, 1979. 346p.	
Second	Van Schilfgaarde, Jan, ed. Drainage for Agriculture. Madison, Wis.: American Society of Agronomy, 1974. 700p.	Second
	Van Wambeke, A. Calculated Soil Moisture and Temperature Regimes of Africa: A Compilation of Soil Climatic Regimes Calculated by Using a Mathematical Model Developed by F. Newhall. Ithaca, N.Y.: Cornell University and Washington, D.C.: Soil Conservation Service, U.S. Dept. of Agriculture, 1982. 1 vol. (U.S. Soil Management Support Services. Technical monog. 3.)	Second
	Van Wambeke, A. Calculated Soil Moisture and Temperature Regimes of Asia: A Compilation of Soil Climatic Regimes Calculated by Using a Mathematical Model Developed by F. Newhall. Ithaca, N.Y.: Cornell University and Washington, D.C.: Soil Conservation Service, U.S. Dept. of Agriculture, 1985. 144p. (U.S. Soil Management Support Services. Technical monog. 9.)	Second
	Van Wambeke, A. Calculated Soil Moisture and Temperature Regimes of South America: A Compilation of Soil Climatic Regimes Calculated by Using a Mathematical Model Developed by F. Newhall. Ithaca, N.Y.: Cornell University and Washington, D.C.: Soil Conservation Service, U.S. Dept. of Agriculture, 1981. 1 vol. (U.S. Soil Management Support Services. Technical monog. 2.)	Second
	Van Wambeke, A. Management Properties of Ferralsols. Rome: Food and Agriculture Organization, 1974. 124p. (FAO Soils Bulletin no. 23)	Second
Second	Vanek, Jan, ed. Progress in Soil Zoology; Proceedings of	Third

Developed
Countries

the 5th International Colloquium on Soil Zoology, Prague,
Sept. 1973. The Hague: Junk, Prague and Academia,
1975. 630p. (English, German or French.)

Third   Vaughan, D., and R.E. Malcolm, eds. Soil Organic Matter   Second
and Biological Activity. Dordrecht and Boston: M. Nijhoff
and W. Junk Pub., 1985. 469p.

Veeger, C., and W. E. Newton, eds. Advances in Nitro-   Third
gen Fixation Research; Proceedings of the 5th International
Symposium on Nitrogen Fixation, Noordwijkerhout, Netherlands, Aug.–Sept. 1983. The Hague, Boston and Wageningen: M. Nijhoff/W. Junk and PUDOC, 1984. 760p.

Second   Veeresh, G. K., and D. Rajogopal, eds. Applied Soil Biol-   Second
ogy and Ecology. New Delhi, India: Oxford University
Press and IBH Pub. Co., 1983. 407p.

Third   Velde, B. Clays and Clay Minerals in Natural and Synthetic Systems. Amsterdam and New York: Elsevier, 1977.
218p.

Third   Vennard, John K., and Robert L. Street. Elementary Fluid
Mechanics. 6th ed. New York: J. Wiley, 1982. 689p. (1st
ed., 1940.) (Available in Spanish as Elementos de Mecánica
de Fluídos. Mexico: Cecsa, 1979. 816p.)

Third   Verma, D. P. S., and Th. Hohn, eds. Genes Involved
in Microbe-Plant Interactions. Vienna, Austria and New
York: Springer-Verlag, 1984. 393p.

Second   Vink, A. P. A. Land Use in Advancing Agriculture. New   Second
York: Springer-Verlag, 1975. 394p.

Vlek, Paul L. G., ed. Micronutrients in Tropical Food   Second
Crop Production. Dordrecht and Boston: M. Nijhof and W.
Junk Pub., 1985. 268p.

W

Second   Waisel, Yoav. Biology of Halophytes. New York: Academic Press, 1972. 395p.

Third   Walker, N., ed. Soil Microbiology. New York: J. Wiley,   Second
1975. 262p.

Second   Wallwork, John A. Desert Soil Fauna. New York: Praeger,   Third
1982. 296p.

Third   Wallwork, John A. The Distribution and Diversity of Soil
Fauna. London and New York: Academic Press, 1976.
355p. (Companion volume to his Ecology of Soil Animals,
1970)

Second   Wallwork, John A. Ecology of Soil Animals. London and
New York: McGraw-Hill, 1970. 283p.

Second   Walsh, Leo M., ed. Instrumental Methods of Analysis of

Developed Countries		Third World
	Soils and Plant Tissue. Madison, Wis.: Soil Science Society of America, 1971. 222p.	
First rank	Walsh, Leo M., and James D. Beaton, eds. Soil Testing and Plant Analysis. Rev. ed. Madison, Wis.: Soil Science Society of America, 1973. 491p. (SSSA Special Publication no. 2) (1st ed. 1967.)	First rank
	Watanabe, I., and Walter G. Rockwood, eds. Nitrogen and Rice. Papers presented at a symposium. Los Baños, Philippines: International Rice Research Institute, 1979. 499p.	Second
Third	Weaver, Charles E., and Lin D. Pollard. The Chemistry of Clay Minerals. Amsterdam and New York: Elsevier, 1973. 213p.	Second
	Webster, C. C. Agriculture in the Tropics. 2d ed. London and New York: Longmans, 1980. 640p. (1st ed. 1966.)	Second
Second	Webster, R. Quantitative and Numerical Methods in Soil Classification and Survey. Oxford, U.K.: Clarendon Press, 1977. 269p.	Second
Third	Weinberg, Eugene D., ed. Microorganisms and Minerals. New York: M. Dekker, 1977. 492p.	Third
	Weiss, Albert, ed. Climate and Agriculture, Systems Approaches to Decision Making. Proceedings. . . . Charleston, S.C., Mar. 1989. No publisher, 304p. (Held in cooperation with the American Meteorological Society.)	Third
	Weiss, Eckehard. Guide to Plants Tolerant of Arid and Semi-arid Conditions: Nomenclature and Potential Uses. Weikersheim, Germany: Margraf Scientific Pub., 1989. 543p.	Third
Third	Westerman, R. L., ed. Soil Testing and Plant Analysis. 3d ed. Madison, Wis.: Soil Science Society of America, 1990. 784p.	Second
Third	Western, S. Soil Survey Contracts and Quality Control. Oxford: Clarendon Press and New York: Oxford University Press, 1978. 284p.	
	Wetselaar, R., J. R. Simpson and T. Rosswall, eds. Nitrogen Cycling in South-East Asian West Monsoonal Ecosystems; Proceedings of a Regional Workshop, Chiang Mai, Thailand, Nov. 1979. Canberra, Australia: Australian Academy of Science, 1981. 216p.	Third
Second	White, Leslie P. Aerial Photography and Remote Sensing for Soil Survey. Oxford, U.K.: Clarendon Press, 1977. 104p.	
Second	White, Robert E. Introduction to the Principles and Practice of Soil Science. 2d ed. Oxford, U.K. and Boston: Blackwell Scientific Publications, 1987. 244p. (1st ed., 1979. 198p.)	Second

Developed Countries		Third World
Third	White, William C., and Donald N. Collins, eds. The Fertilizer Handbook. Washington, D.C.: Fertilizer Institute, 1982. 274p.	Second
	Whitman, C. E. et al., eds. Soil, Water, and Crop/Livestock Management Systems for Rainfed Agriculture in the Near East Region; Proceedings of the Workshop, Amman, Jordan, Jan. 1986. Washington, D.C.: U.S. Agency for International Development, U.S. Dept. of Agriculture, ICARDA, 1989. 343p.	Third
Third	Whittington, William J., ed. Root Growth; Proceedings of the 15th Easter School in Agricultural Science, University of Nottingham, 1969. London: Butterworths, 1969. 450p.	
	Whyte, Robert O. Tropical Grazing Lands: Communites and Constituent Species. The Hague: W. Junk, 1974. 222p.	Third
Third	Wilde, S. A. Soil and Plant Analysis for Tree Culture. 5th rev. ed. New Delhi, India: Oxford University Press and IBH Pub. Co., 1979. 224p. (1st ed. 1941)	
First rank	Wilding, L. P., N. E. Smeck and G. F. Hall, eds. Pedogenesis and Soil Taxonomy. Amsterdam and New York: Elsevier, 1983. 2 vols.	First rank
Second	Wilding, L. P., and R. Puentes, eds. Vertisols: Their Distribution, Properties, Classification and Management. College Station, Tex.: Texas A & M University Press, 1988. 193p. (SMSS Technical Monograph no. 18)	First rank
	Williams, C. N., and K. T. Joseph. Climate, Soil and Crop Production in the Humid Tropics. Rev. ed. Kuala Lumpur, Malaysia and New York: Oxford University Press, 1973, 1976 printing. 177p. (1st ed., 1970.)	Second
Third	Williams, Peter J. The Surface of the Earth: An Introduction to Geotechnical Science. London and New York: Longman, 1982. 212p.	
Third	Wilson, J. R., ed. Advances in Nitrogen Cycling in Agricultural Ecosystems; Proceedings of the Symposium, Brisbane, Australia, May 1987. Wallingford, U.K.: CAB International, 1988. 451p.	Second
Third	Wit, Cornelis T. de and H. van Keulen. Simulation of Transport Processes in Soils. 2d ed. rev. Wageningen, Netherlands: Centre for Agricultural Publishing and Documentation (PUDOC), 1975. 100p. (1st ed., 1972, 1973.)	Third
Second	Wolf, K., W. J. van den Brink and F. J. Colon, eds. Contaminated Soil '88; Proceedings of the 2d International TNO/BMFT Conference on Contaminated Soil, Hamburg, Germany, Apr. 1988. Dordrecht and Boston: Kluwer Academic Pub., 1988. 2 vols.	

Developed Countries		Third World
Third	Wolman, M. G., and F. G. A. Fournier, eds. Land Transformation in Agriculture. Chichester, U.K. and New York: J. Wiley, 1987. 531p.	Third
Third	Wood, Martin. Soil Biology. Glasgow, U.K.: Blackie, 1989. 154p.	Second
	Workshop on Irrigation Management for Diversified Cropping. Paper presented in a Workshop on Irrigation Management for Crop Diversification, IIMI, Nov. 1986. Digana Village via Kandy, Sri Lanka: International Irrigation Management Institute, 1987. 282p.	Third
Third	Wright, Herbert E., and David G. Frey, eds. The Quaternary of the United States: A Review Volume for the VII Congress of the International Association of Quaternary Research. Princeton, N.J.: Princeton University Press, 1965. 922p.	
Third	Wright, Madison J. et al., eds. Plant Adaption to Mineral Stress in Problem Soils. Ithaca, N.Y.: Cornell University Press, 1986. 420p. (Workshop on Plant Adaption to Mineral Stress, NAL/USDA, Beltsville, Md., 1976)	First rank
	Wrigley, Gordon. Tropical Agriculture: The Development of Production. 4th ed. London and New York: Longman, 1982. 496p. (1st ed. 1961.)	Second

Y

Second	Yaalon, Dan H., ed. Aridic Soils and Geomorphic Processes; Proceedings of the International Conference of the International Society of Soil Science, Jerusalem, Israel, Mar.–Apr. 1981. Cremlingen, Germany: Catena Verlag, 1982. 219p.	
First rank	Yaalon, Dan H., ed. Paleopedology: Origin, Nature, and Dating of Paleosols; Symposium on the Age of Parent Materials and Soils, Amsterdam, 1970. Jerusalem: International Society of Soil Science, 1971. 350p.	Second
	Yang, Peter T. Some Special Topics in Soil Science. Taipei, Taiwan: SWAP International, 1989. 490p.	Third
Third	Yaron, Bruno, E. Danfors and Y. Vaadia, eds. Arid Zone Irrigation. New York: Springer-Verlag, 1973. 434p.	Second
	Yawalkar, K. S., J. P. Agarwal and S. Bokde. Manures and Fertilizers. 2d rev. ed. Nagpur, India: Agri-Horticultural Pub. House, 1967. 288p.	Second
	Yawalkar, K. S. Introduction to Agricultural Geology, Soil Physics and Agricultural Climatology. 6th ed. Nagpur, India: Agri-Horticultural Pub. House, 1976. 192p. (2d rev. ed., 1963.) (Cyclostyled Notes no. 1)	Second

Developed Countries		Third World
First rank	Yong, Raymond N., and Benno P. Warkentin. Soil Properties and Behaviour. Amsterdam and New York: Elsevier, 1975. 449p.	Second
	Yoshino, Masatoshi M., ed. Climate and Agricultural Land use in Monsoon Asia. Tokyo: University of Tokyo Press, 1984. 398p.	Third
Third	Young, Anthony. Slopes. London and New York: Longmans, 1975. 288p.	
	Young, Anthony. Tropical Soils and Soil Survey. Cambridge, U.K. and New York: Cambridge University Press, 1976. 468p.	Second
Third	Young, Harry M. No-Tillage Farming and Minimum Tillage Farming by William A. Hayes. Brookfield, Wis.: No-Till Farmer, 1982. 202p. (1st ed., 1973, Milwaukee: Reiman Assoc. by Shirley H. Phillips.)	

Z

Third	Zachar, Dusan. Soil Erosion. Amsterdam and New York: Elsevier, 1982. 547p.	Second
Third	Zelenin, Arkadii N., V. I. Balovnev and I. P. Kerov. Machines for Moving the Earth: Fundamentals of the Theory of Soil Loosening, Modeling of Working Processes and Forecasting Machine Parameters = Mashiny disa Zemlisanykh Rabot. New Delhi, India: Amerind Pub. Co., 1985. 555p.	
	Ziemer, R. R., O'Loughlin and L. S. Hamilton, eds. Research Needs and Applications to Reduce Erosion and Sedimentation in Tropical Steeplands. Proceedings of an International Symposium, Suva, Fiji, June, 1990. Wallingford, U.K.: International Association of Hydrological Sciences, 1990. 396p. (International Association of Hydrological Sciences Publication no. 192)	Third
Third	Zuidam, Robert A. van. Aerial Photo-Interpretation in Terrain Analysis and Geomorphologic Mapping. The Hague: Smits Pub., 1986. 442p.	Third

This list of core monographs was sent for analysis to four university libraries to ascertain the extent of their holdings. Two libraries in the United States were chosen and two libraries in developing countries. The four libraries were: (a) the Albert R. Mann Library at Cornell University; (b) Kansas State University Libraries; (c) Haryana Agricultural University Library in India; and (d) Universiti Pertanian in Malaysia. Table 9.10 gives the percentages of holdings for the four universities.

Table 9.10. Holdings percentages for four universities

	Cornell	Kansas	Haryana	Pertanian
Cited edition	89.6%	61%	36.8%	45.6%
Other edition	2.4%	7%	3.0%	1.6%
No holdings	8.0%	32%	60.2%	52.8%

10. Primary Journals in Soil Science

PETER MCDONALD

Cornell University

As in other scientific disciplines, research results in soil science are first published in journals. Whereas technical reports, conference proceedings, monographs, and government publications also inform and serve in the distribution of research results, journals unquestionably remain the primary outlet for publication.

In these studies, a journal is defined as being serially published, as well as having several articles on different and specific subjects in one issue; these are usually referred to as being "in" a title. Annual proceedings generally belong to the category of journals and are designated as such in the Core Agricultural Literature Project. Similarly, annual reviews, such as *Advances in Agronomy*, are considered to be journals since they fit the criteria above. International and specialized conference proceedings, on the other hand, which are published once or at lengthy intervals, have been categorized as monographs, if they have distinctive titles and subject focus differing with each conference. Most serially published bulletins and handbooks have been considered monographs when each issue is complete in itself and is unrelated to the number which preceded it.

Fifty years ago, the total number of soil-related serials published in English was fewer than twenty-five, including bulletins, newsletters, and annual reports.[1] The bulk of soils material during this period was subsumed in journals dealing with agronomy and general agriculture, such as the *Journal of Agricultural Science* from Cambridge University. Indeed, major foreign language soils serial titles numbered fewer than fifteen. A more thorough analysis of these historical journals is given in Chapter 14. The 1950s and 1960s, however, saw the number of journals dealing with soils-related topics rise dramatically. These years, from 1950 to the present, are the primary focus of this study to arrive at the current core journals in soil science.

1. According to counts from the *Dictionary Catalog of the National Agricultural Library*, the Albert R. Mann Library collection, and the Commonwealth Bureau of Soils, *Bibliography of Soil Science, Fertilizers and General Agronomy* 1931/34–1959/62.

257

A. Source Documents and Methodology

Of the thirty-five monographs and four journal articles used as source documents to identify core soils monographs, twenty-eight were analyzed at a level to obtain data on serially published material (see Table 9.1, Chapter 9). The same methods and caveats outlined for the monographs in Chapter 9 apply for journals, serials, and report series. All journal titles and report series cited in the source documents were tabulated by title and date of publication.

Substantial literature reviews in journals on broad topics related to soil science rarely had the scope or instructional coverage to meet the Core Agricultural Literature Project's criteria, making a comparative analysis between citation patterns in monographs to those in journals difficult to quantify. Journal data reveals that the majority of journals in soil science publish predominantly current results of research. The core project has sought a broader scope, including college-level instructional material, as well as identification of important historical trends of the discipline. Analysis of monographs identified the long-standing core journals of the discipline. The analysis which follows rests primarily on monographic source documents.

B. Journal Literature Findings

Of the 19,642 citations analyzed in the source documents, 14,336, or 72.9%, were to journal articles. The average half-life of these citations was 9.4 years.[2] This is a long half-life in comparison to research-oriented soil journals analyzed by the *Science Citation Index* and elsewhere, where the half-life is more commonly 6.6–8.6 years. The half-life of citations from mongraphic source documents analyzed by the core project is substantially greater than from journal analysis, in part, because monographs take longer to reach publication and deal with more established subjects that do not go out of date so quickly and, thus, are commonly cited over a greater span of years.

As noted, 72.9% of the core project citations are to journals. In AGRICOLA and CAB Abstracts, a similar citation pattern was found. The 27.1% of citations to monographs in the Core Project deserve mention. These included citations both to whole documents and to chapters, which are analo-

---

2. Half-life is the number of years, beginning with the publication year of the source document, which account for 50% of all citations analyzed.

gous to journal counts because the latter are tabulated at the article level. In both cases, parts of the whole are identified. In the major agricultural databases, chapters within books are more likely to be indexed when they are edited works with chapters by different people on distinct subjects.

In the Core Project, source monographs were analyzed to obtain a variety of data, but the percentages in both cases are near 75% for journals. The percentage of citations to journals in AGRICOLA classified as soil-related is 70.7% (1979–91); and in CAB Abstracts it is 74.2% (1984–91). Monographs, pamphlets, and reports comprise the remaining percentages. For AGRICOLA, the latter is heavily weighted by soil surveys. In both online databases and in the Core Project analysis, a clear three-quarters of all soil citations are to literature published in journals.

Analysis of the citations in the source documents shows a surprising number of different journal titles which publish research results appropriate to soils. In all, 819 journals were identified, although 413, or 50.4%, were cited three times or fewer. That still leaves just over 400 journal or serial titles with four citations or more, which is statistically significant. Of all journals cited, 76.4% were cited ten times or fewer (see Figure 10.1). It is difficult to arrive at the total universe of soil journals published worldwide with which to compare the Core Project numbers. The National Agricultural

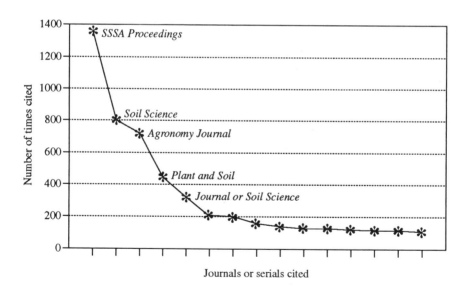

Figure 10.1. Distribution of journal titles and serials.

Library (NAL) lists 419 soil-related journals as received in 1990; about 10–15% can be considered as dead, that is they have ceased publication, whereas *The International Union List of Agricultural Serials* (1990), edited jointly by NAL, FAO, and CAB International, lists 387 current soil titles, including some report series, proceedings, and annual reports. The Albert R. Mann Library at Cornell University, which has a strong soil journal collection, receives just over 100 serials related to the subject. This may more closely reflect a core list than either the NAL or union lists.

Bradford's law of scattering postulates that a small core of journals will garner the majority of references cited, while the remainder will be distributed evenly among a large number of journals.[3] The Core Project analysis found this distribution to hold true in soil science. During the analysis, each time an article in a journal was cited, counts were made for that journal. The top twenty-five journals accounted for 41.3% of all citations, the top 100 journals, for 58.2%, confirming Bradford's law of scattering (see Figure 10.1).

Of the top twenty-five journals, twelve are soil specific, while ten of the remainder deal with aspects of agronomy, which traditionally encompasses soil science. Two titles in the top twenty-five, *Nature* and *Science*, publish substantially outside the discipline of agriculture. *The Journal of Environmental Quality (JEQ)*, ranked eleventh in Table 10.2, serves as a good example of a journal whose subject emphasis straddles many fields not exclusively agricultural. Rather, it addresses a broad range of environmental issues, not just the impact of agriculture on the environment. Soil-related articles predominate; for 1991 alone, they account for 55.5% of the total. *JEQ* is the rule rather than the exception among top-ranked journals. Many of these journals cover several subjects, with agriculture. This is especially true of journals dealing with the physiology and pathology of plants. They are included among the top titles which deal with subjects that are not specific to agriculture. The subject divisions of non-agricultural journals are

five titles cover botanical subjects,
four titles cover microbiology,
two each are devoted to chemistry, hydrology/water research, ecology, and general science, and
one each came from geography, meteorology, and forestry.

These represent 20% of the top 100 journals. the remainder of the top 100 journals remain predominantly within the discipline of agriculture.

3. B.C. Vickery, "Bradford's Law of Scattering," *Journal of Documentation* 4 (1948): 198–203.

As with soil monographs, all types of publishers were represented in the journal counts. Previous subject analyses completed by the Core Agricultural Literature Project have shown that the most important journal publishers are independent organizations.[4] In soils, twelve of the top twenty-five journals are published by professional societies. The first one hundred journals mirror the same pattern, with societies and professional organizations publishing 51% of the titles.

Table 10.1 is a comparison by percentage of the types of publishers for the top 100 soils journals identified by the Core Project, against the twenty-one soils journals identified by the *Science Citation Index (SCI)* ranked by impact factor (which is a measure of the average citation rate per published journal issue), and, finally, the publishers of the 419 soils-related serials acquired by the NAL.

Not surprisingly, the NAL has strong government holdings, since it collects such publications heavily from the United States and abroad. The high count of commercial journals in the *SCI* column reflects a growing trend among publishers to divide subjects into more narrowly focused publications in areas of cutting edge research, although these journals are rarely soils related.[5] Five of the seven commercial soil journals ranked by impact factor in the *SCI* began publication since 1980. In all columns, society and organization publications account for the major share of the total.

C. Ranking of Soil Science Journals

Of the twenty-eight source documents analyzed for data on journals, eight were monographs dealing specifically with tropical soils. Data were

Table 10.1. Soils journal publishers by percent

	SCI	NAL	Core 100 top-ranked journals
Commercial	33.3%	19%	24%
Governments	9.6%	30%	15%
Organizations and societies	47.6%	42%	51%
Universities	9.5%	9%	10%

Sources: SCI Journal Citation Reports, 1988, Section: "Agriculture, Soil Science," p. 363; NAL Listing of Journals Received.

4. Wallace C. Olsen, *Agricultural Economics and Rural Sociology: The Contemporary Core Literature* (Ithaca, N.Y., and London: Cornell University Press, 1991).

extracted to arrive at comparisons between Third World and developed citation patterns. The overall ranking of the journals, as well as ranking by Third World and other criteria, are given in Table 10.2.

Table 10.2. Top journals from core project analysis related to *SCI* impact factors

Overall ranking		A Developed world ranking	B Tropical soil ranking	C ISI citation ranking[a]
1	Soil Science Society of America Journal (O) (1937–)	1	1	7
2	Soil Science (C) (1916–)	3	2	23
3	Agronomy Journal (O) (1907–)	2	6	19
4	Plant and Soil (C) (1948–)	4	3	14
5	Journal of Soil Science (U) (1949–)	5	4	28
6	Pochvovedenie (G) (1899–)	6	—	—
7	Communications in Soil Science and Plant Analysis (C) (1970–)	16–17	5	26
8	Geoderma (C) (1967–)	10	11–12	29
9	Plant Physiology (O) (1926–)	8	36–37	3
10	Journal of Environmental Quality (O) (1972–)	9	31–32	21
11	Nature (C) (1869–)	14–15	13	1
12	Science (O) (1895–)	11	18	2
13–14	Soil Biology and Biochemistry (U) (1969–)	35–36	8	18
13–14	Tropical Agriculture (G)	—	7	39
15	Canadian Journal of Soil Science (O) (1957–)	7	—	27
16	Advances in Agronomy (O) (1949–)	20	16–17	41
17	Australian Journal of Soil Research (G) (1963–)	23	16–17	35–36
18	Journal of Agricultural Science (U) (1905–)	18–19	21	22
19–20	Australian Journal of Agricultural Research (G) (1950–)	38–40	14	30–31
19–20	Agronomie Tropicale (O) (1946–)	—	9	44
21	Journal of the Indian Soil Science Society (O) (1953–)	52–53	11–12	—
22–24	Experimental Agriculture (C) (1933–)	—	10	37
22–24	New Phytologist (U) (1902–)	24–25	26	13
22–24	Transactions of the ASAE (O) (1958–)	21	28–29	8
25	Journal of Soil and Water Conservation (O) (1946–)	28	—	25
26	Ecology (O) (1920–)	14–15	—	5
27	Zeitschrift für Pflanzenernährung und Bodenkunde (O) (1847–)	12	—	32
28–29	Pedologie (O) (1951–)	22	43–44	—
28–29	Pedobiologia (C) (1961–)	13	—	34
30	Agricultural Meteorology (C) (1964–)	18–19	—	—

Overall ranking		A Developed world ranking	B Tropical soil ranking	C ISI citation ranking[a]
31	Phytopathology (O) (1911–)	29	33–35	6
32	Soils and Fertilizers (C) (1938–)	45–46	22	—
33	Revue d'Ecologie et de Biologie du Sol (C) (1946–)	16–17	—	40
34	Turrialba (O) (1950–)	—	15	43
35–36	Australian Journal of Experimental Agriculture (G) (1960–)	—	20	33
35–36	Netherlands Journal of Agricultural Science (O) (1953–)	30	47–48	38
37	Soil Science and Plant Nutrition (O) (1961–)	24–25	—	35–36
38	Sols Africains (O) (1951–)	—	19	—
39–40	Indian Journal of Agricultural Science (O) (1931–)	—	23	42
39–40	Journal of Agricultural and Food Chemistry (O) (1953–)	26–27	—	10
41–42	Agrokhimiya (G) (1964–)	26–27	—	—
41–42	Soil and Tillage Research (O) (1980–)	42–43	36–37	30–31
43	Water Research (U) (1967–)	38–40	—	12
44–45	Journal of Experimental Botany (O) (1950–)	33–34	—	11
44–45	Physiologia Plantarum (O) (1948–)	32	—	4
46	Journal of the Science of Food and Agriculture (O) (1950–)	35–36	—	20
47	Crop Science (O) (1961–)	37	—	16
48–49	Canadian Journal of Botany (G) (1951–)	52–53	38	9
48–49	Journal of Applied Ecology(O) (1964–)	41	—	24
50	Canadian Journal of Microbiology (G) (1954–)	·	24–25	15
51	Agronomie (G) (1981–)	31	—	—
52	East Africa Agriculture & Forestry (O) (1934–)	—	30	—
53	Oikos (C) (1949–)	33–34	—	17

Date: after title is date of initial publication
Publishers: O = Organization; C = Commercial; G = Government; U = University.
[a] *SCI Journal Citation Report 1988.*

Table 10.2 provides these rankings which help explain the value of individual journals. The overall ranking is derived from the Core Project analysis of all source monographs; the final counts were adjusted for skews. Journals with high or exclusive counts from one source only were weighted downward. For example, *Science*, which publishes few articles on soil-related topics, does publish extensively in atmospheric science. The one source document on agroclimatology analyzed by the core project accounted for over 36% of the *Science* citations. Such a heavy count for one journal in a discipline such as atmospheric science, only peripherally related to soils, is a numeric skew. An adjustment downward for this count more closely reflects *Science*'s true importance in the ranking. Similar criteria were used in other cases where apparent skews were identified.

Column A, "Developed world ranking," lists the top journals derived from analysis of the twenty source documents whose primary emphasis was the developed world. Column B, "Tropical soils ranking," lists the same journals by counts from the eight tropical soils documents. Finally, Column C, "ISI citation count ranking," lists the same journals by the number of citation counts they received in the *SCI Journal Citation Reports 1988*. A problem in comparing these rankings is that many journals on the list are not soils specific (such as *Science, Nature, Ecology*, and *Plant Physiology*) and will, therefore, have a higher citation standing in the *SCI* column, which is not soil specific. *SCI* ranks journals according to their entire subject coverage, and some, such as *Science* and *Nature*, may have fewer than 5% of their articles on soil-related topics. Their high standing is, therefore, unrelated to their soil coverage content.

The United States is the predominant country of publication in Table 10.2, with 26.4% of the fifty-three titles. Next is the United Kingdom, at 22.6%; France and the Netherlands are both 9.4%; Australia and Canada are 5.6%; Denmark, Germany, India, and Russia, 3.7%. Countries with one title apiece account for the remaining 6.2%.

Almost all the top journals in Table 10.2 have an established history, with twenty-seven of the top fifty-three journals (50.9%) predating 1950. Only *Soil and Tillage Research* (Amsterdam: Elsevier) began publication since 1980. Compare this with the listing in Table 10.3, where almost 20% of the titles began publication after 1980, reflecting the heavy current-research orientation of *SCI* data. Table 10.3 lists the top-ranked soils journals in the *SCI Journal Citation Reports* by impact factor.

Impact factor is a ratio between the number of citations to a journal in a year and the number of citations it publishes in its articles. The higher the impact factor, the greater the number of citations to it than the number it cited. This usually reflects a "quick turnaround" or current topics which have limited impact over the long term. In choosing its source documents, the Core Agricultural Literature Project sought to find a balance between current trends and literature important to college-level instruction, generally with historical relevance to the development of soil science. What the Core

Table 10.3. SCI ranking by impact factor of thirty-six top journals from core project analysis

Overall ranking by impact factor		Impact factor	Cited half-life (years)
1.	Advances in Agronomy [bc]	1.504	>10.0
2.	Journal of Agriculture and Food Chemistry [b]	1.265	8.3
3.	Soil Biology and Biochemistry [bc]	1.245	6.7
4.	Soil Science Society of America Journal [bc]	1.093	7.3
5.	Journal of Soil Science [bc]	0.974	8.9
6.	Geoderma [bc]	0.921	7.1
7.	Journal of the Science of Food and Agriculture	0.809	>10.0
8.	Biology and Fertility of Soils	0.775	1.8
9.	Agronomy Journal [bc]	0.719	>10.0
10.	Plant and Soil [bc]	0.717	7.6
11.	Journal of Soil and Water Conservation [b]	0.673	5.3
12.	Soil Science [bc]	0.647	>10.0
13–14.	Soil Use and Management	0.642	4.8
13–14.	Soil and Tillage [b]	0.642	4.1
15.	Australian Journal of Soil Research [bc]	0.614	7.5
16.	Crop Science [b]	0.608	9.0
17.	Journal of Agricultural Science [bc]	0.570	>10.0
18.	Canadian Journal of Soil Science [b]	0.555	8.4
19.	Australian Journal of Agricultural Research [bc]	0.543	>10.0
20.	[a]Experimental Agriculture [bc]	0.500	7.2
21.	Soil Science and Plant Nutrition [b]	0.472	7.4
22.	Communications in Soil Science and Plant Analysis [b]	0.422	6.4
23.	Netherlands Journal of Agricultural Science [bc]	0.393	>10.0
24.	Landwirtschaftliche Forschung	0.390	8.3
25.	Transactions of the ASAE [bc]	0.294	7.7
26.	Australian Journal of Experimental Agriculture [bc]	0.279	>10.0
27.	Fertility Research	0.273	4.3
28.	[a]Tropical Grasslands	0.272	7.4
29.	Catena	0.271	5.8
30.	Biocycle	0.264	2.9
31.	Archiv für Acker und Pflanzenbau und Bodenkunde	0.245	4.5
32.	[a]Tropical Agriculture [bc]	0.239	>10.0
33.	Soviet Soil Science (Pochvovedenie) [b]	0.056	>10.0
34.	[a]Turrialba [bc]	0.050	n/d
35.	Soil and Crop Society of Florida	0.025	n/d
36.	[a]Agronomie Tropicale [bc]	0.021	n/d
37.	[a]Indian Journal of Agricultural Science [b]	0.012	>10.0

Source: SCI *Journal Citation Reports,* 1983–88 average.
[a] Tropical agriculture journals.
[b] Titles in Table 10.2. (74.9%)
[c] Titles that match the top 36 in Table 10.2. (47.2%)

Project analysis demonstrates is that ranking journals by impact factor in the *SCI* shows current trends. This is less likely to reveal the long-term relationship between journals whose immediate impact is low and the evolution of the discipline as a whole, where their overall impact may be extensive.

This research slant of *SCI* journals is clearly reflected in their cited half-lives, which are considerably shorter than for more established publications which mix research, analysis, and general overviews. As noted, the half-life of the journal citations analyzed by the Core Project for soils is 9.4 years, compared to the *SCI* (1988) 7.2 mean for soils journals and 5.9 for all science journals. The rule of thumb is the longer the half-life, the less likely the material is to be purely research oriented.

Just over 25% of the *SCI* titles do not appear in the top core list. *Biology & Fertility of Soils* is an example. While its impact factor is high (0.775), its corresponding cited half-life is a brief 1.8 years, an indication that the primary thrust of this journal is popular or quickly outdated research. Indeed, the editors of *Biology & Fertility of Soils* claim "rapid publication" is a "hallmark of the journal."

D. Third World and Tropical Soils in Journals

Table 10.4 ranks the top soils journals by their strength in material published on, or pertinent to, tropical soils. Although Third World countries extend beyond the tropics, it should be noted that almost all tropical soil science is done on farmland in developing countries. While the terms tropical and Third World are not necessarily interchangeable, analysis of tropical soil literature done by the Core Project establishes a clear relationship between them.

Of the top journals identified by the Core Project analysis as being published in the Third World or predominantly tropical in content (see Table 10.5), only eight, or 32%, are covered by *SCI*. These are identified by[a] in Table 10.5. This lack of tropical agriculture coverage in the *SCI* has been noted in the literature and limits comparisons (see Chapter 4).

Top nontropical journals, such as the *SSSA Journal*, dominate tropical soil citation patterns, in part, because they receive worldwide distribution, have a prestigious history as top-quality publications, and are, therefore, commonly cited in the literature. Furthermore, publications such as the *SSSA Journal* out-publish the number of articles in a general agriculture journal, such as *Turrialba*, by a margin of six to one (1990). This means that although the percentage of articles pertaining to the tropics in *Turrialba* may be as high as 96% (1990) and in *SSSA Journal* only 14.6% (1990), in

Table 10.4. Ranking of top journals and serials by tropical soils citation counts in AGRICOLA and core citation analysis

Journals and serials	AGRICOLA (1979–91)	Core analysis	
		Core A	Core B
Advances in Agronomy	1	4	16–17
ªAgronomie Tropicale	2	2	9
ªTropical Agriculture	3	3	7
Communications in Soil Science and Plant Nutrition	4	11	5
Australian Journal of Soil Research	5	7	16–17
ªJournal of the Indian Society of Soil Science	6	9	11–12
ªExperimental Agriculture	7	6	10
ªTurrialba	8	1	15
Soil Science Society of America Journal	9	20	1
Soil Science	10	19	2
Plant and Soil	11	15	3
Journal of Soil Science	12	13	4
Soil Biology and Biochemistry	13	10	8
Pochvovedenie	14	n/d	[79]
Agronomy Journal	15	21	6
Geoderma	16	17	11–12
Canadian Journal of Soil Science	17	n/d	[107]
Australian Journal Experimental Agriculture	18	8	20
Nature	19	14	13
Journal of Agricultural Science	20	18	21
Australian Journal of Agricultural Research	21	12	14
ªSols Africains (dead)	n/d	5	19
Science	n/d	16	17

ª Tropical agriculture journals.

AGRICOLA: Ranking arrived at by dividing all citations in a journal by the number of "tropic? soil?" citations in that journal for the years 1979–91. The percentages were then ranked numerically. The shortfall here is that many pertinent citations to "tropical soils" will not appear, since the terms "tropic?" and "soil?" often do not appear in either the title, the subject field, or the abstract, or only one term will appear and not both. For instance, many articles on tropical volcanic soils may never mention any of these terms, but use "andosols" instead. However, a fair, if approximate, ranking can be derived given that the universe of articles indexed for each journal is in the many hundreds, and that with such large base numbers discrepencies are equalized.

Core A is a ranking arrived at by dividing the total number of citations to a journal from all source documents by the number of citations derived from the eight *tropical soils* source documents. This ranking by percentages reveals, in aggregate, citation patterns to those journals which show the greatest inclusion of articles pertinent to tropical soils research.

Core B is a straight numerical ranking derived from citations tabulated from the eight tropical soils source documents. A journal with more published pages per year and wider distribution will likely come out higher on the list by the sheer number of its articles.

whole numbers, the yearly output of articles on tropical subjects is about even between them, and for soils, the *SSSA Journal* will come out far ahead because of its specific subject scope. Among the top 100 journals ranked by the Core Project using tropical soils source documents, one-fourth were specific to tropical agriculture, though not necessarily soils specific. These are listed in Table 10.5.

Only the the first eight journals in Table 10.5 have sufficient citation counts (27+) to be of statistical value in determining importance in the literature of soil science. The *Journal of Rubber Research*, by comparison, had only seven citations from two of the twenty-eight source documents. Its impact on the literature is, therefore, slight, local, and subject specific. Four of the journals in Table 10.5 are soil specific, the rest are general agriculture.

Table 10.5. Top tropical journals publishing soils material

1.	ªTropical Agriculture (Trinidad) ᵇ
2.	ªAgronomie Tropicale (France) ᵇ
3.	ªExperimental Agriculture (U.K.) ᵇ
4.	Journal of the Indian Society of Soil Science ᵇ
5.	ªTurrialba (Costa Rica) ᵇ
6.	Sols Africains (Nigeria) ᵇ
7.	ªIndian Journal of Agricultural Science ᵇ
8.	Bragantia (Brazil) ᵇ
9.	East African Agriculture and Forestry (Kenya)
10–11.	Philippine Agriculture
10–11.	ªTropical Grasslands (Australia)
12–14.	Nigerian Agricultural Journal
12–14.	Rhodesian Agricultural Journal
12–14.	Cahiers l'ORSTOM: Pédologie (France)
15	JARQ (Japan Agricultural Research Quarterly)
16.	Journal of Tropical Geography (Singapore)
17.	ªPesquisas Agropecuria Brasileira
18.	Agricultura Tropical (Colombia)
19–20.	Indian Farming
19–20.	ªIndian Journal of Agronomy
21.	Revista Brasileira de Ciência do Solo
22.	Malayan Agricultural Journal
23.	Tropical Agriculturist (Sri Lanka)
24.	Suelos Ecuatoriales (Colombia)
25.	Indian Journal of Agricultural Research
26.	Journal of Rubber Research (Malaysia)

Ranking: From project analysis of the eight source documents on tropical soils.
ª Journals indexed in the *SCI 1988*.
ᵇ Ranked Core analysis journals with twenty-seven citation counts or more. (See core recommendations in Table 10.6.)

E. The Core Journals and Serials

The recommended core journals in soil science in publication are top-ranked journals identified by the Core Project's citation analysis. Journals of primary importance are ranked 1, those of lesser rank are coded 2. The five top-ranked journals are the same for both the developed and Third World. Eighteen titles (45%) are on both lists. These recommended forty-six titles represent only 5.4% of all titles identified in the analysis, yet they account for 48.7% of the citation counts.

One title not listed below deserves mention. *African Soils (Sols Africains)*—first published in 1951, in Lagos, Nigeria, by the Organisation de l'Unité Africaine, Commission Scientifique, Technique et de la Recherche—was heavily cited but ceased publication in 1974. During its twenty-three years, it was an important and respected journal and deserves preservation attention (see Chapter 14).

The title *Eurasian Soil Science,* which is listed as core, was formerly *Soviet Soil Science,* but changed title in 1992. Beginnning with the inaugural issue of *Eurasian Soil Science* (vol. 24, no. 2), translations from *Pochvovedenie* and selected translations from *Agrokhimiia* are featured. *Soviet Soil Science,* which began in 1958, is an English translation of the long-lasting Russian serial *Pochvovedenie* (1899–). Both *Pochvovedenie* and *Agrokhimiia* were core journals, but only the translation is being recommended.

Core Journals and Serials in Soil Science

	Third World	Developed countries
Advances in Agronomy. Vol. 1 (1949)+ New York; Academic Press. Annual.	2	2
Agricultural and Forest Meteorology. Vol. 20 (1984)+ Amsterdam: Elsevier Scientific Pub. Co. Bimonthly. Continues: *Agricultural Meteorology.*	–	2
Agronomy Journal. Vol. 1 (1907/09)+ Madison, Wis.; American Society of Agronomy. Bimonthly.	1	1
L'Agronomie Tropicale. Vol. 1 (1946)+ Paris; Institute de Recherche Agronomiques Tropicales et des Cultures Vivieres. Monthly.	1	–
Agronomie. Vol. 1, no. 1 (1981)+ Paris; Institut National de la Recherche Agronomique. Quarterly. Supersedes: *Annales Agronomiques.*	–	2

	Third World	Developed countries
Applied and Environmental Microbiology. Vol. 31, no. 1 (1976) + Washington, D.C.; American Society for Microbiology. Monthly. Continues: *Applied Microbiology.*	2	–
Australian Journal of Agricultural Research. Vo. 1 (1950) + Melbourne; Commonwealth Scientific and Industrial Research Organization. Quarterly, 1950–, Bimonthly, 1974–	2	–
Australian Journal of Experimental Agriculture. Vol. 25, no. 1 (1985) + East Melbourne; Commonwealth Scientific and Industrial Research Organization. Quarterly. Continues: *Australian Journal of Experimental Agriculture and Animal Husbandry.*	2	–
Australian Journal of Soil Research. Vol. 1 (1963) + Melbourne; Commonwealth Scientific and Industrial Research Organization. Quarterly.	2	2
Bragantia. Vol. 1 (1941) + Campinas, Brazil; Instituto Agronomico. Annual. Continues: *Instituto Agronomico Boletim Tecnico.*	2	–
Canadian Journal of Microbiology. Vol. 1 (1954) + Ottawa, Ontario; National Research Council Canada). Monthly.	2	–
Canadian Journal of Soil Science. Vol. 37 (1957) + Ottawa; Agricultural Institute of Canada. Semi-annual, 1957–1963, 3 times a year, 1964-1972, Quarterly, 1973–.	–	1
Communications in Soil Science and Plant Analysis. Vol. 1 (1970) + New York: M. Dekker. Bimonthly.	–	2
East African Agricultural and Forestry Journal. Vol. 26 (1960) + Nairobi; East African Agricultural and Forestry Research Organization. Quarterly. Continues : *East African Agricultural Journal of Kenya.*	2	–
Ecology. Vol. 1 (1920) + Durham, N.C.; Ecological Society of America and the Duke University Press. 6 no. a year. Supersedes: *Plant World.*	–	2
Eurasian Soil Science. Vol. 24 (1992) + Washington; Scripta Publishing. 10 no. a year. Supersedes: *Soviet Soil Science, a translation of Pochvovedenie* (1958–1992)	–	1
Experimental Agriculture. Vol. 1 (1965) + London and New York; Cambridge University Press. Quarterly. Supersedes: *Empire Journal of Experimental Agriculture.* (1933–1964)	2	–
Geoderma. Vol. 1, no. 1 (1967) + Amsterdam; Elsevier Scientific Pub. Co. Monthly.	2	1
The Indian Journal of Agricultural Sciences. Vol. 1 (1931) + New Delhi, India; Indian Council of Agricultural Research. Bimonthly, 1931–47, Monthly. Supersedes: *Agricultural Journal of India; Absorbed: Agricultural Re-*	2	–
Journal of Agricultural and Food Chemistry. Vol. 1 (1953) + Easton, Pa.; American Chemical Society. Biweekly, 1953-54, Monthly, 1955-59, Bimonthly, 1960–.	–	2

	Third World	Developed countries
The Journal of Agricultural Science. Vol. 1 (1905/06)+ Cambridge, Eng.; Cambridge University Press. Bimonthly.	2	2
Journal of Environmental Quality. Vol. 1 (1972)+ Madison, Wis.; Published cooperatively by American Society of Agronomy, Crop Science Society of America, and Soil Science Society of America. Quarterly.	2	1
Journal of the Indian Society of Soil Science. Vol. 1 (1953)+ New Delhi, India; Indian Society of Soil Science. Quarterly.	2	–
Journal of Soil and Water Conservation. Vol. 1 (1946)+ Ankeny, Iowa.; Soil Conservation Society of America. Quarterly, 1946–52, Bimonthly, 1953–.	2	2
Journal of Soil Science. Vol. 1 (1959)+ Oxford; Clarendon Press. Annual, 1959–50, Quarterly, 1951–.	1	1
Nature. Vol. 1 (1869)+ London; Macmillan. Weekly.	2	2
Netherlands Journal of Agricultural Science. Vol. 1 (1953)+ Wageningen; Koninklijk Genootschap voor Landbouwwetenschap. Quarterly.	–	2
The New Phytologist. Vol. 1 (1902)+ London and New York; Academic Press. Frequency varies.	2	2
Pedobiologia. Vol. 1 (1961)+ Germany: Jena Gustov Fischer. Bimonthly.	–	2
Pédologie. Vol. 1 (1951)+ Ghent, Belgium; Societe Belge de Pédologie. Quarterly.	–	2
Phytopathology. Vol. 1 (1911)+ St. Paul, Minn.; American Phytopathological Society. Bimonthly, 1911–17, Monthly 1918–.	2	2
Plant and Soil. Vol. 1 (1948)+ The Hague; M. Nijhoff. Bimonthly, 1976–.	1	1
Plant Physiology. Vol. 1 (1926)+ Bethesda, Md.; American Society of Plant Physiologists. Monthly.	–	1
Revue d'Ecologie et de Biologie du Sol. Vol. 1 (1963)+ Paris; Gauthier-Villars. Quarterly.	–	2
Science. New Series, Vol. 1, no. 1 (1895)+ Washington, D.C.; American Association for the Advancement of Science. Weekly. Supersedes: *Science; A Weekly Record of Scientific Progress*; Absorbed: *Scientific Monthly*, Jan. 1958.	2	2
Soil Biology and Biochemistry. Vol. 1 (1969)+ Oxford; Pergamon Press. Bimonthly.	1	1
Soils and Fertilizers. Vol. 1 (1938)+ Farnham Royal, Eng.; Commonwealth Agricultural Bureaux. Monthly. Supersedes: *Commonwealth Bureau of Soils. Monthly Letter*.	2	1
Soil Science. Vol. 1 (1916)+ Baltimore, Md.: Williams and Wilkins. Monthly.	1	1

	Third World	Developed countries
Soil Science and Plant Nutrition. Vol. 1 (1955)+ Tokyo, Japan; Society of the Science of Soil and Manure. Quarterly.	1	2
Soil Science Society of America Journal. Vol. 1 (1936)+ Madison, Wis.; Soil Science Society of America. Bimonthly. Continues: *Soil Science Society of America. Proceedings.*	1	1
Transactions of the ASAE. New Series, Vol. 1 (1958)+ St. Joseph, Mich.; American Society of Agricultural Engineers. Bimonthly. Supersedes: *American Society of Agricultural Engineers. Transactions 1907–1934/35.*	2	2
Tropical Agriculture. Vol. 1 (1924)+ Trinidad: Imperial College of Tropical Agriculture. Quarterly.	1	–
Turrialba. Vol. 1 (1950)+ Turrialba, Costa Rica; Inter-American Institute of Agricultural Sciences. Quarterly.	2	–
Zeitschrift für Pflanzenernahrung und Bodenkunde. = Journal of Plant Nutrition and Soil Science Vol. 1 (1851)+ Weinheim/Bergstraat; VCH. Bimonthly.	–	2

Of the forty-four serials listed as core, forty-three were indexed by CAB Abstracts, and twenty-six by AGRICOLA, in 1990. AGRICOLA is less inclined to index tropical or foreign language journals, concentrating instead on English-language publications.

Two other journals also require special mention. *Nature* and *Science*, while only peripherally soil related, have consistently ranked high in the counts in this and many other Core Project subject disciplines. They are also of interest because, to date, they are the only scientific journals readily available, full-text, in computer format. Both are listed on UMI's full-text CD-ROM database, *ProQuest: General Periodicals Ondisc*, where the last five years of both journals are accessible from computer terminal workstations. Although few science journals have been transferred into machine-readable formats, this is expected to change soon. The availability of these two distinguished publications presages the beginning of a trend to access full articles of many science serials online, either at library workstations or remotely, via telecommunication systems from terminals in the home or office. A clear example of this trend comes from the tri-societies: ASA, CSSA, and the SSSA. These prestigious scientific bodies have created a Peer Reviewed Electronic Journal Subcommittee in order to explore the feasibility of an electronic professional journal for publication on computer software and supporting databases.

Thus, the identification of the most important journals for research and instruction, as the Core Agricultural Literature Project is doing, in order to transfer this literature to CD-ROM technology, can be seen as the vanguard of a growing trend. This effort will facilitate the intellectual framework for the transfer of this technology.

F. New Serial Publications, 1983 +

The following is a selected listing of recently begun journals in soil science, including titles in agronomy with broad soil coverage. This is not an evaluative list, and most popular and newsletter titles are excluded.

Recently Begun Journals in Soil Science (1983 +)

Advances in Soil Science. Vol. 1. (1985) + New York: Springer Verlag. Annual.
Arid Soil Research and Rehabilitation. Vol. 1. (1988) + New York: Taylor and Francis. Quarterly.
Better Farming in Salt Affected Soils. Vol. 1. (1985) + Karnal, India: Central Soil Salinity Research Institute. Irregular.
Biology and Fertility of Soils. Vol. 1. (1985) + Berlin: Springer International. Quarterly.
Ciencia del Suelo: Revista de la Asociacion Argentina de la Ciencia del Suelo. Vol. 1. (1983) + Buenos Aires: La Asociacion. Biannual.
Ecology and Farming: International IFOAM Magazine. Vol. 1. (1990) + Tholey Theley, Germany: International Federation of Organic Agriculture Movements. Quarterly.
European Journal of Agronomy. Vol. 1 (1992) + Montrouge, France: Gauthier-Villars, European Society of Agronomy. Quarterly.
Journal of Production Agriculture. Vol. 1. (1988) + Madison, Wis: American Society of Agronomy, Crop Science Society of America, Soil Science Society of America. Quarterly.
Journal of Soil Contamination. Vol. 1 (1992) + Boca Raton, Fla: Lewis Publishers. Quarterly.
Land Degradation and Rehabilitation. Vol. 1. (1989) + Chichester, UK: Wiley. Bimonthly.
Pedosphere. Vol. 1. (1990) + Beijing, China: Science Press. Quarterly.
Sahara Review. Vol. 1. (1989) + Alexandria, Egypt: Balba Group for Soil and Water Research. Quarterly.
Soil Technology. Vol. 1. (1988) + Cremlingen, Germany: CATENA Verlag. Quarterly.

Soil Use and Management. Vol. 1. (1985)+ Oxford: British Society of Soil Science. Quarterly.

South African Journal of Plant and Soil = *Suid Afrikaanse Tydskrif vir Plant en Ground*. Vol. 1. (1984)+ Pretoria: Bureau for Scientific Publications.

11. Soil Surveys and Maps

RALPH J. MCCRACKEN

North Carolina State University

DOUGLAS HELMS

Soil Conservation Service, U.S. Department of Agriculture

The first identifiable record of soils as independent natural bodies important for plant growth and, therefore, worthy of study apparently is found in China a little more than 4,000 years ago.[1] Nine broad classes of soils were established, given descriptive names, and used as a basis for taxation and establishing sizes of land holdings.

According to Micha Strzemski, Aristotle (384–22 B.C.) and his student Theophrastes differentiated "earth" (soils) into a set of qualities (e.g., dry vs. wet).[2] Theophrastes named soils as edaphs (from which the term edaphology is derived) and recognized layers in the soil, now called horizons. He distinguished between six kinds of soil, each with varying properties that affected plant growth.

A group of Roman philosophers—Cato the Elder (234–149 B.C.), Varro (116–27 B.C.), and especially Columella (about 45 A.D.)— discussed soil differences which caused variations in plant growth.[3] For example, Columella ranked soils from best to worst, according to their suitability for plant growth.

As pointed out by Roy W. Simonson, the recognition of geology as a science, and its use in studies of the soil in the late eighteenth and early nineteenth centuries, was a landmark for soil classification.[4] This develop-

---

1. (a) Yun-shen Wang, "Soil Classification in Ancient China," *Acta Pedologica Sinica* 18 (1) (1979): 1–8 (English abstract). (b) Roy W. Simonson, *Soil Classification in the Past: Roots and Philosophies* (Wageningen, Netherlands: International Soil Reference and Information Centre, 1985).

2. Micha Strzemski, *Ideas Underlying Soil Systematics = Mysli Przewodnie Systematyki Gleb*, trans. Jerzy Bachrach (Warsaw: Foreign Scientific Publ., Dept. of the National Center for Scientific, Technical, and Economic Info., 1975). Available from the National Tech. Info. Service.

3. S. W. Buol, F. D. Hole, and R. J. McCracken, *Soil Genesis and Classification*, 3d ed. (Ames, Iowa: Iowa State University Press, 1989).

4. Simonson, "Historical Aspects of Soil Survey and Soil Classification, Part I. 1899–1910," *Soil Survey Horizons* 27 (1) (1986): 3–11.

ment led to the methodology for mapping bodies of soils, a narrowly conceived view which saw soils as nothing more than the results of weathering of the underlying rocks. Reliance on this concept delayed the development of the idea of soils as independent natural bodies. German scientists, especially Emil Ramann[5] and F. A. Fallou,[6] began to develop agrogeology: soil as more than weathered rock—a medium for plant growth.[7] Fallou first proposed the use of the term pedology, which he regarded as a theoretical, geologically oriented soil science (a term widely used today, referring to the science of soil genesis and classification), in contrast to agrology—the applied and practical, agronomically oriented aspect of soil science.

A. Beginnings of Soil Surveys and Associated Soil Classification and Mapping

People involved in current soil surveying, mapping, and classification generally regard soils as independent natural bodies formed under the influence of varying combinations of five soil-forming factors, which differ from place to place. Soil survey, as used in the United States, refers to the process of field mapping of soils, followed by preparation of a report describing the soil units shown on the map and presentation of information about the soils with respect to their agricultural and nonfarm uses. Soil classification is the placing of soils in a system in such a way as to help record and remember their properties and to serve as a vehicle for transferring research and experimental results from one site to others.

One of the earliest efforts at mapping natural resources, including some types of soil information, was carried out in the eighteenth century in England.[8] The British Board of Agriculture published the first set of maps with a specific soil legend in the 1790s. These were followed by a series of county maps containing colored soil maps. But these were not based on any formal soil classification system. They mainly depicted the distribution of relatively homogeneous areas with approximately similar soil properties relevant for land use.

In the United States, A. Eaton and T. R. Beck appear to have made the first geological survey map designed to improve agriculture, for Albany

---

5. Emil Ramann, *Bodenkunde* (Berlin: Springer, 1911).

6. F. A. Fallou, *Pedologie oder Allgemeine un Besondere Bodenkunde* (Dresden: 1862).

7. Buol, *Soil Genesis and Classification.*

8. D. H. Yaalon, "The Earliest Soil Maps and Their Logic," *Newsletter,* International Society of Soil Science Work Group on History, Philosophy and Sociology of Soil Science (1990): 2–3.

County, New York.[9] They used "alluvial formations" for what we now call soils and devised a two-category classification system.[10]

Vasilii V. Dokuchaev in Russia[11] and Eugene W. Hilgard in the United States[12] first wrote about the concept of soils as independent bodies, and seemingly proposed the concept independently.[13] Dokuchaev published soil maps in 1883 which distinguished soils as natural bodies differing according to their genesis. Apparently, the first systematic soil map based on a formal genetic soil classification was produced by N. M. Sibertzev in Russia.[14] However, D. N. Yaalon stated that the first "pedologically oriented soil maps were probably prepared in 1856 by A. L. Grossul-Tolstoy of the Crimea region some two to five years before Dokuchaev formally conceptualized the approach."[15]

While Dokuchaev's ideas influenced soil surveying and mapping in Russia, for several years after the initiation of soil mapping in the United States, most American soil scientists held to the vestiges of an earlier concept of soils as a type of geologic material at the earth's surface suitable for plant growth. It was decades into the twentieth century before the Russian concepts influenced soil surveying, mapping, and classification in the United States, and several more years before Hilgard received some credit for his pioneering work.[16]

B. Early Soil Classification, Survey and Mapping in the United States

Hilgard, who served as state geologist of Mississippi in the mid-nineteenth century and later continued his work at the University of California, contributed immensely to new approaches for differentiation of soils on some basis other than geology. As author of *Report on the Geology and Agriculture of Mississippi*, published in 1860, he presented new ideas about soils. Hilgard distinguished between the surface soils and subsoils in the

9. A. Eaton and T. R. Beck, *A Geological Survey of the County of Albany* (Albany, N.Y.: S. Southwick, 1820).

10. Board of Agriculture, State of New York, *Geological Survey of the County of Albany* (New York: New York Board of Agriculture, Vol. I, 1820).

11. Vasilii V. Dokuchaev, *Russian Chernozem = Russkii Chernozem*, trans. N. Kaner (Jerusalem: Israel Program for Scientific Translations, 1967). Available from the U.S. Dept. of Commerce, Clearinghouse for Federal Scientific and Technical Information, Springfield, Va.

12. Eugene W. Hilgard, *A Report on the Relation of Soils to Climate* (1892 [USDA Weather Bureau Bulletin no. 3]).

13. Buol, *Soil Genesis and Classification*.

14. N. M. Sibertzev, "Genetic Classification of Soils," *Aleksandr. Agri. Inst.* 9 (1895): 1–23. (In Russian).

15. Yaalon, *The Earliest Soil Maps and Their Logic*, p. 2.

16. Buol, *Soil Genesis and Classification*.

soil profile based on root depth. His emphasis on examining virgin soils in their natural sites and his recognition of ongoing processes in soil formation led his biographer, Hans Jenny, to conclude that Hilgard definitely viewed the soil as a natural body.[17] He apparently produced the first true soils map in the United States. Later, he became the first in the United States to recognize climate as an important soil-forming factor, noting that several soil properties varied with climate, based on studies in California.[18]

In the latter part of the nineteenth century, the U.S. Department of Agriculture (USDA) became involved in the development and support of upcoming sciences in agriculture.[19] There is probably no greater illustration of this than the development of soil survey and mapping, and soil classification, which is sometimes called the science of pedology. The federal government was and is the chief supporter of and leader in this area. In 1894, USDA established a Division of Agricultural Soils in its Weather Bureau, with Milton Whitney as the director. In 1897, the Division of Soils was made an independent unit within the USDA; Congress raised its stature to that of the Bureau of Soils on July 1, 1901.

Whitney is credited with developing the first soil classification system related to soil survey used as a basis for soil mapping.[20] As the first Chief of the USDA Bureau of Soils, he initiated the first federally supported soil survey in 1899.[21] The main basis of his system was the difference in geological materials and formations in which the soils had formed. He had previously conducted soil fertility and soil chemistry research in the embryonic soils program at North Carolina State College, now North Carolina State University. Later, while affiliated with the state agricultural experiment station in Maryland, he studied the physical properties of soils. In his capacity in the USDA, Whitney saw the need to be of immediate assistance to farmers. In 1896, several years before the soil mapping program began, he stated the importance of soil mapping in his plans for the division: "One of the most important objects of the work of the division should be to investigate and map the important soil areas in accordance with the geologic relations and the agriculture value."[22] By 1900, after the soil survey work had

17. Hans Jenny, *E. W. Hilgard and the Birth of Modern Soil Science* (Pisa: Agrochimica, 1961).

18. (a) Simonson, *Historical Aspects of Soil Survey and Soil Classification.* (b) Hilgard, *A Report on the Relation of Soils to Climate.*

19. A. H. Dupree, *Science in the Federal Government: A History of Policies and Activity to 1940* (Cambridge, Mass.: Belknap Press of the Harvard University Press, 1957).

20. Milton Whitney, *The Physical Properties of Soils in Relation to Moisture and Crop Distribution* (1892 [USDA Weather Bulletin no. 4]).

21. USDA, *Field Operations of Division of Soils* (1899 [USDA Report no. 64]).

22. Whitney, *Report of the Chief of the Division of Agricultural Soils* (Washington, D.C.: USDA, 1896).

actually started, Whitney more precisely defined the purpose of the work: "to construct maps in the field showing the area and distribution of the soil types: to explain as fully as possible from geologic considerations the origin of the soil, and to have the soil chemist and physicist study the differences in soil types."[23]

A trial soil survey was made of about 250 square miles near Hagerstown, Maryland, in 1898, recorded in an annual report of the Division of Agricultural Soils and in correspondence about the fiftieth anniversary of the soil survey program in 1949. The results of the 1898 trial survey have disappeared. Mapping in Cecil County, Maryland, also started in 1898, was completed in 1899, and the results of that were published later.[24] The first volume of field operations, published by what was then the Division of Soils (*USDA Bulletin* 64), includes maps and texts for surveys of the Connecticut Valley, the Pecos Valley, and the Salt Lake Valley. Publication of the soil survey of Cecil County, Maryland, was held over to the following year so that the topographic maps from the U.S. Geological Survey could be used as the base.

The first published soil survey, containing both maps and soil descriptions, started with the survey of a small area in Virginia.[25] Among the early soil surveys published by this new Division of Soils was the *Raleigh to New Bern Area* in eastern North Carolina. This survey provided a basis for the systematic investigation of the fertilizer requirements of crops, especially tobacco. Thirteen soil types were recognized in this initial survey and the scale of mapping was one inch per mile.[26]

These early Bureau of Soils surveys reflected a strong geologic influence. For example, soil groups in that system were conceived and defined as being a series of soil materials from a particular geologic formation, hence the origin of the term "soil series." Soil surveys in the United States still use the soil series as the basic soil map unit. In the currently used *Soil Taxonomy* soil series are the taxa, or classes, in the lowest category.[27]

In the early days, criticisms of these soil surveys focused on two points. Whitney placed a very strong emphasis on soil texture as the property he thought best indicated the value of soils for agricultural uses. The embry-

23. Whitney et al., *Field Operations of the Division of Soils, 1900.* 2nd report (Washington, D.C.: USDA, 1900), p. 6.

24. Simonson, Personal communication.

25. G. A. Weber, *The Bureau of Chemistry and Soils: Its History, Activities and Organization* (Baltimore: Johns Hopkins Press, 1979).

26. W. D. Lee, *The Early History of Soil Survey in North Carolina* (Raleigh, N.C.: N.C. Soil Science Society, North Carolina State University Soil Science Dept., 1984 [Spec. Publ.]).

27. USDA, Soil Survey Staff, *Soil Taxonomy: A Basic System of Soil Classification for Making and Interpreting Soil Surveys* (Washington, D.C.: U.S. Govt. Print. Off., 1975 [Soil Cons. Serv., USDA Handbook no. 436]).

onic soil classification system relied too heavily on geologic formations in classifying broad groups of soils. One of Whitney's chief critics was Hilgard, who actually sought to have Whitney replaced as head of the bureau because of his strongly held views on these points.[28] Cyril G. Hopkins, of Illinois, was another severe critic of Whitney. Cooperative relations in soil surveys between Illinois and the Bureau of Soils were discontinued because of the conflict between the two men. Afterward, the Illinois Station conducted its own independent survey program for many years.[29]

Use and application of these early Bureau of Soils soil surveys began to reveal flaws in Whitney's theories on the importance of soil texture. It was soon agreed that geological formations were not an adequate basis for classifying and mapping soils. For example, an early soil province map, made in 1909, displayed a glacial and loessial province extending from the Atlantic coast to the state of Montana.[30] Soil surveyors began to realize that there were significant differences among soils in close proximity due to factors other than the nature of the geologic material from which they were formed.

Looking backward, it is tempting to find fault with Whitney's narrowness of view and with what some saw as his autocratic methods.[31] The significance of this period is the development of the detailed soil surveys and the finding that soil mapping and soil classification are mutually dependent activities. The development of *Soil Taxonomy*, which was made possible by the ability to test the theories in the field, is an important example of this relationship. When revisiting the missteps of the early history of the soil survey, it is well to recall Charles E. Kellogg's description of the problem: "These early soil scientists were faced with a most difficult dilemma: they could not classify and map soils without knowing their characteristics; and they could not know their characteristics until they had examined representative examples in the field and in the laboratory. Not until they had classified and mapped could they understand the significance of combinations of soil characteristics."[32]

Soils exist in nature. Soil classifications and concepts of soil mapping units are the methods whereby humans attempt to organize knowledge so as to understand nature better. Soil scientists and those who use soil survey information can benefit from an understanding of this evolving history of soil science. Historians of science have learned a lesson that should be

28. Jenny, E. W. Hilgard and the Birth of Modern Soil Science.
29. R. Simonson, Personal communication.
30. Whitney, Soils of the United States (Washington, D.C.: U.S. Govt. Print. Off., 1909 [USDA Bureau of Soils Bulletin no. 55]).
31. Jenny, E. W. Hilgard and the Birth of Modern Soil Science.
32. Charles E. Kellogg, Soil Classification and Correlation in the Soil Survey (Washington, D.C.: USDA Soil Cons. Serv., 1959), p. 17.

valuable to soil scientists looking for antecedents of their discipline. Stephen J. Gould warns against seeking heroes whose views conform with ours, while condemning other scientists as being wrong, narrow in views, or lacking objectivity. Gould has shown that this warning is particularly relevant to discussions of taxonomic and other classification systems that rely upon theoretical constructs. Theories of classification are products of their time and often make valuable contributions in some particulars and not in others. Gould writes: "Changes in theory are not simply the derivative results of new discoveries but the work of creative imagination influenced by contemporary social and political forces."[33] Latter-day soil scientists have had a tendency to regard Whitney's approach as an impediment and claimed he placed too great an emphasis on soil texture. Part of Whitney's motivation was to counteract what he saw as Justus Liebig's overemphasis on the depletion of soil minerals as the basis for decline in productivity of cropland. C.F. Marbut apparently believed this was Whitney's main contribution.[34] Whitney's charge from Congress was to use the study of soils to promote agriculture; soil textures and the geologic origin of the soil materials were the best-known and most convenient ways to carry out this charge at the time.

In 1912, G.N. Coffey, of the USDA, proposed and used a soil classification system based on soil genetic factors, somewhat similar to the genetic concepts of Dokuchaev.[35] He was influenced by the works and concepts of several Russian soil scientists, including Dokuchaev, K.D. Glinka, and Sibirtsev, and may have been influenced by the work of Hilgard in the United States.[36] However, this approach did not quite take hold in the United States and was overshadowed by the continuing concept of soil as primarily a geologic material. To his credit, Coffey was first in the United States to describe soils as independent natural bodies that should be classified according to their own internal properties, based on differences due to climatic, vegetational, and parent material differences.[37] However, the geological bias was still strong in his system. In his committee report, he stated that the light-colored forested soils should be separated from the dark, prai-

33. Stephen J. Gould, "On Heroes and Fools," in *Ever Since Darwin: Reflections in Natural History* (New York: Norton, 1977), p. 7.

34. C. F. Marbut, "Contribution of Soil Surveys to Soil Science," *Society Promotion of Agricultural Science Proceedings* 41 (1921): 116–142.

35. G. N. Coffey, *A Study of Soils of the United States* (Washington, D.C.: U.S. Govt. Print. Off., 1912 [USDA Bureau of Soils Bulletin no. 85]).

36. Simonson, "Lessons from the First Half Century of Soil Survey, I. Classification of Soils," *Soil Science* 74 (4) (1952): 249–57.

37. (a) Coffey, *A Study of Soils of the United States*. (b) Coffey et al., "Progress Report of the Committee on Soil Classification and Mapping," *Journal of the American Society of Agronomy* 6 (1914): 284–87.

rie soils at the highest category to make the classification and mapping similar to that in Russia.[38] The committee did not accept Coffey's recommendation.[39] The Coffey proposal was an idea whose time had not yet arrived, being developed only fifteen years after the American soil survey program, with its strong geological basis, had begun.

C. The Maturing of U.S. Soil Survey and Classification, 1910–1937

It remained for Marbut (director of the USDA Soil Survey Division, 1910–34) to introduce fully in the United States the idea of classifying soils according to their inherent characteristics related to genetic factors (climate, vegetation, time, landscape position), in addition to Whitney's parent material (nature of the geologic material from which they have formed). Marbut, a geologist by training with interests and research experience in geomorphology, was appointed as a soil scientist in the USDA Bureau of Soils soil survey program in 1909, after fifteen years as professor of geology at the University of Missouri. In 1910, Marbut was placed in charge of the federal soil survey program in the USDA Bureau of Soils. Thus started a distinguished twenty-five year career. It included the development of the first comprehensive scientific soil classification system. Marbut oversaw a great period of growth in soil surveys, and he placed much emphasis on soils as independent natural bodies owing their genesis to an array of soil-forming factors. Marbut described the basic soil map unit of the United States soil survey, the soil series, as a "group of soils having the same range in color, the same character of subsoil, particularly as regards color and structure, broadly the same type of relief and drainage and a common or similar origin."[40]

Although initially oriented to a geologically based soil classification and survey approach, Marbut was heavily influenced by reading and translating a book by the Russian soil scientist Glinka, *The Great Soil Groups of the World and Their Development*,[41] in which he described soil zones and soil types recognized by distinctive properties of their soil profiles.[42] Soil types

38. Coffey, "Progress Report of the Committee on Soil Classification and Mapping."

39. (a) Simonson, *Historical Aspects of Soil Survey and Classification*. (b) Coffey, "Progress Report."

40. Marbut, H. H. Bennett, J. E. Lapham, and M. H. Lapham, "Soils of the United States," *U.S. Bureau of Soils Bulletin* 96 (1913): 1–791.

41. K. D. Glinka, *The Great Soil Groups of the World and Their Development*, translated from German by Marbut (Ann Arbor, Mich.: Edwards Bros., 1927).

42. (a) Marbut, "Contribution of Soil Surveys to Soil Science." (b) Glinka, *The Great Soil Groups of the World*. (c) Glinka, *Die Typen der Bodenbildung, ihre Klassifikation un Geographische Verbreitung* (Berlin: Gebruder Borntraeger, 1914).

in the Russian system are roughly analogous to great soil groups in the nomenclature used in the United States.[43] Glinka was a student of Dokuchaev and introduced in soil science the concept of the soil profile, with its genetic layers, or horizons—the A,B,C formulation. Glinka's proposal to use the A-B-C notation was adopted in Russia just prior to the Second International Congress of Soil Science in 1930. The purpose was to have uniform conventions for descriptions of the profiles to be shown to visitors from other countries in the excursions conducted at the time of the congress. This information came from C. C. Nikiforoff, who was a student of Glinka.[44] Glinka relied on certain key properties of soil profiles to distinguish soil types.[45] This window to the Russian soil classification-soil survey principles and procedures influenced Marbut to use similar concepts and methods in developing a soil classification system for the United States soil survey.

In a statement prepared in 1928 for a monograph on the history of the U.S. Bureau of Chemistry and Soils, the history of United States soil survey to that time was described by Marbut as follows:

> Soil survey in the United States was started by Whitney. When Whitney started the mapping program in the United States, he was not aware of the Russian and German work with soil surveys. Because of Whitney's concept of soil texture as an over-riding soil property, the United States federal soil survey work was begun in 1899, on the basis of differentiating soils on basis of soil texture, mainly in the surface layer of soil. The initial approach was to differentiate soils in detail rather than by differentiation into broad groups and general characteristics. This initial soil survey work was done on basis of geological formations, grouping soils in a "province map"—based on geological rather than soil features. It was a little later concluded that soil province differentiation "had no scientific value." Information about the work of Russian soil scientists, especially Glinka, reached the United States in 1914, including descriptions of great soil groups of the world. From this it was concluded in the United States soil survey program that emphasis in soil survey should be on study of the soil itself rather than geologic materials below it.[46]

With wider understanding of Russian and other European soil studies, it became clear to Marbut and his colleagues that "investigation of soils in the field and the resulting determination of their anatomy and structure constitutes the absolute fundamental basis for the orientation for all other kinds of

43. (a) Glinka, *The Great Soil Groups of the World*. (b) Glinka, *Die Typen der Bodenbildung*.
44. Simonson, Personal communication.
45. (a) Marbut, "Contribution of Soil Surveys to Soil Science," p. 128. (b) Glinka, *The Great Soil Groups of the World*. (c) Glinka, *Die Typen der Bodenbildung*.
46. Weber, *The Bureau of Chemistry and Soils*, pp. 91–92.

investigations concerning the soil."[47] Marbut concluded his 1928 statement with the words "Soil science has become soil science with its own methods, its own point of view and generalizations based on its own facts."[48]

On the other hand, Marbut and colleagues made a wide distinction between American methods of soil mapping and that of Russia and other predecessors.[49] They claimed a unique creation—the development of the soil unit for detailed mapping purposes was a unique American creation. In 1921, Marbut wrote:

> The Russian workers have presented a well worked-out basis for the differentiation of soils, having done this by means of studies carried on over a wide area of the country without attempting to go into detail anywhere. It has remained for the American workers to determine the criterion on which the definition of the ultimate soil unit is based and by which it may be identified and mapped. The American work began with the determination of characteristics in detail, the Russian work with the determination of broad general characteristics persisting over large areas. The two groups in this work, therefore, have complemented each other. The American details will be fitted and are being fitted into the broad Russian scheme with such adjustments as the more thorough study of details show to be advisable.[50]

In 1922, Marbut laid out the morphological properties necessary for a soil profile to be classed as a distinct taxonomic and mapping unit.[51] These included number and kind of horizons in the soil profile; texture, structure, chemical composition, and the arrangement and thickness of the horizons of the soil profile; plus, the "geology of the soil material."

The resulting classification was used as a guide in the making of soil maps, which is illustrated in Marbut's 1935 landmark monograph *Soils of the United States*.[52] Soils were divided into pedalfers, with aluminum and iron as the main substances in the soil profiles, and pedocals, with calcium carbonate the main substance. This more detailed and final version was among Marbut's greatest contributions to soil science. It served as the underlying basis for United States soil surveys at detailed and generalized levels for many years.[53] It also served as the starting point and basis for later refinements. The Marbut system contained six categories: orders, sub-

47. Ibid., p. 94.
48. Ibid., p. 98.
49. Ibid.
50. Marbut, "Contribution of Soil Surveys to Soil Science," p. 136.
51. Marbut, "Soil Classification," *American Soil Survey Workers Report* III (1922): 24–32.
52. Marbut, "Soils of the United States," in *Atlas of American Agriculture, Part III*, ed. O. E. Baker (Washington, D.C.: U.S. Govt. Print. Off., 1935).
53. Glinka, *Die Typen der Bodenbildung*.

orders, great soil groups (approximately equivalent to Russian soil types), families, series, and types. These last two categories provided the basis for the detailed county soil surveys used for transferring research results and experiences from one location to another. Features of Marbut's 1935 classification which have not carried over to later classifications are the concept of mature, or normal, soils as a reference point and the recognition of only two classes, or taxa, at the order, or highest category, level. These two classes are the pedocals, in which calcium carbonate has accumulated as a result of limited leaching in less humid climates, and pedalfers, in which iron and aluminum compounds have accumulated under humid conditions.[54] There still is great use and acceptance of the concepts of great soil groups and subgroups (categories 5 and 4 of the Marbut scheme) and of the soil series (category 2). The former are widely used for generalized soil maps of large areas, whereas the series and the type are the basis, together with the phase, for map units of the standard county soil surveys. These are the resulting main units actually used in the United States now. Some modifications were made in developing the new soil classification system, *Soil Taxonomy*.[55] The other categories of the 1935 Marbut system were never completely defined, thus, there were, in effect, two classifications in the United States until adoption of *Soil Taxonomy* (e.g., one was based on great soil groups and the other on the series and types).[56]

Marbut died while on a soil study excursion in China in 1934.[57] Kellogg succeeded him as director of the USDA Soil Survey in 1935 and served as director of the USDA Division of Soil Survey from 1935 to 1971. Kellogg continued Marbut's emphasis on principles of soil genesis and properties of soils as the guides for classification and mapping of soils. He was primarily responsible for the modifications of the Marbut soil classification that were published in the 1938 USDA Yearbook, *Soils and Men*, including introduction of the Russian concepts of zonal, intrazonal, and azonal soils.[58] Kellogg also led the preparation of the *Soil Survey Manual*, a guide to the making of soil surveys.[59] Later, he was a prime mover in the decision to

54. G. D. Smith, "Historical Development of Soil Taxonomy—Background," in *Pedogenesis and Soil Taxonomy*, ed. L. P. Wilding, G. F. Hall, and N. E. Smeck (Amsterdam and New York: Elsevier, 1983), pp. 23–49.

55. USDA, Soil Survey Staff, *Soil Taxonomy*.

56. Wilding, Smeck, and Hall, eds., *Pedogenesis and Soil Taxonomy* (Amsterdam and New York: Elsevier, 1983).

57. Simonson, "Historical Aspects of Soil Survey and Classification. Part IV. 1931–1940," *Soil Survey Horizons* 27 (1986): 3–10.

58. M. Baldwin, Kellogg, and James Thorp, "Soil Classification," in *Soils and Men* (Washington, D.C.: U.S. Govt. Print. Off., 1938 [USDA 1938 Yearbook]), pp. 979–1001.

59. (a) Kellogg, *Development and Significance of the Great Soil Groups of the United States* (Washington, D.C.: USDA, 1936 [USDA Misc. Publ. no. 229]). (b) USDA, Soil Cons. Serv., *Soil Survey Manual* (Washington, D.C., 1951 [USDA Agriculture Handbook no. 18]).

prepare a new soil classification system, called soil taxonomy, which incorporated new information and replaced the concepts of soil zonality and intrazonality, pedocals and pedalfers.[60] It had not always been possible to distinguish clearly between pedocals and pedalfers or zonal and intrazonal soils over a range of conditions. Experience, recent research, and new ideas convinced proponents of a new classification system that the system should be based on quantitative measurement of soil properties related to soil genetic processes. Guy D. Smith, of the Soil Survey Division, was designated by Kellogg to take the lead in the effort to prepare the new soil classification system. The process involved many seminars, conferences, and work sessions; seven approximations, or drafts; and comments from soil scientists worldwide. This system, complete with a new nomenclature for all except the series at lowest level, was published in 1975 as *Soil Taxonomy*, a major milestone in soil survey.[61] Since that date, some changes have been made, including the recognition of a new order of andosols, soils formed from volcanic ash, and thereby differing in some key properties from other soils.[62]

Another landmark publication in this period was *Factors of Soil Formation* by Jenny.[63] Jenny defined and described, in a quantitative and analytical fashion, the five soil-forming factors which interact in varying degrees to produce the differing types of soils. Jenny further developed his ideas on soil formation in a more clearly depicted ecological setting in the book *The Soil Resource, Origin and Behavior*.[64] It, too, has had a strong impact on the research, analyses, and classification supporting soil surveys and mapping.

D. Two Soil Survey Schemes in the United States, 1938–1975

Following the establishment of the Soil Conservation Service, with H. H. Bennett as chief, in 1935, soil conservation surveys were performed on farms and ranches for which erosion control planning was requested. Some were also made in watershed protection areas and areas where reservoir sedimentation studies were being carried out. These soil conservation surveys included mapping of four factors: erosion, land use, slope, and soil type.[65] The land capability system was developed as an aid to those receiv-

60. USDA, Soil Survey Staff, *Soil Classification, A Comprehensive System, 7th Approximation* (Washington, D.C.: USDA Soil Cons. Serv., U.S. Govt. Print. Off., 1960).

61. USDA, Soil Survey Staff, *Soil Taxonomy*.

62. Buol, *Soil Genesis and Classification*.

63. Jenny, *Factors of Soil Formation: A System of Quantitative Pedology* (New York: McGraw Hill, 1941).

64. Jenny, *The Soil Resource, Origin and Behavior* (New York: Springer Verlag, 1980).

65. W. C. Boatright, *Erosion and Related Land Use Conditions on the Froid Demonstration Project, Montana* (Washington, D.C.: Montana USDA Soil Cons. Serv., 1938).

ing technical assistance. In this system, soils were placed in eight classes, according to the degree of erosion hazard and capability for varying degrees and types of land use and management practices. The class depended on the soil properties and the degree and length of the slope of the land.[66] These land capability surveys were a type of technical mapping made for a specific, applied purpose, thereby contrasting with the detailed standard soil surveys, which were based on a scientific soil classification system and which could be interpreted for a great number of purposes.

With the increase in availability of the standard soil survey maps and the merging of the Division of Soil Survey with the Soil Conservation Service in 1952, the soil conservation surveys were discontinued, and the standard soil survey maps were used for interpretation of land capability classes. Specific guidelines were developed for the assigning of the capability classes, and the county soil survey reports also began to carry a section on land capability classifications.

A significant landmark of soil surveys and classification in the United States was the publication of the *7th Approximation* to a new, comprehensive soil classification system in 1960.[67] Following trials and comments, including international participation, this approximation was replaced by *Soil Taxonomy* in 1975.[68]

Guy D. Smith, chief architect and principal author of *Soil Taxonomy*, has described the background, history, and rationale of the development of this system.

(1) The changing concepts of the soil series—due to the new knowledge about soil properties based on research in soil chemistry, soil physics and soil clay mineralogy. Farming practices were changing rapidly and adjustments in boundary lines and ranges of soil series were needed to make the knowledge transfer process and agricultural advisory system work (functions of a classification system) more effective.

(2) Need for increasingly more quantitative interpretations of soil properties—for farming, forestry, and rangeland and for nonagricultural uses of soils.

(3) The great expansion of soil survey activity and applications, including support of soil conservation programs.

(4) There was (and still an increasing need for generalization of soil properties and soil classification for several purposes, including nonagricultural uses of soils. However, the 1938 system did not lend itself well to these purposes.[69]

66. E. A. Norton, *Soil Conservation Survey Handbook* (Washington, D.C.: USDA, Soil Cons. Serv., 1939 [USDA Misc. Pub. no. 352]).

67. USDA, Soil Survey Staff, *Soil Classification*.

68. USDA, Soil Survey Staff, *Soil Taxonomy*.

69. Guy D. Smith, "Historical Development of Soil Taxonomy—Background," in *Pedogenesis and Soil Taxonomy*, ed. Wilding.

Several leading American pedologists, in addition to Smith, contributed significantly to the development of soil classification and soil survey concepts and procedures in the period from the publication of the 1938 classification to the publication of *Soil Taxonomy* in 1975. These were soil scientists affiliated with the USDA soil survey program and with the state agricultural experiment stations at the land-grant universities. (Several of these contributions were published in a special soil classification issue of *Soil Science*, vol. 67.) Simonson, of the USDA Division of Soil Survey, contributed studies of theories of soil genesis[70] and described the use of the soil series in the United States soil surveys.[71] He also described the lessons learned from the first half-century of soil survey and soil classification in the United States.[72] M. G. Cline, professor of agronomy at Cornell University, set forth some basic principles in soil classification.[73] The logic used in the development of *Soil Taxonomy* was also described by Cline,[74] as well as historical highlights in soil genesis, morphology, and classification during the period of installation of the new *Soil Taxonomy*.[75] F. F. Riecken, professor of agronomy at Iowa State University, with Smith, analyzed the lower categories of the 1938 soil classification system as part of the preparation for *Soil Taxonomy*.[76] Riecken also described the use of the 1938 soil classification in farming.[77] The higher categories of the 1938 classification system and its modifications, as a precursor to *Soil Taxonomy*, were discussed by James Thorp, USDA soil scientist, and Smith in a special issue of *Soil Science* in 1949.[78]

E. The Post-*Soil Taxonomy* Period

A study of the international usage and reception of *Soil Taxonomy* was conducted by Cline, who reported that, as of 1980, this system was used as

70. Simonson, "Outline of a Generalized Theory of Soil Genesis," *Soil Science Society of America Proceedings* 23 (1959): 152–58.

71. Simonson, "The Soil Series as Used in the U.S.A.," *Transactions of the 8th International Congress of Soil Science* 5 (1964): 17–24.

72. (a) Simonson, "Lessons from the First Half Century of Soil Survey: II. Mapping of Soils," *Soil Science* 74 (5) (1952): 323–30. (b) Simonson, "Lessons from the First Half Century of Soil Survey: I."

73. M. G. Cline, "Basic Principles of Soil Classification," *Soil Science* 67 (1949): 81–91.

74. Cline, "Logic of the New System of Soil Classification," *Soil Science* 96 (1963): 17–22.

75. Cline, "Historical Highlights in Soil Genesis, Morphology and Classification," *Journal of the Soil Science Society of America* 41 (1977): 250–54.

76. F. F. Riecken and Smith, "Lower Categories of Soil Classification: Family, Series, Type and Phase," *Soil Science* 67 (1949): 107–15.

77. Riecken and Smith, "Lower Categories of Soil Classification."

78. Thorp and Smith, "Higher Categories of Soil Classification: Order, Suborder and Great Soil Groups," *Soil Science* 67 (1949): 117–26.

the primary soil classification in support of soil surveys in twelve countries.[79] Soil scientists in nineteen other countries commonly used it for international communications. Some elements of its principles and concepts are being used in many other countries. It has caused adjustment from a qualitative to a quantitative approach in many places, but the system is still evolving, especially in respect to soils of the tropics.

International committees have been established by the U.S. National Co-operative Soil Survey, in cooperation and consultation with non-United States soil scientists and groups, to work on unfinished problems and new concerns arising from use of *Soil Taxonomy*. The Soil Management Support Services (SMSS) project of the USDA Soil Conservation Service was entrusted with the coordination of these international committees. The SMSS is an international soils project funded by the U.S. Agency for International Development.

These international committees and their chairpersons are[80]

International Committee on Soils with Low Activity Clays, F. R. Moorman, Netherlands, chairman (has completed its work);

International Committee on Oxisols, S. W. Buol, United States, chairman (has completed its work);

International Committee on Andosols, M. L. Leamy, New Zealand, chairman (completed its work before the untimely death of Leamy in 1990);

International Committee on Soil Moisture Regimes, Armand R. Van Wambeke, United States, chairman;

International Committee on Aridisols, A. Osman, Syria, chairman;

International Committee on Vertisols, J. Comerma, Venezuela, chairman;

International Committee on Soils with Aquic Soil Moisture Regimes, J. Bouma, Netherlands, chairman;

International Committee on Spodosols, T. Miller, United States, chairman;

International Committee on Soil Families, B. Hajek, United States, chairman.

These committees are composed of volunteers who are especially interested in the topic. They make recommendations to the Soil Survey Staff of the USDA Soil Conservation Service, which refers the recommendations to the Soil Taxonomy Committee of the Soil Science Society of America for additional review and comment. Review comments are also solicited from scientists around the world who have had experience with the soils under

79. Cline, "Experience with Soil Taxonomy of the United States," *Advances in Agronomy* 33 (1980): 193–226.

80. (a) Strzemski, *Ideas Underlying Soil Systematics*. (b) Buol, *Soil Genesis and Classification*. (c) Smith, *The Guy Smith Interviews: Rationale for Concepts in Soil Taxonomy* (Washington, D.C.: Soil Cons. Serv., 1986 [SMSS Technical Monograph no. 1]).

consideration. Comments are also received from the National Work Planning Conference of the National Cooperative Soil Survey.[81] The USDA SCS Soil Survey Staff then implements the consensus of the recommendations.

Three results of these committee studies have been executed in *Soil Taxonomy*. A new (11th) order of andosols has been formed for placement of soils which are formed mostly from volcanic ash and which have distinctive and unique properties, including the presence of large amounts of poorly crystalline nearly amorphous clay minerals. Studies of the low activity clays resulted in the establishment of new subgroups within the alfisols and ultisol orders to accommodate soils with clays of very low exchange capacities. These suborders incorporate the prefix element "kandic" to denote the presence of the kandic horizon, defined as possessing significant amounts of low-activity clays.[82] The oxisol order has been completely revised as a result of recommendations of the International Committee on Oxisols.

Additions to Soil Taxonomy

After opportunity for comment, amendments which have been accepted are announced in the USDA SCS *National Soil Taxonomy Handbook*. The amendments are also published in *Keys to Soil Taxonomy*.[83] The plan is to revise and reissue *Keys to Soil Taxonomy* about every two years, recording the amendments which have been made and which are maintained in the Soil Survey Division of the USDA Soil Conservation Service. Such amendments have, in the past, been published in *Soil Survey Horizons* (Winter 1983 and Winter 1986 issues), a publication of the Soil Science Society of America.[84]

F. Emerging Concerns and Problems

Among the concerns, dilemmas, and problems with which soil survey and classification specialists must deal in the future are

(1) There is a need for delimiting, defining, and quantitatively characterizing the basic soil body (now known as the pedon), which will reconcile this concept

81. (a) H. Eswaran, USDA Soil Cons. Ser., Div. of Soil Survey, Washington, D.C., Personal communication. (b) USDA Soil Cons. Serv., Div. of Soil Survey, Soil Management Support Services, Washington, D.C., Final Report to U.S. Agency for International Development, Washington, D.C.

82. Buol, *Soil Genesis and Classification*.

83. USDA, Soil Survey Staff, *Keys to Soil Taxonomy*, 4th ed. (Blacksburg, Va.: Crop and Soil Envir. Science Dept., Virginia Polytechnic Institute and State University, 1990 [SMSS Technical Monograph no. 19]).

84. Amendments and Revisions to Soil Taxonomy, Part I, Soil Survey Horizons Winter 1983; Amendments and Revisions to Soil Taxonomy, Part II, *Soil Survey Horizons*, Winter 1986.

with that of soils as a continuum on the landscape, rather than as a quilt-like pattern of independent natural bodies.

(2) What is the spatial variability of key soil properties and the relative impurity (quantitative amount of differing soil bodies within a delineated mapping unit)? How can this mapping requirement be reconciled with the need for precise site information about soils and high interpretative accuracy?

(3) What is the significance of the landscape position of a soil body with respect to its hydrology, erosional history, nutrient content, and the pedogenesis of its properties (all of which can affect the soil's productivity)?

(4) In the realm of environmental concerns, there is a need for greater understanding of the significance of soils as sources for greenhouse gases, which contribute to global warming, especially carbon dioxide, methane, and nitrous oxide. Conversely, we need to understand the effects of resultant global climate changes on soil properties and productivity.

(5) What is the degree of susceptibility to depletion by erosion among soil types, with potential permanent loss of productivity?

(6) A present and future concern with soils with a high erosion hazard is the extent of damage from off-site sedimentation in streams and harbors and on floodplains, as a result of soil erosion from agricultural fields.

(7) There is a need for improved methods of integrating soils data into geographic information systems, expert systems, and computer mapping programs.

(8) A major long-term concern is whether the regional and global supply of soils and their productivity will be sufficient to maintain a satisfactory food supply in the face of the rapidly growing world population.

These points of concern give rise to problems and tensions when applying soil classification systems to soil surveys. Classification systems help us organize our knowledge, remember properties, and transfer our experience and knowledge about soils. Yet, the variability and complexities of nature do not yield easily to our desire for neat classes and categories well expressed in soil map units. These difficulties surface when contrasting the basic soil body, or soil cover unit, concept, which is so useful in mapping, to the concept of a precisely and quantitatively defined discrete soil individual, which is ideal for classification purposes. Soil surveys need concepts of approximately homogeneous physical bodies of soil to establish and describe soil map units. Soil taxonomists need concepts of individual soils which can be aggregated to produce soil taxa—the soil classification units. These difficulties are compounded by the necessity of accepting the natural conditions of soil as a continuum on the landscape, rather than sets of discrete natural bodies.[85] There has been a search for a basic geographic unit of

85. R. W. Arnold, "The Future of the Soil Survey," *Future Developments in Soil Science Research*, ed. L .L. Boersma et al. (Madison, Wis.: Editorial Committee, Soil Sci. Soc. Am., 1987), pp. 261–68.

soil; at least nine proposals have been made according to R. W. Arnold.[86] Two different approaches have been used.[87] One method is to recognize broad landscape units, then subdivide them to produce elementary units of soil cover. A second approach is to start with elementary geographic units (smallest individuals complete in themselves), and recognize broad clusters and combinations of these.

The soil pedon and polypedon concepts (one of the nine approaches listed by Arnold) attempt to overcome these problems, in order to have definable mapping units for soil surveys and workable concepts of the basic soil individual for use in the classification process.[88] These concepts have been adopted for use in the National Cooperative Soil Survey.[89] The pedon is the smallest volume that can be called a soil, yet is large enough for recognition of all of the horizons present in the soil profile.[90] It may range from one to approximately ten square meters. The polypedon is a collection of contiguous pedons alike in the properties used to define a soil series.[91] Soil geographers study the soil continuum, or soil cover, and soil cover patterns. The Russian soil scientist V. M. Fridland described the soil cover pattern as "the general form of spatial distribution of soil expressed in a definite manner by repetition."[92] Fridland[93] and F. D. Hole and J. B. Campbell[94] described the elementary unit of soil cover as a natural soil body with its own natural boundaries. It differs from a soil map unit by containing no inclusions of similar soils. These concepts of the soil cover pattern and the soil body appear to be useful in geographic and geologic studies. The lack of purity (the high percentages of dissimilar pedons in some map units) diminishes the value of these map units and the polypedon concepts for soil surveys. However, such impurity seems to be a practical necessity in soil surveys.

86. Arnold, "Concepts of Soil and Pedology," in *Pedogenesis and Soil Taxonomy*, ed. Wilding pp. 1–21.

87. F. D. Hole and J. B. Campbell, *Soil Landscape Analysis* (Totawa, N.J.: Rowman and Allanheld, 1985).

88. (a) Boersma, ed., *Future Developments in Soil Science Research*, A Collection of SSSA Golden Anniversary Contributions Presented at the Annual Meeting, New Orleans, La., Nov.-Dec. 1986. (b) Arnold, "Concepts of Soil and Pedology."

89. (a) USDA, Soil Survey Staff, *Soil Taxonomy*. (b) USDA, Soil Survey Staff, *Soil Classification*. (c) Simonson and D. R. Gardner, "Concepts and Functions of the Pedon," *Transactions of the 7th International Congress of Soil Science* 4 (1960): 127–31. (d) W. M. Johnson, "The Pedon and the Polypedon," *Soil Science Society of America Proceedings* 27 (1963): 212–15.

90. USDA, Soil Survey Staff, *Soil Classification*.

91. Johnson, "The Pedon and the Polypedon."

92. (a) V. M. Fridland, *Struktura Pochvennogo Pokrova* (Moscow: Nauka, 1972). Trans. N. Kaner, *Pattern of the Soil Cover* (Moscow: Dokuchaev Inst. Acad. of Sci.; and Jerusalem: Israel: Israel Program for Scientific Translations, 1976). Available from U.S. National Tech. Info. Serv., Springfield, Va. (Original Russian version published in Moscow, 1972.) (b) Fridland, *Soil Combinations and Their Genesis* (New Delhi: Amerind, 1976), p. 43.

93. Fridland, *Pattern of the Soil Cover*.

94. Hole, *Soil Landscape Analysis*.

Spatial Variability and Purity of Soil Mapping

Concerns about the geographic purity of soil mapping units, the desire to improve their interpretive accuracy, and the precision of statements that can be made about them has recently turned attention to the spatial variability of mapping units.[95] The recent trend of increased quantification of soil information, and the desire to gain a better understanding of soil distribution patterns, has also drawn attention to the variability within soil mapping units. L. P. Wilding and L. R. Drees listed the kinds of information that measures of soil spatial variability will provide.[96]

Field checks of the taxonomic purity of mapping units of detailed soil maps indicate that it may range from 50% to 65%.[97] Later studies, after the adoption of *Soil Taxonomy*, with its quantitative definitions, found that mapping units representative of specified soil series seldom comprised more than 45–50% of the soils indicated in the mapping unit name.[98] P. H. T. Becket and R. Webster reported that the percentage of purity of soil mapping units ranges from about 50%, for soil series, to 75%, for soil orders.[99] However, F. P. Miller et al. believe that the taxonomic purity of the soil map units is not a proper measure of the quality or precision of a soil survey.[100] In the future, greater attention should be paid to interpretive accuracy of soil mapping units. Users of soil surveys need quantification of soil spatial variability and the assignment of confidence limits for soil mapping unit composition with respect to specific soil properties and soil performance.[101]

Because of these concerns, a number of geostatistical procedures have been adapted from mineral prospecting and from techniques for quantitative characterization of plant species present in vegetated areas. These include kriging, semivariograms and isarithmic mapping.[102] However, kriging and other geostatistical methods can be very costly and time consuming. Gener-

95. (a) Buol, *Soil Genesis and Classification.* (b) Wilding and L. R. Drees, "Spatial Variability and Pedology," in *Pedogenesis and Soil Taxonomy*, ed. Wilding, Smeck, and Hall, pp. 83–116.

96. Wilding, Smeck, and Hall, eds., *Pedogenesis and Soil Taxonomy.*

97. Wilding, R. B. Jones, and G. M. Schafer, "Variations of Soil Morphological Properties Within Miami, Celina, and Crosby Mapping Units in West-Central Ohio," *Soil Science Society of America Proceedings* 29 (1965): 711–17.

98. Wilding, Smeck, and Hall, eds., *Pedogenesis and Soil Taxonomy.*

99. P. H. T. Beckett and R. Webster, "Soil Variability: A Review," *Soils Fertilization* 34 (1971): 1–15.

100. F. P. Miller, D. E. McCormack, and J. R. Talbot, "Soil Surveys: Review of Data Collection Methodologies, Confidence Limits and Uses," in *The Mechanics of Track Support, Piles and Geotechnical Data* (Transactions of the Research Board, National Academy of Science, 1979 [Transaction Research Record 733]).

101. Miller, "Soil Survey Under Pressure: The Maryland Experience," *Journal of Soil and Water Conservation* 33 (1978): 104–11.

102. Buol, *Soil Genesis and Classification.*

ally, the technique is used only for special research studies.[103] R. B. Daniels and L. A. Nelson believe that a better understanding of the random (not predictable from current knowledge) and systematic (now partially predictable) spatial variability of soils must be reached before we can successfully apply statistical analyses to quantification of the soils composition of geographic areas.[104] They believe that the extent of unpredictability depends upon the researchers' understanding of processes, present and past, that have operated, or are now operating, on the soil landscape. Few studies have been designed to obtain data needed for an understanding of the soil variability. They attribute the current interest in the convenience of geostatistical methods, such as kriging, to the lack of understanding among many soil scientists of the basic causes of soil spatial variability. Many also fail to recognize the potential misapplication of geostatistical methods which were developed for interpolation between points. Daniels and Nelson propose sample surveys that measure soil properties and soil performance along transects in key areas. They urge a multidisciplinary approach to studies of soil spatial variability and of the relative variations in soil performance.

Soils Modeling and Systems Analysis to Meet Site-Specific Environmental Problems

Research to support soil surveys has generally concentrated on the effects of the soil-forming factors (climate, vegetation, general landscape position, parent material, and time) on internal soil-forming processes and properties, as well as the geographic distribution of the various types of soils.[105] Soil science has been well served by these activities. The functional factorial framework of state factors for predicting and analyzing origins of differing soils made possible the quantitative basis for *Soil Taxomony*.[106] The capability for quantification has enhanced the interpretive value of soil surveys for soil conservation programs and for increasing the efficiency of crop and forestry production. Simonson reminds us that many of these pedogenic

103. Ibid.
104. R. B. Daniels and L. A. Nelson, "Soil Variability and Productivity: Future Developments," in *Future Developments in Soil Science Research*, ed. Boersma et al., pp. 279–91.
105. (a) Buol, *Soil Genesis and Classification*. (b) Jenny, *Factors of Soil Formation*. (c) Jenny, *The Soil Resource, Origin and Behavior*. (d) Smeck, E. C. A. Runge, and E. E. Mackintosh, "Dynamics and Genetic Modeling of Soil Systems," in *Pedogenesis and Soil Taxonomy*, ed. Wilding, Smeck, and Hall, pp. 1–51. (e) R. B. Bryant and C. G. Olson, "Soil Genesis: Opportunities and New Directions for Research," in *Future Developments in Soil Science Research*, ed. Boersma et al., pp. 301–14.
106. USDA, Soil Survey Staff, *Soil Taxonomy*.

processes are simultaneously or sequentially in action in soils, and the resulting soil profiles thus reflect the relative strength of these pedogenic processes.[107]

Wilding et al. saw the "conceptual value" of these soil genesis models, but also perceived some concerns and needs.[108] They called for more adequate qualitative and quantitative data on soil-forming processes. The problems they cited with the currently available soil genesis models include: lack of continuity in their application; inconsistency with the basic principles of hierarchies, in that hypotheses, laws and principles developed at one level are not necessarily applicable at other levels; and these models are "conceptual in nature and cannot be rigorously tested experimentally."[109] Also, most of such models lack mathematical solutions.

R. B. Bryant and C. G. Olson saw an increasing need for more soil survey-genesis information to meet growing site-specific environmental problems, especially those caused by human activities.[110] Current models and analytical approaches do not provide all the needed answers. They cite examples of the need for soil system modeling techniques for solute transport and chemical equilibrium which have a high potential for studies of natural soil-forming processes.

Both sets of authors urge greater application of rigorous systems and simulation modeling concepts to soil analyses. They also cite urgent needs for more qualitative and quantitative soil data acquired by interdisciplinary groups. As pointed out by Bryant and Olson, these new kinds of data, techniques, and technologies should not supplant field studies of soil genesis and classification.[111] Instead, they will provide the setting and means for using these data and relationships for a fuller understanding of soils, hence better soil classification and soil surveys.

D. Holt has discussed application of that portion of the field of artificial intelligence known as knowledge-based or expert systems in soil science, including soil genesis and classification.[112] He sees three successive levels of sophistication in development of expert systems for soils. The first level is that of computer-based if-then rules: if a soil has a certain specific set of characteristics, then it falls into a certain taxon of a specific category. The second level consists of providing a computer with a number of examples,

107. Simonson, "Lessons from the First Half Century of Soil Survey."
108. Wilding, Smeck, and Hall, eds., *Pedogenesis and Soil Taxonomy.*
109. Ibid., p. 68.
110. Bryant, "Soil Genesis."
111. Ibid.
112. D. Holt, "Potentials for Artificial Intelligence and Supercomputers in Soil Science," in *Future Developments in Soil Science Research*, ed. Boersma et al., pp. 459–68.

then allowing it to infer its own decision rules, based on patterns it detects in the examples. Ralph J. McCracken and R. B. Cate studied the use of rule-based systems in soil classification and concluded that applications of principles of cognitive science and measurement theory will provide opportunities to improve soil taxonomic systems.[113] The third level of complexity described by Holt includes incorporation of one or more simulation models (similar to those described by Bryant and Olson[114] and Wilding et al.[115]) into a soil expert system. Usefulness of soil-related expert systems with included simulation models will be proportional to their closeness to reality, according to Holt.[116]

Significance of Soils in Landscape Positions

Arnold stated in his articles on the future of soil survey that "soil landscapes are significant hydrologic and pedogenic entities" and that "emphasis on soil properties of landscape positions and some combinations and predicting interactions with catenary units would change guidelines for the soil survey."[117] Daniels and Nelson indicate that the site history of how a soil's landscape position came to be, through geologic processes and hydrology, "[is] not fully appreciated by most soil scientists."[118] Some examples from research illustrate the influence of landscape position of soils: The geomorphology (landform and landscape position) of local areas has a major influence on the location and distribution of soils with specific properties.[119] Data on soil hydrologic properties from the United States Midwest show changes associated with distance from the summit of a landscape position.[120] In a survey of climate and topographic relations with respect to the morphology of United States Great Plains soils, C. W. Honeycutt et al. found that the depth of the clay maximum in soils increased with increasing precipitation on summit positions of the landscape.[121] Also, the depth of the

113. McCracken and R. B. Cate, "Artificial Intelligence, Cognitive Science and Measurement Theory Applied in Soil Classification," *Soil Science Society of America Journal* 50 (1986): 557–61.

114. Bryant, "Soil Genesis."

115. Wilding, Smeck, and Hall, eds., *Pedogenesis and Soil Taxonomy.*

116. Holt, "Potentials for Artificial Intelligence and Supercomputers in Soil Science."

117. Arnold, "Concepts of Soil and Pedology," p. 3.

118. Ibid., p. 3.

119. R. V. Ruhe, R. B. Daniels, and J. G. Cady, *Landscape Evolution and Soil Formation in Southwestern Iowa* (1967 [USDA Technical Bulletin no. 1349]).

120. D. D. Malo, B. K. Worcester, D. K. Cassel, and K. O. Matzdorf, "Soil-Landscape Relationships in a Closed Drainage System," *Soil Science Society of America Proceedings* 41 (1921): 116–42.

121. C. W. Honeycutt, R. D. Heil, and C. V. Cole, "Climate and Topographic Relations of Three Great Plains Soils. I. Soil Morphology," *Soil Science Society of America Journal* 54 (1990): 469–75.

clay increased from the summits to lower landscape positions, within a given climatic zone. They suggest that this greater downslope is due more likely to the hillslope erosional history or subsurface lateral flow than to any increase in precipitation at the lower ends of the slopes.

These examples tend to support the need for more attention to landscape position in soil surveys and classifications.

Soil Surveys, Classification, and Environmental Concerns

Two current environmental concerns present the need and opportunity for greater use of soil maps and soil classification. First, there is the possibility of reduced productivity in some soils should significant global warming occur. These soils are potentially significant sources of carbon dioxide as a "greenhouse gas", as soil carbon is converted to carbon dioxide. Some plant-soil systems can sequester carbon against emissions into the atmosphere. Soil carbon is the largest of the carbon pools interacting with the atmosphere.[122] Soil carbon in active exchange with the atmosphere constitutes approximately two-thirds of the carbon in terrestrial ecosystems.[123] Therefore, it is important to have information on the carbon content of various soils. Some of this carbon will be emitted to the atmosphere, contributing to the greenhouse effect if global warming should be significant. The amount will be determined by the total carbon present in the soil and the soil temperatures. Second is the question of soil erosion as a source of off-site sedimentation, whose detrimental effects and costs may be greater than those of on-site, on-farm soil erosion. In addition to the sediment itself, the soil particles may carry pollutants into the water.[124] Soils differ in their contribution to off-site sedimentation damages according to their erodibility and landscape location. Soil surveys and their interpretations can be quite useful in meeting these problems.

A new erosion prediction technology in the United States, the Water Erosion Prediction Program (WEPP), is being developed and will rely on local and national soil databases, including data from the National Cooperative Soil Survey and the SCS Pedon Data Base. Increased attention will be given to how parameter values in hydrologic and erosion equations are re-

122. A. F. Bouwman, ed., *The Role of Soils and Land Use in the Greenhouse Effect* (Chichester, U.K.: Wiley, 1990).

123. W. M. Post, W. R. Emanuel, P. J. Zinke, and A. G. Stangenberger, "Soil Carbon Pools and World Life Zones," *Nature* 298 (5870) (1982): 156–59.

124. (a) G. R. Foster and L. G. Lane, "Beyond the USLE: Advancements in Soil Erosion Prediction," in *Future Developments in Soil Science Research*, ed. Boersma et al., pp. 319–26. (b) E. H. Clark, II, "The Offsite Costs of Soil Erosion," *Journal of Soil and Water Conservation* 49 (1985): 19–22.

lated to fundamental properties of soils as provided by the soil survey program.[125]

S. W. Buol et al. used soil survey data to compute soil mean annual temperatures and the mesic-frigid soil temperature boundary.[126] They then predicted shifts in cropping patterns in the temperate regions of the Northern Hemisphere at a three-degree carbon increase in mean annual temperature, due to global warming, and adjusted the present mesic-frigid soil family boundary northward, on the three-degree basis. The results suggest a 26% decrease in the area of corn belt land in the United States. The wheat production area in the USSR was projected to increase 142%, in Canada, 46%. No significant increases in wheat area were projected for the United States.

Soils Data for Future World Food Needs

The maintenance of production and the environment over time (sustainability) is mostly dependent upon soil resources and climate.[127] B. A. Stewart et al. believe that careful analysis of these two factors can provide significant information for formulating guidelines and policies. From the beginnings of agriculture until about 1950, increased food production came mostly from expanding the crop land. The large increases in production per unit area of land from major crops between 1950 and 1985 are mostly due to a 6.9 times increase in energy used (fertilizers, fuel, machinery). Also, the world's irrigated area increased from 94 million hectares, in 1950, to 271 million, in 1985. The current annual increase in irrigated area is less than 1%. These authors present information on area, climate, and food production constraints of the six main soil orders, which are the main food producers about which significant information is available.

Total area of all orders of *Soil Taxonomy* are reported by Stewart et al. in Table 11.1 below (andosols from volcanic ash are a recent addition and are not computed).

These authors cite the mollisols as being the most potentially productive, with fewer constraints on achieving sustainability, and alfisols as next in potential productivity, but with more constraints. Vertisols are important in semiarid and arid areas, but have many constraints on sustained production.

125. Boersma, et al., eds., *Future Developments in Soil Science Research*.

126. Buol, P. A. Sanchez, J. M. Kimble, and S. B. Weed, "Predicted Impact of Climatic Warming in Soil Properties and Use," in *Impact of Carbon Dioxide Trace Gases and Climate Change on Global Agriculture* (Madison, Wis.: American Society of Agronomy, 1990 [Special Publication no. 53]).

127. B. A. Stewart, R. Lal, S. A. El-Swaify, and Eswaran, "Sustaining the Soil Resource Base of an Expanding World Agriculture," *Proceedings of the 14th International Congress of Soil Science* 7 (1990): 296–301.

Table 11.1. Total area of all orders of *Soil Taxonomy*

Soil order	Area, billion hectares (ha)
Alifisols	1.71
Aridisols	2.44
Entisols	1.11
Histosols	0.12
Inceptisols	1.17
Mollisols	1.11
Oxisols	1.11
Spodosals	0.56
Ultisols	0.78
Vertisols	0.23
TOTAL	9.28 billion ha

Source: B. A. Stewart el al., "Sustaining the Soil Resource Base of an Expanding World Agriculture," *Proceedings of the 14th International Congress of Soil Science* 7 (1990): 296–301.

The soil order with the most production constraints on potential productivity is the aridisol taxon of the arid regions.

The area of potentially arable soils in the world was estimated as 3.1 billion hectares by Kellogg and A. C. Orvedal, and the total soil area was estimated as 13.16 billion hectares.[128] These estimates were made by use of soil maps of the world and estimating the suitability of the soils for cultivation based on their descriptions and properties. P. Buringh estimated that the total area of cultivable soils was 3.3 billion hectares, of which 1.5 billion was under cultivation at the time.[129] These estimates were derived by study of soil maps prepared by the Food and Agriculture Organization and by various countries.

The world soil inventory, or surveys, were only about one-fourth completed at a level of detail adequate for analyses of productivity level as of 1979, therefore, one must be cautious in predicting the sufficiency of soils to support the growing world population.[130] Buringh called for completion of soil surveys of potential agricultural land as soon as possible.[131]

128. Kellogg and A. C. Orvedal, "Potentially Arable Soils of the World and Critical Measures for Their Use," *Advances in Agronomy* 221 (1969): 109–70.
129. P. Buringh, "The Capacity of the World Land Area to Produce Agricultural Products," in *Organic Production: The Relationship Between Agricultural and Natural Vegetation Production Rates,* ed. Frank B. Gollev and June H. Cooley. (Athens, Ga.: International Association for Ecology, 1985), pp. 9–13.
130. Armand Van Wambeke, "Land Resources and World Food Issues," in *World Food Issue* (Ithaca, N.Y.: Center for Analysis of World Food Issues, NYS College of Agriculture and Life Science, Cornell University, 1984).
131. Buringh, "The Capacity of the World Land Area."

Buol and Pedro A. Sanchez concluded that there is ample area of infertile soil that can be converted to high levels of food production by fertilizer use, but infrastructure, including dependable markets, is needed to convince farmers that the risky investments have a reasonable chance of paying off at harvest time.[132] F. H. Sanderson[133] and K. R. Farrell et al.[134] projected that most of the future increases in food production will be due to improved technology and increased production of currently cultivated soils, rather than from bringing additional soils not now in cultivation into production.

G. The Content of Soil Maps

As previously described, a soil survey is a systematic field study of the soils of a designated geographic area, including the recording of kinds of soils present and showing their distribution on a map. The kinds of classified soils are identified on the maps by symbols, and their locations are delineated by drawing boundaries around them. The legend correlates the map symbols with the soil class name. Each kind of soil displayed has a set of unique characteristics, or properties, as a result of its being formed as a natural body by a specific array of soil-forming properties. Thus, a soil map shows areas occupied by different kinds of soils.[135] If the soil map is part of a program of systematic study of a given area, it is ordinarily incorporated in a soil survey report or accompanied by a technical or scientific paper presenting information on the properties of and interpreting use suitabilities for the soil classes shown on the map.

For cartographic purposes, separate and apart from the classification and taxonomic aspects, a soil map unit may be considered a collection of areas defined and named in terms of or in relation to their soil taxa components.[136] Map units bearing the same identification symbol on the map differ in some way from all others shown on the map. The kinds of units used to identify and map soils depend on the purposes of the soil survey program, the soil pattern, and the scale of mapping.

132. Buol and Sanchez, "Soils of the Tropics and the World Food Crisis," *Science* 188 (4188) (1975): 598–603.

133. F. H. Sanderson, "An Assessment of Global Demand for U.S. Agricultural Products to the Year 2000: Economic and Policy Dimensions," *American Journal of Agricultural Economics* 66 (5) (1984): 577–84.

134. K. R. Farrell, F. H. Sanderson, and T. T. Vo, "Feeding a Hungry World" (Washington, D.C.: Resources for the Future, 1984), *Resources* 76: 1–20.

135. USDA, Soil Survey Staff, *National Soil Taxonomy Handbook*.

136. Ibid.

H. Generalized Types of Soil Maps

Derived Soil Maps. Though most soil maps published by the U.S. National Cooperative Soil Survey, with the USDA Soil Conservation Service as lead agency, are based on field investigations and direct firsthand placement of soil boundary lines by field soil surveyors, there are circumstances in which it is necessary to locate the soil map units other than by direct observation by the soil surveyor. These derivations are generalized and schematic soil maps.[137]

Schematic maps are made from knowledge of effects on soil properties by soil-forming factors, not by field soil surveyors.[138] They may be derived from previously existing geologic, vegetation, and climatic maps, or by inferences about soil properties based on knowledge of the impacts of various climates, parental materials, and landscape position over time and in varying landform situations.[139] These maps are generally of a regional nature and are estimates of the patterns of soil distribution.[140]

Generalized maps are also designed to show broad regional views at smaller scales than the primary soil maps from which they are derived.[141] These generalized maps are prepared either by graphic generalization, which retains the original detailed legend, or by taxonomic generalization, in which boundaries are removed between neighboring delineations that are taxonomically similar.[142] Buol and colleagues[143] present several examples of generalized soil maps, which are usually regional, continental, and global.[144]

I. Base Maps for Soil Map Preparation

The quality of the cartographic base material affects the accuracy of map unit boundaries and soil identification, rate of progress of the field surveys,

137. (a) Buol, *Soil Genesis and Classification.* (b) USDA, Soil Survey Staff, *Soil Taxonomy.* (c) Hole, *Soil Landscape Analysis.*

138. (a) Hole, *Soil Landscape Analysis.* (b) USDA, Soil Survey Staff, *National Soil Taxonomy Handbook.*

139. Orvedal, M. Baldwin, and A. J. Vessel, "Soil Classification and Soil Maps: Compiled Maps," *Soil Science* 67 (1949): 177–81.

140. Hole, *Soil Landscape Analysis.*

141. Ibid.

142. Ibid.

143. Buol, *Soil Genesis and Classification.*

144. (a) Glinka, *Die Typen der Bodenbildung.* (b) R. Dudal, *Definitions of Soil Units for the Soil Map of the World* (Rome: FAO, 1968 [World Soil Resource Report no. 33]). (c) I. P. Gerasimov, *Soil Map of the World: Approaches to Soil Classification—World Soil Maps Compiled by Soviet Soil Scientists* (Rome: FAO/UNESCO, 1968 [World Soil Resource Report no. 32]). (d) USDA, Soil Cons. Serv., *Distribution of Principal Kinds of Soils: Orders, Suborders and Great Groups* (Reston, Va.: U.S. Geol. Surv., 1967 [Sheet 85, National Atlas of the USA]).

and the costs and quality of the final published map. At present, in the National Cooperative Soil Survey, aerial photographs are the most common mapping base for soil surveys. The cost of these airphotos is about 1–2% of the total soil survey cost, if extant photography can be used, and about 5%, if new aerial photography must be flown for the survey.

Prior to the use of aerial photography as a soil mapping base, planimetric maps were prepared in the field by the soil surveyors, as the soil mapping proceeded, using plane table, alidade, compass, and odometers, attached to buggy wheels in the early days and later to cars or pickup trucks. A very graphic description of the trials and tribulations of the plane table methods for preparing base maps is presented by W. D. Lee.[145] Among the most significant advances in soil survey was the introduction of airphotos as base maps.[146] Airphotos replaced the old planimetric map method starting in 1923. By 1935, it was the standard in the United States.[147] The first known use of airphotos in the early 1920s for soil mapping was by William Battle Cobb, a professor of agronomy at what is now North Carolina State University. Tom Bushnell, professor of agronomy at Purdue University, also pioneered the use of airphotos for soil surveys and apparently was responsible for the first county soil survey in the United States prepared with the use of airphoto base maps. This work was done in Jennings County, Indiana, in 1929 and 1930.[148] The American Soil Survey Association endorsed the use of airphotos in soil mapping in 1930.[149]

The use of airphotos as the base for published soil maps contained in county soil survey reports did not become common until about 1956, due to costs and extensive cartographic adjustments required for this technology. The development of technology for using airphotos as base maps for published detailed soil maps proceeded rapidly after 1956.[150] Controlled mosaic airphoto bases were the main means for soil map publication for a number of years. This initial use of controlled airphoto mosaics was abandoned in the late 1970s for more satisfactory procedures. Rectified enlarged air pho-

145. Lee, *The Early History of Soil Survey in North Carolina.*

146. (a) W. U. Reybold, "The Use of Aerial Photography in the U.S. National Cooperative Soil Survey Program," USDA Soil Cons. Serv., Washington, D.C., Unpublished report. (b) Simonson, "Use of Aerial Photographs in Soil Surveys," *Photogrammetric Engineering* 16 (1950): 308–15.

147. (a) Simonson, "Use of Aerial Photographs in Soil Surveys." (b) D. R. Gardner, "The National Cooperative Soil Survey of the U.S.," Thesis, Grad. School Public Admin., Harvard University, 1957.

148. (a) Tom M. Bushnell, "Aerial Photography and Soil Survey," *American Soil Survey Association Bulletin* X (1929): 23–28. (b) Bushnell, "A New Technique in Soil Mapping," *American Soil Survey Association Bulletin* XIII (1932): 74–81.

149. T. D. Rice, "Report of the Committee on Resolutions," *American Soil Survey Association Bulletin* XII (1931): 197–98.

150. Reybold, "The Use of Aerial Photography."

tos have been used for publication of areas of low relief, with no hills or mountains. The use of orthphotoquads is especially important for the rapidly growing practice of digitizing the soil maps for computer storage and retrieval.

By the late 1950s, aerial photographs of most of the agricultural areas of the United States were used in monitoring federal government crop allotments and in soil conservation programs. These airphotos were mainly at a scale of 1:20,000. In the late 1960s, the USDA Soil Conservation Service initiated a program of smaller scales (e.g. 1:24,000 feet). The U.S. Geological Survey and the USDA, in the early 1970s, started using the more satisfactory high-altitude photography for airphoto bases for field mapping. In 1980, a national program of high-altitude photography was begun to coordinate acquisition of airphoto coverage of the conterminous United States. This program used two cameras, one producing photos at 1:58,000 scale with color infrared film, and the second producing orthphotoquad photography at 1:80,000 scale. Field soil survey parties are furnished black-and-white reproductions for their field mapping use and alternate color infrared prints for use in airphoto interpretations and stereo viewing for premap preparation for the field surveys.[151]

In 1986, the National High Aerial Photography (NHAP) program was revised to a one-camera, 1:40,000 scale program with a choice of using either color infrared or black and white film. The larger scale provided the detailed additional image resolution desired by soil surveyors and soil conservation planners. The new program was named the National Aerial Photography Program (NAPP).[152]

Timing of aerial photography is important for soil mapping. The best periods for aerial photography for soil mapping are when trees are leafless and wet soils are drying, because soil and landscape distinctions are clearer then. B. D. Kunze and G. D. Lemme describe in some detail types and uses of aerial photography for soil surveys.[153] Procedures and techniques for use of airphotos in soil surveys are presented in the *Soil Survey Manual*,[154] *Guide for Soil Map Compilation on Photo Base Map Sheets*,[155] the *Soil Conservation Service National Soils Handbook*,[156] and in *USDA Agriculture*

151. Ibid.
152. G. W. Teselle, director, Cartography and Geographic Information Systems Division, USDA SCS, Washington, D.C., Personal communication.
153. B. O. Kunze and G. D. Lemme, "Photography for Order 2 Soil Surveys," *Soil Survey Horizons* 27(3) (1986): 10–17.
154. USDA, Soil Survey Staff, *Soil Survey Manual*.
155. USDA, Soil Cons. Serv., *Guide for Soil Map Compilation on Photobase Map Sheets* (Washington, D.C.: USDA Soil Cons. Serv., 1970).
156. USDA, Soil Cons. Serv., *National Soils Handbook* (Washington, D.C., 1983).

Handbook 294: Aerial Photo Interpretation in Classifying and Mapping Soils.[157]

Types of aerial photography now available for use in soil mapping are panchromatic, color, infrared, color infrared, and both high-altitude and low-altitude photography.[158] Remote sensing by earth-orbiting satellites and high-flying planes is being investigated for use in producing base maps for soil surveys, and as a complementary aid to soil identification for mapping.[159] Digital elevation data and remotely sensed Landsat and other spectral data have been processed into slope, aspect, elevation, and spectral class maps, which can then be combined with a digitized soil "premap" to produce computerized statistics to aid in producing field soil survey maps.[160]

The USDA Soil Survey Division has recently agreed to a policy that requires the use of orthophotography made from NAPP imagery for all future soil survey base maps, using primarily scales of 1:12,000 and 1:24,000.[161]

J. Soil Classification and Soil Maps

Soil cover, the soil continuum or soilscape, is the succession over space of bodies of differing soils together with bodies which are not soil.[162] Taxonomic units (or taxa) make up the classes in a multiple-category classification system: soil orders, suborders, great groups, subgroups, families, and series are used to identify groups of taxa at varying levels of generalization (from highest to lowest, in the groups listed here).[163] The pedon is the space-occupying representation in nature of a soil taxon and is the minimum sampling and study unit of soil taxa, averaging a meter or two square. A polypedon is a taxonomic landscape unit composed of two or more contig-

157. USDA, Soil Cons. Serv., *Aerial Photo Interpretation in Classifying and Mapping Soils* (Washington, D.C., 1966 [USDA Agriculture Handbook no. 294]).

158. (a) Simonson, "Historical Aspects of Soil Survey and Classification." (b) Ramann, *Bodenkunde.*

159. (a) Simonson, "Historical Aspects of Soil Survey and Classification." (b) Ramann, *Bodenkunde.* (c) M. F. Baumgardner, "Remote Sensing for Resource Management: Today and Tomorrow," in *Remote Sensing for Resource Management*, ed. C. J. Johannsen and J. L. Sanders (Ankeny, Iowa: Soil Conservation Society of America, 1982), pp. 16–29. (d) J. R. Anderson, "Land Resources Map Making from Remote Sensing Products," in *Remote Sensing for Resource Management*, ed. Johannsen and Sanders, pp. 63–67.

160. E. H. Horvath, E. A. Fosnight, A. A. Klingebiel, D. G. Moore, and J. E. Stone, *Using a Spatial and Tabular Database to Generate Statistics from Terrain and Spectral Data for Soil Surveys* (Madison, Wis.: Soil Science Society of America, 1987), pp. 91–98.

161. McCracken, "Artificial Intelligence."

162. Hole, *Soil Landscape Analysis.*

163. Ibid.

uous pedons, a taxonomically homogeneous unit at the series level in *Soil Taxonomy*.[164] A soil mapping unit is a delineation, or area, within a boundary on a map which is used to represent the locations of polypedons on landscapes.[165] Van Wambeke and T. R. Forbes describe a soil mapping unit as "an aggregate of all soils that are identified by a unique symbol, color or name or other representation on a map."[166]

In the real outdoor world, it is possible for a particular soil landscape to be composed of pedons belonging to more than one taxon, due to complex geologic parent material and local variability in other soil-forming factors. If differing pedons are side by side on a landscape, a taxonomic boundary would divide this landscape unit into two sets of pedons, each one assigned to different soil taxon. If different pedons are intermixed on a given landscape, due to local variations in one or more of the soil-forming factors, the result is a complex pattern of differing pedons.

In soil mapping, different types of mapping units are recognized; each type is given one of the following names:[167]

Consociations—Mapping units made up mainly of phases (such as differing slope classes or degrees of erosion) of a single taxon.

Complexes—Soil mapping units composed of two or more different and contrasting taxa in a recurring pattern on the local soilscape so complex this distribution cannot be shown at the map scale used.

Associations—Composed of mapping units in which contrasting taxa are present, as in a complex, but the patterns of these differing soils are coarse and can be resolved at map scales commonly used in standard county soil surveys. The association is commonly used in defining and describing mapping units for small-scale generalized maps.

Taxadjuncts—These consist of polypedons with soil properties outside the range of any recognized taxa at the soil series level (lowest taxonomic level), but with use and management requirements differing only slightly from those of established soil series taxa. This general topic is discussed for a different perspective in a later section describing some of the emerging problems and concerns about local spatial variability of soils and purity of soil mapping.

164. USDA, Soil Survey Staff, *Soil Taxonomy*.

165. Buol, *Soil Genesis and Classification*.

166. Van Wambeke and T. R. Forbes, *Guidelines for Using "Soil Taxonomy" in the Names of Soil Map Units* (Washington, D.C., 1986 [USDA Soil Cons. Serv., Soil Management Support Service, Technical Monograph no. 10]).

167. (a) Hole, *Soil Landscape Analysis*. (b) USDA, Soil Survey Staff, *Soil Taxonomy*.

K. Landmark Soil Maps

Soil surveys are field studies of the soils of a specific area, including soil maps, reports, or other publications, which give information about land use potential, agricultural forestry, conservation, community and regional planning, land valuation, and nonagricultural land uses. These surveys can be related to one of five classes, or orders, of intensity, according to the degree of detail, map scale and relative suitability for one or more of the land uses mentioned. A first order soil survey is one with maps of very large scale for highly detailed planning for specialty crops, very intensive nonagricultural land use, or research plots. Fifth order surveys have maps of very small scale and are highly generalized and suitable only for very broad and nonintensive land use planning or agricultural uses. Second, third, and fourth order surveys fall between these extremes.

Other categories of soil maps and their accompanying reports are those for detailed use at county or parish levels; general regional maps with smaller map scales and generalized soil maps; those which are suitable for regional multistate use; and those which are for use at country, continental, and global levels. The county and parish maps are the standard soil maps in the United States, usually at order two or three level, done by the National Cooperative Soil Survey. Scales may range from 1:12,000 to 1:31,680; all current issues are airphoto base maps, but the older ones are on planimetric bases.

Landmark County Soil Surveys and Maps

The following are landmark maps at the county or parish level on the older planimetric bases which have relatively small scales (1:50,000 and higher):

Tama County, Iowa—series 1938, and published in 1951 by Division of Soil Survey, USDA. Map scale 1:63,360, planimetric base. Illustrates soil patterns of mostly mollisols in the heart of the United States corn belt. Among the first to give detailed soil management and conservation practices by phases of soil series.

McKenzie County, North Dakota—series 1933, published 1942. Map scale 1:63,360. In area of more ustic (drier) mollisols than Tama County map. Wheat production and cattle ranches. Among the first in the United States used as the basis for rural land tax assessment which displayed soil productivity ratings and provided a native plant list.

Benton County, Tennessee—completed in 1941. Main soils are ultisols and alfisols. Was one of the first uses of soil surveys for regional agricultural and nonagricultural land use and development programs, as part of the Tennessee Valley (TVA) regional agricultural development program. Financed in part by TVA.

More recent county and parish soil surveys, published on airphoto bases, with larger scales, which were among the first to reflect the use of *Soil Taxonomy* in soil classification as recommended by the USDA Soil Survey Staff (1984) are

Franklin Parish, La. (1981)—Maps and soil classification in an area with many entisols.

Rawlins County, Kans. (1981)—Maps in areas of mollisols.

Victoria County, Tex. (1981)—Soil maps in areas of vertisols and mollisols.

All of these maps contain the relatively new feature of data and interpretations on the uses of soils for nonagricultural purposes.

Landmark United States State and Regional Soil Maps

According to G. A. Weber,[168] the first map of any U.S. state which displayed the nature and location of the soils was a geologic map of Massachusetts.[169] The first state soil map which showed the distribution of soils classified according to their physical characteristics[170] was prepared by Whitney and associates and published by the U.S. Weather Bureau.[171] In this map and accompanying publication, Whitney emphasized the importance of soil texture and other physical characteristics and related them to moisture availability for crop plants. This publication had at least two consequences: it led Whitney into a career-long controversy with other soil scientists over the relative importance of soil moisture related to texture, compared to the importance of plant nutrient content of soils; and it led to the establishment by the USDA of a permanent organization on the mapping and study of soils. Some examples of older soil maps of states and regions:

Soil Survey of Hawaii (Cline and party)—Completed in 1939 and published in 1946. Soils predominantly oxisols. The first to show grouping of mapping units into taxonomic families, and among the first to display extensive laboratory data of the main soils. Used as the basis for developing revised classification of tropical soils from basic rocks and for improvements in the knowledge of the genesis of these soils.

168. Weber, *The Bureau of Chemistry and Soils.*
169. E. Hitchcock, *Final Report on the Geology of Massachusetts*, Massachusetts Geological Survey (Northampton, Mass.: J. H. Butler, 1841).
170. Weber, *The Bureau of Chemistry and Soils.*
171. Whitney, *The Physical Properties of Soils.*

Table 11.2. Information on scales of soil maps, sizes of real polypedons, soilscapes, and map units representing them

Intensity of soil survey (Smyth 1977) (map scale)	Representative map scales and corresponding area (ha of minimum size [0.4 cm² on the map]) of delineation (Eswaran et al. 1977)		Usual ranges of areas (ha) of soil bodies and common map scales at which they are delineated (Hole 1978)		
	(scale)	(ha)	Polypedon (ha)	Multi-poly-pedonic units (ha)	<1:50,000,000 (scale)
	1:5,000,000	100,750	—	>300,000,000 (global units)	<1:50,000,000
Syntheses (Smaller than 1,000,000)				250,000,000–300,000,000 (continental units)	1:10,000,000–1:50,000,000
	1:1,000,000	4,030	—	250,000,000 (megasociations)	1:1,000,000–1:10,000,000
Exploratory (1:250,000– 1:1,000,000)	1:500,000	1,008	—	250,000–2,500,000	1:300,000–
	1:250,000	252	—	(macrosociation)	1:1,000,000–
Low intensity (1:100,000– 1:250,000)	1:200,000	161	—	2,500–250,000	<1:100,000–
	1:100,000	40.3	—	(mesociation)	1:300,000

Intensity	Scale		Polypedon size		Scale range
Medium intensity (1:25,000–1:100,000)	1:50,000	10.1		>2,500 (microsociation)	>1:100,000
	1:25,000	2.52	>100 (megapolypedon)	—	1:32,000–1:62,500
High intensity	1:20,000	1.61	10–100 (macropolypedon)		1:16,000–1:32,000
(1:10,000–1:25,000)	1:10,000	0.40	1–10 (mesopolypedon)		1:8,000–1:16,000
Very high intensity (larger than 1:10,000)	>1:10,000	<0.40	0.1–1 (micropolypedon)		1:4,000–1:8,000
			<0.1 (nanopolypedon)		Larger than 1:4,000

Source: S. W. Buol, *Soil Genesis and Classification* (Ames: Iowa State University Press, 1989). By permission of the publisher.

Soil Survey of Puerto Rico (R. C. Roberts and party)—Completed in 1936, published in 1942. Scale 1:50,000, planimetric base. Uses classification of these tropical soils prior to the development of *Soil Taxonomy*.

Soils of the Southern States and Puerto Rico (1975)—Published by Agricultural Experiment Stations of the Southern States and Puerto Rico. Southern Cooperative Service Bulletin 174 (Buol, ed.) Illustrates the use of the higher categories of *Soil Taxonomy* to produce a meaningful map.

Soils of the Great Plains (1982, A. R. Aandahl, University of Nebraska Press, Lincoln, Neb., 282p.)—Portrays predominantly mollisols classed in the higher categories of *Soil Taxonomy*, and much about the use, management, and productivity of these soils.

Distribution of Principal Kinds of Soils: Orders, Suborders and Great Groups (Soil Conservation Service, 1967. Sheet 85, *National Atlas of the United States,* A. Gerlach, ed., published by U.S. Geological Survey, Washington, D.C.)—Reflects the use of the new *Soil Taxonomy* and contains much more information than the 1935 United States soil map.

C. F. Marbut, 1935,"Soils of the United States, Part III," in *Atlas of American Agriculture,* USDA, Washington, D.C. Maps reflect the soil classification system devised by Marbut.

Landmark Global Soil Maps

S. W. Buol, F. D. Hole, and R. J. McCracken, *Soil Genesis and Classification,* 3d ed., Ames, Ia.: Iowa State University Press, 1989. 446p. Map on pp. 350–353. Contains 1:50,000,000 scale map of the soils of the world using language and concepts of *Soil Taxonomy.* Shows global distribution of world soils by orders and suborders, based on information from World Soil Geography Unit, USDA Soil Conservation Service, Washington, D.C.

I. P. Gerasimov, "World Soil Maps Compiled by Soviet Soil Scientists," in *World Soil Resources Reports 32* (Rome; FAO, 1968); pp. 22–77. Reflects Russian system of soil classification, with emphasis on genesis of the soils.

R. Dudal, "Definitions of Soil Units for the Soil Map of the World," in *World Soil Resources Reports 33* (Rome; FAO, 1968). Reflects hybridization of concepts and terminology of *Soil Taxonomy,* with classifications used by Russians and other European countries. Atlas sheets contain the maps of the various parts of the world. Also see Dudal, *Key to Soil Units for the Soil Map of the World,* Vol. 1, Legend (Rome: FAO/UNESCO, 1974).

C. E. Kellogg and A. C. Orvedal, "Potentially Arable Soils of the World and Critical Measures for Their Uses," *Advances in Agronomy* 21 (1969):

109–170. Contains small-scale map of the world, using concepts and terminology of *Soil Taxonomy*.

H. Eswaran, N. Bliss, D. Lytle, and D. Lammers, *The 1:30,000,000 Map of "Major Soil Regions of the World"* (Washington, D.C.: U.S. Geological Survey, in press). This soil resources of the world map is based on a FAO map which was completed in 1990. Map units are based on suborders of *Soil Taxonomy* and soil temperature regimes. Map is equal area projection. Total area of each soil unit is presented.

Information on the scales of soil maps, the sizes of polypedons, and the map units representing them is shown in Table 11.2. Note that only those soil maps of medium or higher intensity are suitable and practical for detailed farm level and small site nonagricultural and environmental interpretations.[172]

172. Buol, *Soil Genesis and Classification*.

12. Major Soil Maps of the World

PETER McDONALD

Cornell University

Quantitative and qualitative appraisals of soil resources on a global basis
have engaged the minds of soil specialists from the beginning of the century. But it was during the 1950s and early 1960s that researchers in the
field moved toward a comprehensive soil map of the world in order to
correlate soil units from different regions and compare them on a global
scale. Following a recommendation of the International Society of Soil Science, a joint FAO/UNESCO project was undertaken in 1961 to prepare a
soil map of the world.

A. FAO/UNESCO *Soil Map of the World*

This project was under the authority of an international advisory panel,
within the framework of FAO and UNESCO programs. Many countries had
carried out extensive national soil surveys already, but globally there was
no cohesive nomenclature or survey methods; legends and systems of classification varied widely from region to region. The first task was a comparative study of soil maps from around the world, based upon the compilation
of available soil survey material and field correlations. Approximately 600
soil maps of different scales and legends were compiled to form this map.
Compilation was followed by corroborating laboratory work and the organization of international expert meetings and study tours to correlate material
and for filling in gaps where maps and observations were lacking. The map
was planned at a scale of 1:5,000,000.

In 1968, at the Ninth Congress of the International Society of Soil Science in Adelaide, Australia, the first draft of the *Soil Map of the World* was
presented. The congress approved the outline of the legend, the common
nomenclature, and the definitions of soil units. With this approval, the first
sheets of the map were printed in 1970. Successive drafts of regional por-
tions of the soil map and legend were prepared and printed over the next

decade. Major soil correlations were undertaken in South America, Mexico and Central America, North America, Europe, Africa, South and Southeast Asia, North and Central Asia, and Australasia. The final *Soil Map of the World* comprised a ten-volume set. Twenty years passed from inception to completion. The map is now available from FAO in digitized form, using ARC/INFO software, in geographical projection.

As noted elsewhere, the FAO/UNESCO map received the highest number of citations of any work in the Core Agricultural Literature Project analysis of source documents. This map obviously remains an invaluable resource for scientists throughout the world. With better satellite photography, such as Landsat, and the growing sophistication of regional research, with its attendant needs for greater map detail at smaller scales in diverse localities, many countries have begun extensive soil mapping of their own, at scales of 1:1,000,000 or less.

B. Surveys and Maps

Ralph J. McCracken and Douglas Helms give an extensive description of soil surveys and maps in Chapter 11. The distinction between surveys and maps is not completely defined, but maps tend to cover regional, country, or continental levels (scale 1:100,000 to 1:5,000,000) and, thus, lack great detail, describing instead major soil classes. Surveys, on the other hand, serve more detailed needs at county and township levels (scale 1:12,000 to 1:50,000) and can often describe soil conditions at the farm level.

C. Soil Classification

Several major systems of soil classification are used as the basis for soil maps. These divergent systems can cause confusion. At the larger map scale, the less detailed FAO/UNESCO system has shown itself to be both workable and adequate at the "great group" level. However, because international consensus of the weight which each soil unit should have within a unified system is difficult to apply at a global level, the FAO/UNESCO map is, therefore, "a monocategorical classification of soils, and not a taxonomic system subdivided into categories at different levels of generalization." The *Soil Map of the World* was not meant to replace any particular classification scheme, but to serve as a common denominator, "to improve understanding between different schools of thought." Some of the maps listed below follow the FAO/UNESCO classification system.[1]

Another major classification system used internationally on soil maps is that of *Soil Taxonomy* of the U.S. Soil Survey Staff (1975). The taxonomic system of the Soil Survey Staff differs from the FAO/UNESCO system in its ability to describe soil types in much greater detail, including their zonality, evolution, morphology, and ecology. Its inception was contemporaneous to that of the FAO/UNESCO map. Here is how McCracken describes it in Chapter 11: "A significant landmark of soil surveys and classification in the United States was the publication of the *7th Approximation* to a new, comprehensive soil classification system in 1960. Following trials and comments, including international participation, this approximation was replaced by *Soil Taxonomy* in 1975." By 1980, thirty-one countries used the *Soil Taxonomy* system. Today, that number has almost doubled.[2]

Both the FAO/UNESCO system and the system of *Soil Taxonomy*, when known to apply to a particular map, are given as a data element on the entries below. Little information was available on other map classification systems, such as the Russian and French, but presumably many of the maps listed in this chapter use these or local systems.

D. List Compilation

The Core Agricultural Literature Project is indebted to M. Purnell and his colleagues at FAO, who supplied the bulk of the map citations in the list. The Land and Water Development Division Map Library of the FAO has a large collection of soil maps, as well as other related maps, such as vegetation, land use, geology, climate, and subnational boundaries. All are available to visitors of this facility at Via delle Terme di Caracalla, Rome, Italy. Searches of the collection can be made, and citations by country or subject can be provided. The map collection and database are constantly updated.

FAO has a history of leadership in the cataloging and distribution of map information. The first list of world soil maps by country was deposited in the FAO library in 1961. A more thorough listing appeared under the directorship of the World Soil Resource Office the following year. Finally, in 1965, FAO published its third edition under the title *Catalogue of Maps*, which was used as the basis for work on its *Soil Map of the World* in 1970.

1. FAO/UNESCO *Soil Map of the World*, Volume 1, "Legend" (Paris: UNESCO, 1970), pp. iv, v.

2. M. G. Cline, "Experience with Soil Taxonomy in the United States," *Advances in Agronomy* 33 (1980): 193–226.

Until the publication of the listing in this chapter, the 1965 work was the only comprehensive listing of world soil maps readily available to scientists. Only a few maps from the 1965 publication are named in the current list, generally because newer maps are not readily available, e.g., Iceland and Iran. The list which follows supersedes that of 1965.

Three sources were used for compilation of the list. The basic FAO-supplied catalog was augmented with South American listings supplied by the library of the Centro Internacional de Agricultura Tropical (CIAT) in Cali, Colombia. Special thanks go to Hans Van Baren, acting director of the International Soil Reference and Information Center (ISRIC) in Waeningen, Netherlands. ISRIC has been very helpful in coordinating aspects of this map project and in reviewing the list.

E. The Map Catalog

The catalog is divided into three parts: (a) world; (b) continents; and (c) alphabetical by country. Only those maps dealing predominantly with soils were included, at a maximum scale of 1:1,000,000. A few maps are at scales smaller than 1:250,000, but the bulk fall between these ranges.

Maps dealing mainly with land use, vegetation, geology, climate, and rainfall are excluded. Although this map catalog is not comprehensive, efforts were made to gather information on the most current map material available for each country at scale 1:1,000,000 to 1:250,000. In some instances, the best country map available was in the FAO/UNESCO *Soil Map of the World*.

Countries such as the United States pose a special problem, since they have been mapped at levels of 1:24,000 down to 1:15,840. The number of maps and soil surveys available for the continental land-mass of the United States runs into the thousands. Indeed, most published soil surveys cover one or more counties in each state. For over thirty years, the soil surveys have been printed on a photomosaic basis. The latest *List of Published Soil Surveys*, issued by the Soil Conservation Service, explains how surveys that are still in print can be obtained: "Land users in the area surveyed and professional workers who have use for the survey can obtain a free copy from the state or local office of the Soil Conservation Service, from their county agent, or from their congressman. Many libraries keep published soil surveys on file for reference. Also, soil conservation district offices and county agricultural extension offices have copies of local soil surveys that can be used for reference."[3]

3. Soil Conservation Service, *List of Published Soil Surveys*, 1991 (Washington, D.C.: USDA, Soil Cons. Serv., March 1992), p. 1.

Countries with soil surveys similar to that of the United States, such as Australia, Canada, Brazil, and some countries of the European Economic Community, also have excellent map series, some with accompanying resource guides. It is beyond the scope of this listing to describe each map individually, so only the major national map series have been listed for these countries.

Finally, soil maps of the former Union of Soviet Socialist Republics are listed separately, under the individual countries of the Commonwealth of Independent States, but the list of existing maps for these countries is very incomplete.

Section C lists the country soil maps in alphabetical order with this information, when known:

Title:	[Either the individual map or the series]
Scale:	[e.g. relational, 1:250,000]
Authors:	[Generally the cartographer or compiler, if known]
Publisher:	
Classification:	
Type of map:	[Digitized, Landsat, hand-tinted, black & white, etc. No indication for printed maps.]
No. of sheets:	[Generally for series maps]
Coordinates:	[Global longitude and latitude]
Coverage:	[Complete or partial]
Notes:	[Additional information, distinguishing characteristics, etc.]

Catalog of Major Soil Maps

A: World

1. *A Physical Resource Base (main climatic and soil divisions in the developing world); Base des Ressources Physiques; Base de Recursos Fisicos.*
   Scale: 1:25,000,000
   Publisher: FAO, Land and Water Development Division, 1983.
2. *FAO/UNESCO Soil Map of the World.*
   Scale: 1:5,000,000
   Publisher: UNESCO, Paris/FAO Land and Water Development Division, 1972–77.
   No. of sheets: 19
   Classification: FAO/UNESCO (1974).
   Coverage: Vol. I and map I: Legend; Vol. II and maps II-1, II-2: North America; Vol. III and map III: Mexico and Central America; Vol. IV and maps IV-1, IV-2: South America; Vol. V and maps

V-1, V-2: Europe; Vol. VI and maps VI-1, VI-2, VI-3: Africa; Vol. VII and maps VII-1, VII-2: South Asia; Vol. VIII and maps VII-1, VIII-2, VIII-3: North and Central Asia; Volume IX and map IX: Southeast Asia; Vol. X and maps X-1, X-2: Australasia.

3. *World Map on Status of Human-Induced Soil Degradation; Mapa Mundial de la Degradacion de Suelo Inducida por el Hombre; Degradation du Sol Resultant des Activites Humaines.*
   Scale: 1:15,000,000
   Authors: Oldeman, L.R.; Hakkeling, R.T.A.
   Publisher: Winand Staring Centre; UNEP/ISRIC, Nairobi, Kenya, 1990.
   No. of sheets: 3
   Type of map: Digitized

4. *World Soil Resources; Ressources en Sols du Monde; Recursos de los Suelos del Mundo.*
   Scale: 1:25,000,000
   Publisher: FAO, Land and Water Development Division, 1990.
   Classification: FAO/UNESCO, revised legend.
   Notes: The map is accompanied by an explanatory text (*World Soil Resources Report* no. 66, FAO, Rome).

B: Continents

Africa

5. *Provisional Map of Present Degradation Rate and Present State of Soil, Based on an Interpretation of Major Environmental Parameters.*
   Scale: 1:5,000,000
   Authors: Riquier, J., et al.
   Publisher: FAO/UNEP/UNESCO, 1980.
   No. of sheets: 6

Asia

6. *Bassin Inférieur du Mékong, Carte Pédo- Géomorphologique.*
   Scale: 1:1,000,000
   Countries: Laos; Thailand; Cambodia.
   Publisher: Bureau pour le développement de la production agricole (BDPA)/Comité pour la coordination des études sur le bassin inférieur du Mékong, IGN, 1977.
   No. of sheets: 2 Classification: FAO/UNESCO.

7. *Pochvennaia Karta Azii: K Proektu FAO/UNESCO Pochvennoi Karty Mira = Soil Map of Asia: To the Project of the FAO/UNESCO World Soil Map.*
   Scale: 1:6,000,000
   Authors: Kovad, V. A.; Lobova, E. V., eds.
   Publisher: Moscow, GUGK, 1971.
   No. of sheets: 6 (59x99 cm)
   Size of map: 1 map, 177x198 cm

Europe

8. *Atlas der Donaulaender; Atlas of the Danubian Countries; Atlas des pays Danubiens.* Sheet 161, part B: Soils (by J. Fink, 1984).
   Scale: 1:2,000,000
   Countries: Austria; Yugoslavia; Czechoslovakia; Hungary; Albania; Bulgaria; Romania.
   Authors: Breu, J.
   Publisher: Oesterreichisches Ost- und Suedosteuropa Institut, Wien (Publisher), Bundesamt fuer Eich- und Vermessungswesen (Landesaufnahme) und Eugen Ketterl (Printer); F.Deuticke (Distributor), 1988.
   Classification: FAO/UNESCO.

9. *Soil Map of Europe; Carte des Sols de l'Europe; Mapa de Suelos de Europa.*
   Scale: 1:2,500,000
   Countries: Albania; Austria; Belgium; Denmark; Finland; France; Germany; Greece; Iceland; Ireland; Italy; Luxembourg; Netherlands; Norway; Poland; Portugal; Spain; Sweden; Switzerland; United Kingdom; Yugoslavia.
   Authors: Tavernier, R.; Dudal, R.
   Publisher: Working Party on Soil Classification and Survey of the European Commission on Agriculture, FAO, 1963.
   No. of sheets: 6

10. *Soil Map of Middle Europe 1:1,000,000.*
    Scale: 1:1,000,000
    Countries: Switzerland; Austria.
    Authors: Tavernier, R., et al.
    Publisher: International Society of Soil Science, Wageningen, 1986.
    Classification: FAO/UNESCO.

11. *Soil Map of the European Communities.*
    Scale: 1:1,000,000
    Countries: Ireland; United Kingdom; France; Belgium; Netherlands; Germany; Spain; Greece; Italy; Spain; Portugal; Denmark; Luxembourg.
    Authors: Tavernier, R., et al.
    Publisher: Commission of the European Communities, Directorate-General for Agriculture, Coordination of Agricultural Research, Luxembourg, EEC, 1985.
    No. of sheets: 9
    Classification: FAO/UNESCO.

Near East

12. *Provisional Map of Present Degradation Rate and Present State of Soil, Based on an Interpretation of Major Environmental Parameters.*
    Scale: 1:5,000,000
    Authors: Riquier, J., et al.
    Publisher: FAO/UNEP/UNESCO, 1980.
    No. of sheets: 6

South America

13. *Land in Tropical America = La Tierra en America Tropical = A Terra na America Tropical.* 3 vols.
    Author: Cochrane, T.
    Publisher: Centro Internacional de Agricultura Tropical (with assistance from EMBRAPA, Brazil), Cali, Columbia, 1985.
    Notes: In English, Spanish & Portuguese. Vol. 1: A guide to climate, landscape and soils in Amazonia, the Andean Piedmont, Central Brazil and the Orinoco. Vol. 2: A Legend to the land system map.

14. *Land Systems Map. Physiography, Climate, Vegetation, Topography, Soils of the Central Lowlands of Tropical South America* (Vol. 2).
    Scale: 1:5,000,000
    Countries: Brazil; Ecuador; Colombia; Venezuela; Bolivia.
    Authors: Cochrane, T.T.; Sanchez, L.G.
    Publisher: In: *Land in Tropical America*, Vol. 1, 2 (parts 1 & 2), and 3, CIAT/EMBRAPA-CPAC, EMBRAPA, 1985. (See Map #12a)
    No. of sheets: 46

C: Country Soil Maps

Afghanistan

15. *Esquisse Sédimentologique et Pédologique de l'Afghanistan Méridional (Seistan, Regestan, Margo).*
    Scale: 1:500,000
    Authors: Pias, J,
    Publisher: ORSTOM-CNRS, Paris, 1979.
    Coverage: Partial coverage.

Albania

16. Scale: 1:2,000,000
    Classification: FAO/UNESCO.
    Coverage: Complete coverage in *Atlas of the Danubian Countries*. (See Map #8)

Algeria

17. *Carte des Sols de l'Algérie.*
    Scale: 1:500,000
    Authors: Durand, J.H.
    Publisher: Inspection générale de l'agriculture, Gouvernement général de l'Algérie, 1954.
    No. of sheets: 7
    Coverage: Partial coverage. Also local coverage at larger scales (1:100,000, 1:50,000, etc.).

Angola

18. *Carta Generalizada dos Solos de Angola.*
    Scale: 1:3,000,000
    Publisher: Junta das missoes geograficas e de investigacoes do ultramar,
                    Ministerio do Ultramar, 1965.
    Coverage: Complete coverage.
19. *Carta General dos Solos de Angola. 4. Distrito de Cabinda.*
    Scale: 1:500,000
    Publisher: Missao de pedologia de Angola e Mocambique, Junta de
                    Investigacao do Ultramar No 57, Lisboa, 1968.
    Coverage: Partial coverage.
    Notes: #9–Huila, 1:1,000,000, 1959; #27–Huambo, 1:500,000,
                    1961; #45–Mocamedes, 1:1,000,000, 1963; #63–Uizae Zaire,
                    1:1,000,000, 1972.

Antigua and Barbuda

20. *Soil and Land Use Surveys No. 19A and 19B, Antigua (1:25,000) and Barbuda
    (1:50,000).*
    Scale: 1:50,000
    Authors: Hill, J.D.
    Publisher: University of the West Indies, Trinidad, 1966.
    No. of sheets: 2
    Coverage: Complete coverage.

Argentina

21. *Atlas de Suelos de la Republica Argentina.*
    Scale: 1:500,000
    Publisher: Instituto Nacional de Tecnologia Agropecuaria, Centro de
                    Investigaciones de Recursos Naturales, Bueños Aires; PNUD
                    ARG/85/019, UNDP, 1990.
    No. of sheets: 38
    Classification: Soil Taxonomy.
    Coverage: Complete coverage.
22. *Mapa de Suelos de la Provincia de Bueños Aires.*
    Scale: 1:500,000
    Publisher: INTA, Secretaria de Agricultura, Ganaderia y Pesca; CIRN
                    Instituto de Evaluacion de Tierras; Proyecto PNUD
                    ARG/85/019, UNDP, 1989.
    No. of sheets: 11
    Coverage: Partial coverage.
23. *Geomorfologia de la Pampa Deprimida: Base para los Estudios Edafologicos y
    Agronomicos.*
    Authors: Tricart, Jean
    Publisher: Instituto Nacional de Technologia Agropecuaria, Librart,
                    Bueños Aires, Argentin, 1973.

Notes: Resource map of Argentina. 202pp. of text. 10 folded leaves of plates. Maps 23cm. Includes bibliography.

Australia

24. *A Soil Map of Australia (Handbook of Australian Soils).*
     Scale: 1:10,000,000
     Authors: Campbell, R.G.; Northcote, K.H.
     Publisher: CSIRO Division of Soils, Adelaide, 1968.
     Coverage: Complete coverage.

Austria

25. *Soil Map of Middle Europe.*
     Scale: 1:1,000,000
     Publisher: International Society of Soil Science, Wageningen, Netherlands, 1986.
     Classification: FAO/UNESCO
     Coverage: Complete coverage. Also at 1:2,000,000 in: *Atlas of the Danubian Countries.* (See Map #8)

Bahrain

26. *Bahrain Surface Materials Resources Survey; Soils (Provisional).*
     Scale: 1:50,000
     Publisher: Ministry of Works, Power and Water, Bahrain, 1974–75.
     Type of map: Black & White
     Coverage: Complete coverage.

Bangladesh

27. *Land Resources Inventory.*
     Scale: 1:250,000
     Publisher: FAO-UNDP BGD/81/035 Agricultural Development Adviser Project/Ministry of Agriculture, Bangladesh, 1985.
     No. of sheets: 17
     Coverage: Complete coverage.

Barbados

28. *Soil Map, Barbados.*
     Scale: 1:25,000
     Authors: Vernon, K.C.; Carroll, D.M.
     Publisher: Imperial College of Tropical Agriculture, Trinidad, W.I, 1965.
     No. of sheets: 3
     Coverage: Complete coverage.

Belgium

29. Scale: 1:1,000,000
    Coverage: Complete coverage in *Soil Map of the European Communities*.
                   (See Map #11)

Belize

30. *British Honduras* (Provisional Soil map).
    Scale: 1:250,000
    Authors: Wright, A.C.S.
    Publisher: Directorate of Overseas Surveys, U.K, 1958.
    No. of sheets: 2
    Coverage: Complete coverage.

Benin

31. *Carte Pedologique de Reconnaissance.*
    Scale: 1:200,000
    Publisher: ORSTOM, Paris, 1976–78.
    No. of sheets: 9
    Coverage: Complete coverage.
32. *Carte Pedologique de Dahomez.*
                   Scale:1:1,000,000
    Authors: Williams, P.; Volkoff, B.
    Publisher: Paris, ORSTOM, 1967.

Bhutan

33. *Soils and Land Capability Maps, Tashigang and Mongar Districts.*
    Scale: 1:50,000
    Publisher: Sinclair Knight & Partners PTY Ltd/FAO RAS 79/123 Soil and
                   Land Capability Survey, Tashigang & Mongar Districts, FAO,
                   1983.
    No. of sheets: 4
    Type of map: Inserted in report.
    Coverage: Partial coverage.

Bolivia

34. *Estudio de Suelos de la Provincia de Vallegrande, Resumen de la Clasificacion
    taxonomica de Suelos (Soil Taxonomy).*
    Scale: 1:250,000
    Authors: Encinas Guzman, M.
    Publisher: Corporacion Regional de Desarrollo de Santa Cruz
                   (CORDECRUZ)-PNUD-FAO BOL 86/011 Proyecto Desarrollo
                   Agropecuario, 1989.

No. of sheets: 6
Classification: Soil Taxonomy.
Type of map: Black & White
Coverage: Partial coverage.
35. *Experimental Digitized Map.*
    Scale: 1:50,000
    Publisher: Programa del Satelite Tecnologico de Recursos Naturales, ERTS-Bolivia, La Paz, 1977.
    Coverage: States of Huayllamarca North and South, Eucaliptas North and South.
    Size: 56x36 cms.
    Coordinates: 17.53S 67.32W 18.04S 67.48W
    Notes: Experimental Digitized Map from Landsat Data.
36. *Potencial Agricola del Uso de la Tierra de Bolivia: Un Mapa de Sistema de Tierra = The Agricultural Land Use Potential of Bolivia.*
    Authors: Cochrane, T.
    Publisher: Editorial Don Bosco, La Paz, Bolivia, 1973.
    Notes: 826pp. of text. Illustrated with maps, 2 folded in pockets 26cm. Includes bibliography.

Botswana

37. *Soil Map of the Republic of Botswana.*
    Scale: 1:1,000,000
    Authors: De Wit, P.; Nachtergaele, F.
    Publisher: Ministry of Agriculture, Gaborone, FAO BOT/85/011 Soil Mapping and Advisory Services Project, 1990.
    No. of sheets: 2
    Classification: FAO/UNESCO (Revised legend).
    Coverage: Complete coverage. Also covered partially at 1:250,000 scale.
38. *Botswana: Soils Map.*
    Scale: 1:250,000
    Publisher: Caborone, Dept. Field Services, Ministry of Agriculture, 1985.
    No. of sheets: 41

Brazil

In recent years, two major soil map series have been published. They are listed here in abbreviated form. Most maps cover large areas, usually states of Brazil.
39. *Mapa de Solos do Brasil.*
    Scale: 1:5,000,000
    Authors: Camargo, M.N.
    Publisher: Empresa Brasileira de Pesquisa Agropecuaria (EMBRAPA), Servico Nacional de Levantamento e Conservacao de Solos, Brazil, 1981.
    Coordinates: 6.00N 34.30W 35.00N 74.00W.
    Coverage: Complete coverage.

Size: 121x93 cms.
Notes: Impr. by Editora Grafica Barbero. Inset: Archipielago de
               Fernando de Noronha, 1:100,000; and Ilha de Trinidade,
               1:100,000.
40. Estado de Bahia. 1976. 1:1,000,000. 72x147 cms.
     18.00S 43.50W 8.45S 37.00W
     *Mapa Exploratior—Reconocimiento de Sols Margen Derecha del Rio San
     Francisco del Estado de Bahia.* Folha Sul.
41. Baxia Intensidade. 1983. 1:500,000. 53x118 cms.
     19.00S 49.15W 14.30S 46.30W
     Map is an annex to the book: *Levantamento de Reconhecimento de Baixa
     Intensidade dos Solos e Aptidao Agricola das Terras de parte da Regiao Geo-
     Economica de Brasilia.*
42. Estrada de Ferro Carajas. 1:444,444. 88x56cms.
     6.30S 50.30W 2.30S 43.45W.
     Worked performed and published with Companhia Vale do Rio Doce. Map is
     an annex to the book: *Levantamento de Reconhecimento de media Intensidade
     dos Solos e Avaliacao da Aptidao Agricola das terras da Area da Estrada de
     Ferro Carajas.* In three sheets: Sao Luis (No.1), Maranhao (No.2), Para (No.3)
43. Estado de Goias. 1:300,000. 62x114cms.
     15.30S 47.30W 12.30S 46.00W.
     Published with SUDESCO. Map is an annex to the book: *Levantamento do
     Reconhecimento de Media Intensidade dos Solos e Avaliacao da Aptidao
     Agricola das Terras da margem direita do Rio Parana—Estado de Goias.*
44. Estado do Maranhao. 1986. 1:100,000. 78x103cms.
     00.00S 49.00W 1.00S 42.00W
     Published with SUDENE/DRM.
45. Estado de Mato Grosso. 1982. 1:500,000. 66×87cms.
     17.45S 59.45W 14.00S 56.30W
     Map is an annex to the book: *Levantamento de Reconhecimento de Media
     Intensidade, Avaliacao da Aptidao Agricola das terras e Indicacao da Culturas
     em Areas Homogeneas de Solos de alguns municipios do Sudoeste do Estado de
     Mato Grosso*
46. Estado do Parana. 1981. 1:600,000. 112×96cms.
     27.30S 54.30W 22.30S 48.00W
     Map is accompanied by the book: *Levantamento de Reconhecimento dos Solos
     do Estado do Parana.* Vol. 1
47. Estado do Piaui. 1983. 1:1,000,000. 96×85cms.
     11.00S 46.00W 2.30S 38.00W
     Map is accompanied by the book: *Levantamento Exploratorio—Reconhecimento
     de Solos do Estado do Piaui.* Vol. 1
48. Polo Tapajos. 1984. 1:500,000. 54×45cms.
     4.05S 56.15W 2.20S 5.40W
     Map is an annex to the book: *Levantamento de Reconhecimento de Media
     Intensidade dos Solos e Avaliacao da Aptidao Agricola das Terras da Area do
     Polo Tapajos.*
49. Polo Pre-Amazonia: Maranhense. 1982. 1:500,000. 28×43cms.
     5.45S 46.42W 3.45S 45.45W
     Map is an annex to the book: *Levantamento de Reconhecimento de Media*

Intensidad dos Solos e avaliacao da Aptidao Agricola das terras da Area do Polo Pre-Amazonia Maranhese.

50. Triangulo Mineiro. 1980. 1:500,000. 103 × 55cms.
    20.15S 51.00W 18.00S 47.00W
    Map is an annex to the book: *Levantamento de Reconhecimento de Media Intensidade dos Solos e Availiacao de Aptidao das Terras do Triangulo Mineiro.*

51. *Mapas Exploratorios de Solos, Projeto Radambrasil.*

Scale:	1:1,000,000
Publisher:	Departamento Nacional da Producao Mineral (DNPM), 1974–87.
Coverage:	Partial coverage at FAO (9 sheets).
Notes:	This series was issued by DNPM of the Ministerio de Minas y Energia. All bear the title: *Mapa Exploratorio de Solos,* then subtitled by the region mapped. Scale in all cases is 1:1,000,000.
Subtitles:	Amazonas; Aracaju/Recife; Araguaia/Tocantins; Belem; Boa Vista/Roraima; Campo Grande; Corumba; Cuiba; Fortaleza; Goiania; Goias; Guapore; Jaguaribe/Natal; Javari/Contamana; Jurua; Juruena; Macapa; Manaus; Pico da Neblina; Porto Velho; Purus; Rio Branco; Rio de Janeiro/Vitoria; Rio Doce; Santarem; Salvador; Tapajos; Tocantins; Tumucumaque.

52. *Levantamento de Reconhecimento dos Solos.*

Scale:	1:
Publisher:	Ministerio de Agricultura, Geocarta, S.A.

53. *Mapa Suelos Margen Izq. del Rio San Francisco. Estado de Bahia.* Ministerio de Agricultura, 1973. 1:1,000,000. 68 × 83 cms.
    15.30S 46.30W 8.45S 40.30W
    Notes: Mapa Exploratorio-Reconocimiento de Suelos Margen izquierda del Rio San Francisco, Estado de Bahia.

54. *Mapa Exploratorio—Reconhecimento de Solos. Estado do Bahia.* Ministerio de Agricultura, 1973. 1:1,000,000. 68 × 83 cms.
    15.30S 46.30W 9.15S 40.00W
    Notes: Mapa de suelos, anexo del libro: "Levantamento Exploratorio—Reconhecimento de Solos da Margem Esquerda do Rio Sao Francisco Estado da Bahia."

55. *Suelos del Area Bajo Influencia del Reservorio del Tres Marias.* Minas G. Ministerio de Agricultura. EMBRAPA - EPAMIG, 1976. 1:1,600,000. 46 × 52 cms.
    19.00S 45.30W 18.15S 44.50W
    Notes: Mapa de Levantamiento de Reconocimiento Detallado de Suelos del Area bajo la influencia del Reservorio de Tres Marias. Estado de Minas Gerais.

56. *Levantamento de Reconhecimento dos Solos.* Ministerio de Agricultura. 54 × 49 cms.
    24.00S 51.30W 22.30S 49.30W
    Notes: Mapa de suelos, anexo del libro: "Levantamento de Reconhecimento dos solos do Noroeste do estado do Parana."

57. *Levantamento de Reconhecimento dos Solos do Sudoeste—Parana*. Ministerio de Agricultura. 1974. 1:300,000. 83 × 55 cms.
    26.30S 55.00W 25.00S 52.30W
    Notes: Mapa de suelos, anexo del libro: "Estudo Expedito de Solos nas partes Central e Sul do Estado do Bahia para Fins de Classificacao, Correlacao e Legenda Preliminar."
58. *Levantamento de Reconhecimento dos Solos. Sudeste do Parana*. Ministerio de Agricultura. 1974. 1:300,000. 35 × 19 cms.
    25.30N 50.00W 25.00N 49.00W
    Notes: Mapa de suelos, anexo del libro: "Levantamento de Reconhecimento dos solos do Sudoeste do Estado do Paraná 1a. Parte."
59. *Levantamento de Reconhecimento dos Solos—Oeste de Parana*. Ministerio de Agricultura, 1974. 1:300,000. 67 × 37 cms.
    25.00S 54.30W 24.00S 52.30W
    Notes: Mapa de suelos, anexo del libro: "Levantamento de Reconhecimento dos solos do oeste do Estado do Paraná."
60. *Levantamento de Reconhecimento dos Solos. Parana*. Ministerio de Agricultura, 1977. 1:300,000. 36 × 61 cms.
    Transversal Mercator
    26.00S 49.00W 24.30S 48.00W
    Notes: Mapa de suelos, anexo del libro: "Levantamento de Reconhecimento dos Solos do Litoral do Estado do Paraná (Area 11)."
61. *Levantamento de Reconhecimento dos Solos. Estado do Parana*. Ministerio de Agricultura, 1971. 1:313,000 (Aprox.). 64 × 50 cms.
    24.00S 51.30W 22.30S 49.30W
    Notes: Mapa de suelos, anexo del libro: "Levantamento de Reconhecimento dos solos do Nordeste do Estado do Paraná."
62. *Mapa Exploratorio—Reconhecimento de solos. Estado de Alagoas*. Ministerio de Agricultura, 1972. 1:400,000. 96 × 60 cms.
    10.30S 38.30W 8.30S 35.00W
    Notes: Mapa de suelos, anexo del libro: "Levantamento Exploratorio-Reconhecimento de Solos do Estado de Alagoas."
63. *Mapa Exploratorio—Reconhecimento de Solos do Estado de Sergipe*. Ministerio de Agricultura, 1973. 1:400,000. 79 × 68 cms.
    11.45S 38.30W 9.00S 36.00W
    Notes: Mapa de Suelos, anexo del libro: "Levantamento Exploratorio-Reconhecimento de Solos do Estado de Sergipe."
64. *Carta de Suelos del Estado de Sao Paulo*. Ministerio de Agricultura CNEPA—SNPA, 1960. 1:500,000. 2 Hojas de 93 × 122 cms. 185 × 122 cms.
    25.00S 53.00W 20.00S 44.00W
    Notes: Mapa de Suelos.
65. *Mapa Exploratorio—Reconhecimento do Solos. Estado da Paraiba*. Ministerio de Agricultura, 1971. 1:500,000. 94 × 56 cms.
    8.30S 39.00W 6.00S 34.30W
    Notes: Mapa de Suelos, anexo del libro: "Levantamento-Reconhecimento de Solos do Estado da Paraiba.

66. *Mapa Exploratorio—Reconhecimento de Solos. Rio Grande do Norte.*
    Ministerio de Agricultura, 1968. 1:500,000. 88 × 55 cms.
    5.00S 38.40W 5.00S 34.30W
    Notes: Mapa de suelos, anexo del libro: "Levantamento Exploratorio-
                    Reconhecimento de solos do estado do Rio Grande do Norte."
67. *Mapa de Suelos de Region Bajo Influencia de la Cia. Del Valle.* Rio Doce,
    Geocarta S.A. Ministerio de Agricultura, 1970. 1:500,000. 93 × 77 cms.
    21.00S 44.00W 17.30S 39.30W
    Notes: Mapa de Levantamiento Exploratorio de Suelos de la Region
                    bajo la influencia de la compania del Valle del Rio Doce, entre
                    los estados de Minas Gerais y Espiritu Santo.
68. *Levantamiento de Reconhecimento de Suelos del sur de Mato. Grosso.* Geocarta
    S.A. Ministerio de Agricultura, 1971. 1:600,000. 104 × 73 cms.
    24.00S 58.00W 20.00S 52.00W
    Notes: Mapa de suelos.
69. *Mapa Exploratorio—Reconhecimento de Solos. Pernambuco.* Ministerio de
    Agricultura, 1969. 1:600,000. 121 × 41 cms.
    9.30S 41.30W 7.00S 34.45W
    Notes: Mapa de suelos, anexo del libro: "Levantamento Exploratorio-
                    Reconhecimento de solos do Estado do Pernambuco."
70. *Mapa Exploratorio—Reconhecimento de Solos, Estado do Ceara.* Ministerio de
    Agricultura, 1972. 1:600,000. 82 × 106 cms.
    8.30S 41.30W 2.45S 37.00W
    Notes: Mapa de suelos, anexo del libro: "Levantamento Exploratorio-
                    Reconhecimento de solos do Estado do Ceará."
71. *Levantamento de Reconhecimento dos Solos. Estado do Parana.* Ministerio de
    Agricultura, 1981. 1:600,000. 111 × 96 cms.
    27.30S 54.45W 22.30S 48.00W
    Notes: Mapa de Suelos, anexo del libro: "Levantamento de
                    Reconhecimento dos Solos do Estado do Paraná" Tomo II.
72. *Levantamiento Edafologico del Estado del Rio Grande do Sul.* Division de
    Pedología y Fertilidad del Suelo, 1967. 1:1,000,000. 56x30 cms.
    29.40S 55.30W 27.00S 50.00W
    Notes: Mapa sobre Levantamiento edafologico, anexo del libro:
                    "Levantamento de Reconheçimiento dos Solos do Estado do
                    Rio Grande do Sul. Primeira Etapa, Planalto Rio-Grandense."
73. *Delineamento Esquematico dos Solos do Brasil.* Camargo, M.N & Bennema,
    J., 1966. 1:10,000,000. 45 × 45 cms.
    30.00S 78.00W 4.00N 34.00W
    Notes: Mapa sobre delineamiento de suelos de Brasil, anexo del libro:
                    "Delineamento Esquematico dos solos do Brasil."

Bulgaria

74. *Soil Map of Bulgaria.*
    Scale: 1:400,000
    Authors: Koinov, V.
    Publisher: 1968.

Coverage: Complete coverage. Covered also at 1:2,000,000 scale in *Atlas of the Danubian Countries*. (See Map #8)
Notes: In Cyrillic.

Burkina Faso

75. *Carte Pédologique de Reconnaissance de la République de Haute-Volta (Centre-Nord, Est, Ouest-Nord, Ouest-Sud, Centre- Sud).*
     Scale: 1:500,000
     Authors: Boulet, R.; Leprun, J.C.
     Publisher: ORSTOM, Centre de Dakar, 1969.
     No. of sheets: 5
     Coverage: Complete coverage.

Burundi

76. *Carte des Associations des Sols du Rwanda et du Burundi.*
     Scale: 1:1,000,000
     Authors: Van Wambeke, A.
     Publisher: Institut National pour l'Etude Agronomique du Congo, Institut Geographique Militaire, Bruxelles, 1962.
     Coverage: Complete coverage.
77. *Carte des Sols du Burundi.*
     Scale: 1:250,000
     Authors: Sottiaux, G.
     Publisher: Institut des Sciences Agronomiques du Burundi (ISABU), Bruxelles, 1980.
     Coverage: Complete coverage.
78. *Soil Map of IGADD countries, 1:1,000,000 (in prep.).*
     Scale: 1:1,000,000
     Publisher: FAO-ISRIC, Wageningen, FAO-GIS, Rome, 1992.
     Classification: FAO/UNESCO, Rev. legend.
     Type of map: Digitized.
     Coverage: Complete coverage.

Cambodia

79. *General Soil Map; Carte Générale des Sols.*
     Scale: 1:1,000,000
     Authors: Crocker, C.D.
     Publisher: USAID/Secrétariat d'Etat ǎ l'Agriculture, 1963.
     Coverage: Complete coverage.

Cameroon

80. *Carte Pédologique du Cameroun Occidental ǎ 1/1,000,000.*
     Scale: 1:1,000,000

Authors: Vallérie, M.
Publisher: ORSTOM, Centre de Yaounde, Bondy, Notice Explicative No. 45, 1970.
Coverage: Partial coverage; completed by other 1:1,000,000 scale map.

81. *Carte Pédologique du Cameroun oriental ă l'Echelle de 1:1,000,000.*
Scale: 1:1,000,000
Authors: Segalen, P.; Martin, D.
Publisher: ORSTOM, Centre de Yaounde, 1965.
No. of sheets: 2
Coverage: Partial coverage; completed by other 1:1,000,000 map.

82. *Les Sols et les Ressources en Terres du Nord- Cameroun.*
Scale: 1:500,000
Authors: Brabant, P.; Gavaud, M.
Publisher: ORSTOM/MESRES, Yaounde, 1985.
No. of sheets: 6
Coverage: Partial coverage.

Canada

83. *Canada Land Inventory: Land Capability Analysis.*
Scale: 1:250,000
Author: Canada Dept. of Agriculture.
Publisher: Canada Dept. of Agriculture, Ottawa, 1969–.
Coverage: Most of Canada.
Notes: Map sizes vary. Explanation of analysis on each map in English and French.

84. *Soils of Canada.*
Scale: 1:5,000,000
Publisher: Soil Research Institute, Department of Agriculture, Ottawa, 1972.
Coverage: Complete coverage.

Cape Verde

85. *Mapa dos Solos da Ilha de Sao Nicolau.*
Scale: 1:30,000
Authors: Nunes, M.
Publisher: Missao de Estudos Agronomicos do Ultramar, Lisboa, 1962.
No. of sheets: 5
Type of map: Black & White
Coverage: Partial coverage.

Central African Republic

86. *Carte Pédologique de la République Centrafricaine.*
Scale: 1:1,000,000
Authors: Boulvert, Y.

Publisher: ORSTOM, Bondy, France, 1983.
No. of sheets: 2
Coverage: Complete coverage. Also partial coverage at 1:200,000 scale.

Chad

87. *Carte Pédologique de Reconnaissance*.
Scale: 1:200,000
Publisher: Ministére de l'Agriculture et des Eaux et Forêts, Direction de
                l'Agriculture-ORSTOM, 1962–74.
No. of sheets: 22
Coverage: Partial coverage.
88. *Carte Pédologique du Tchad a l'Echelle de 1:1,000,000*.
Scale: 1:1,000,000
Authors: Pias, J.
Publisher: ORSTOM, Bondy, 1968.
No. of sheets: 2
Coverage: Complete coverage.

Chile

89. *Suelos Volcanicos de Chile, Mapas Geologicos y Carta de Suelos (Carta de
    Suelos Derivados de Materiales Volcanicos)*.
Scale: 1:250,000
Publisher: Instituto de Investigaciones Agropecuarias, Ministerio de
                Agricultura, 1985.
Coverage: Complete coverage.

China

90. *The Soil Atlas of China*.
Authors: Institute of Soil Science, Academia Sinica
Publisher: Cartographic Publishing House, Beijing, China, 1986.
Coverage: Country coverage at 1:14,000,000. Other local coverage at
                1:1,000,000.
Notes: 128pp. of text in English and Chinese. 31pp of photos, 32
                maps ranging in scale from 1:4,000,000 to 1:20,000,000 on all
                aspects of soils in China. (Can be purchased by writing:
                Cartographic Publishing House, P.O .Box 821, Nanking,
                China. Price $190.)
91. *Soil Map of China* (1st edition).
Scale: 1:10,000,000
Authors: Li Jin; Zhou Minzong.
Publisher: Institute of Soil Science, Academia Sinica, Beijing, 1988.
Coverage: Complete coverage.
92. *Soil Map of the People's Republic of China*. Revised version (1990) based on
    the FAO/UNESCO Soil Map of the World.

Scale: 1:5,000,000
Publisher: CPR/87/029 Land Resources, Use, Productivity Assessment
                  Project; FAO-GIS, 1990.
Classification: FAO/UNESCO.
Type of map: Digitized
Coverage: Complete coverage.

Colombia

93. *Map of Atlantic Coast.*
     Scale: 1:100,000
     Publisher: Instituto Geografico Agustin Codazzi, Bogota, 1981.
     Coverage: "Vias de Comunicacion, Relieve y Rios."
     Size: 70 × 102 cms.
     Coordinates: 10.15N 75.20W 11.10N 74.40W
94. *Map of States Sucre and Bolivar.*
     Scale: 1:100,000
     Publisher: Instituto Geografico "Agustin Codazzi," Bogota, 1976.
     Size: 94 × 73 cms
     Notes: Accompanies the volume: "Estudios General de Suelos de los
                      Municipos de Buenavista, Since, Galeras, San Pedro, Los
                      Palmitos, San Juan de Betulia y Malangue."
95. *Mapa de Suelos.*
     Scale: 1:1,500,000
     Publisher: Instituto Geografico Agustin Codazzi, Subdireccion Agrologica,
                      1983.
     Coverage: Complete coverage.
96. *Mapa del Suelos del Cauca.*
     Scale: 1:100,000
     Publisher: F.N.C./Departemento de Investigacion y Programacion
                      Prodesarrollo, 1977.
     Coverage: State of Cauca.
     Size: 4 maps, 94 × 67 cms
     Notes: Accompanies the volume: "Uso Potencial del Suelo en la Zona
                      Cafetera del Dpto. del Cauca."
97. *Mapa de Suelos y su Aptitud de Uso.*
     Scale: 1:5,000,000
     Authors: Cortes Lombana, A.
     Publisher: Instituto Geografico Agustin Codazzi, Colombia, 1976.
     Coverage: Complete coverage.
98. *Carta de Clasificacion de Tierras: Republica de Colombia.*
     Scale: 1:500,000
     Authors: Ministerio de Hacinenda y Credito Publico, Instituto
                      Geografico Agustin Codazzi, Direccion Agrologica
     Publisher: Bogotá, Instituto Geografico Agustin Codazzi, Direccion
                      Agrologica, 1973.
     Size: 1 map in 19 sections, each 42 × 83 cm (336 × 269 cm)

Comoros

99. *Carte Pedologique d'Anjouan.*
    Scale: 1:100,000
    Publisher: Institut de Recherche Scientifique de Madagascar, 1951.
    Coverage: Partial coverage.

Congo

100. *Carte Pedologique du Congo.*
     Scale: 1:200,000
     Publisher: ORSTOM, Centre de Brazzaville, 1974.
     No. of sheets: 4
     Coverage: Partial coverage.
101. *Pedologie.*
     Scale: 1:2,000,000
     Authors: Boissezon, P. de; Martin, G.
     Publisher: ORSTOM, Centre de Brazzaville, 1969.
     Coverage: Complete coverage.

Cook Islands

102. *Soil Maps of Mauke, Mitiaro, Mangaia, Aitutaki, Atiu, Rarotonga.*
     Scale: 1:15,000
     Publisher: New Zealand Soil Bureau, Department of Scientific and
                    Industrial Research, Wellington, N.Z., 1980.
     No. of sheets: 6
     Coverage: Partial coverage.

Costa Rica

103. *Mapa de Suelos de Costa Rica Y Mapa de Capacidad de Uso del Suelo de
     Costa Rica.*
     Scale: 1:200,000
     Authors: Vasqez Morera, A.
     Publisher: FAO GCP/COS/OO9/ITA Apoyo al Servicio Nacional de
                    Conservacion de Suelos y Aguas, FAO, 1989.
     No. of sheets: 18 Classification: Soil Taxonomy.
     Type of map: Blue Print
     Coverage: Complete coverage.
104. *Asociacion de Sub-Grupos de Suelos de Costa Rica: Mapa Preliminar.*
     Scale: 1:200,000
     Authors: Perez, S.; et al.
     Publisher: San José, Oficina de Planificacion, Sectorial Agropecuaria,
                    1978.
     Size: 9 maps, 54 × 69 cm. or smaller

Cuba

105. *Mapa Genetico de Suelos.*
      Scale: 1:250,000
      Publisher: Academia de Ciencias de Cuba, Instituto de Suelos; Instituto
                      Cubano de Geodesia y Cartografia, 1971.
      No. of sheets: 19
      Coverage: Complete coverage.

Cyprus

106. *General Soil Map of Cyprus.*
      Scale: 1:200,000
      Authors: Soteriades, C.G.; Grivas, G.C.
      Publisher: Section of Soils and Plant Nutrition, Department of
                      Agriculture, Nicosia, 1970.
      Classification: FAO/UNESCO.
      Coverage: Complete coverage.

Czechoslovakia

107. *Pudni Mapa CSSR.*
      Scale: 1:500,000
      Authors: Hrasko, J.; Nemecek, J.
      Publisher: Vydala Slovenska Kartografia, n.p. Bratislava, 1973.
      No. of sheets: 2
      Coverage: Complete coverage. Also covered at 1:2,000,000 scale in *Atlas
                      of the Danubian Countries.* (See Map #8)

Denmark

108. Scale: 1:1,000,000
      Classification: FAO/UNESCO.
      Coverage: Covered by *Soil Map of the European Communities.* (See Map
                      #11) Also 1:500,000 in *Soil Map of Denmark* (1970, reprint of
                      1935) by C.H. Bornebusch et al.

Djibouti

109. *Soil Map of IGADD Countries, 1:1,000,000* (in prep.).
      Scale: 1:1,000,000
      Publisher: FAO-ISRIC, Wageningen, FAO-GIS, Rome, 1992.
      Classification: FAO/UNESCO, Rev. legend.
      Type of map: Digitized
      Coverage: Complete coverage.

Dominican Republic

110. *Asociaciones de Suelos de la Republica Dominicana.*
    Scale: 1:250,000
    Authors: Pretell, O.F.; Soto. G.
    Publisher: Organizacion de los Estados Americanos, Washington, D.C,
                  1967.
    Coverage: Complete coverage.

Ecuador

111. *Mapa general de Suelos del Ecuador.*
    Scale: 1:1,000,000
    Authors: Mejaia Vallejo, Luis.
    Publisher: Sociedad Ecuatoriana de la Ciencia del Suelos/I.G.M. Instituto
                  Geografico Militar, Quito, Ecuador, 1986.
    Size: 111 × 83 cms
    Coordinates: 6.00S 81.30W 1.30N 72.00W
    Notes: With data on temperature vegetation, and geology. Includes 4
                  ancillary maps and 2 schematic cross sections.
112. *State Maps of Ecuador MAG—PRONAREG—ORSTOM*
    Scale: 1:200,00
    Coverage: States of Macara, Tulcan, Ibarra, Latacunga, Ambato, Salinas,
                  Riobamba, Zamora, Canar, Jipijapa, Porto Viejo, Quito and
                  Cuenca, 1984.
    Size: 56 × 37 cms   Coordinates: 4.00S 79.30W 4.40S 80.30W
    Notes: Specially published for the Banco Central del Ecuador.

Egypt

113. *Soil Map of Egypt.*
    Scale: 1:100,000
    Publisher: Academy of Scientific Research and Technology, 1986.
    No. of sheets: 43
    Coverage: Complete coverage. Includes also local coverage at 1–250,000
                  scale.

El Salvador

114. *Mapa Pedologico de El Salvador.*
    Scale: 1:300,000
    Authors: Rico, M.
    Publisher: Universidad de El Salvador, Departamento de Suelos, 1974.
    Coverage: Complete coverage.

Estonia

115. *Soil Map of the European Part of USSR.*
     Scale: 1:2,500,000
     Authors: Lobova, E.V.; Rozov, N.N.
     Publisher: USSR Academy of Sciences, Dokuchaev Institute of Soils, 1947.
     No. of sheets: 4
     Coverage: Partial coverage of ex-U.S.S.R.
     Notes: In Cyrillic.

Ethiopia

116. *Geomorphology and Soils.*
     Scale: 1:1,000,000
     Authors: Henricksen, B.L.; Ross, S.; Tilimo, S.; Wijnte-Bruggeman, H.Y.
     Publisher: Land Use Planning and Regulatory Department, FAO/UNDP Assistance to Land Use Planning Project (ETH 78/003), Tech. Rep. 1, Maps, 1983.
     No. of sheets: 8   Classification: FAO/UNESCO (1974).
     Coverage: Complete coverage.
117. *Soil Map of IGADD Countries, 1:1,000,000.*
     Scale: 1:1,000,000
     Publisher: FAO-ISRIC, Wageningen, FAO-GIS, Rome, 1992.
     Classification: FAO/UNESCO, Rev. legend.
     Type of map: Digitized
     Coverage: Complete coverage.

Fiji

118. *The Soil Resources of the Fiji Islands (Soil Maps and Land Classification Maps).*
     Scale: 1:125,000
     Authors: Twyford, I.T.; Wright, A.C.S.
     Publisher: Soil Bureau, Department of Scientific and Industrial Research, New Zealand, 1961.
     No. of sheets: 8 at 1:126,720 and 1 at 1:760,320
     Coverage: Complete coverage.

Finland

119. Scale: 1:1,000,000
     Coverage: Covered at 1:1,000,000 scale (1984).
120. *Suomen Maaperäkartta/Suomen Maanteellinen Seura.*
     Scale: 1:2,000,000
     Publisher: Helsinki, Topografikunta, Suomen Maanteellinen Seura, 1969.
     Size: 60×41 cm

France

121. *Carte Pédologique de la France a l'Echelle du Millionieme.*
    Scale: 1:1,000,000
    Authors: Dupuis, J.
    Publisher: Commission de la Carte Pedologique de France, Institut
                    National de la Recherche Agronomique, 1966.
    No. of sheets: 2
    Coverage: Complete coverage. Also covered at 1:1,000,000 by *Soil Map
                    of the European Communities* and at larger scale (1:250,000),
                    by INRA. (See Map #11)

French Guiana

122. *Atlas des Départments d'Outre Mer: La Guyane.*
    Scale: 1:350,000
    Publisher: Centre d'Etudes de Geographie Tropicale (CNRS)/ORSTOM,
                    1979.
    Coverage: Complete coverage at 1:1,000,000 scale. Partial coverage at
                    1:350,000.

Gabon

123. *Carte Pédologique du Gabon.*
    Scale: 1:2,000,000, part at 1:200,000
    Authors: Martin, D.
    Publisher: ORSTOM, Bondy, France, 1981.
    No. of sheets: 2
    Coverage: Complete coverage.

Gambia

124. *Soil Associations, Maps 1 to 4.*
    Scale: 1:125,000
    Authors: Dunsmore, J.R.; Blair Rains, A.
    Publisher: Land Resources Division, Land Resource Study 22 (The
                    Agricultural Development of the Gambia: An Agricultural,
                    Environmental and Socioeconomic Analysis), 1976.
    No. of sheets: 4
    Coverage: Complete coverage.

Germany

125. Scale: 1:1,000,000
    Classification: FAO/UNESCO.
    Coverage: Covered by *Soil Map of the European Communities*. (See Map
                      #11) Also covered at 1:1,000,000 scale by BGR (1986).

126. *Boden, Atlas DDR 6.*
     Scale: 1:750,000
     Authors: Haase, G.; Schmidt, R.
     Publisher: Akademie der Wissenschaften der DDR, 1984.
     Coverage: Partial coverage (ex-German Democratic Republic).

Ghana

127. *Soil Map of Ghana.*
     Scale: 1:1,500,000
     Authors: Obeng, H.B.
     Publisher: Survey Department, Accra, 1971.
     Coverage: Complete coverage.

Greece

128. Scale: 1:1,000,000
     Classification: FAO/UNESCO.
     Coverage: Covered by *Soil Map of the European Communities.* (See Map
                     #11)

Grenada

129. *Soil Map of Grenada.*
     Scale: 1:18,000
     Authors: Vernon, K.C.; Payne, H.
     Publisher: The Regional Research Centre of the British Caribbean,
                     Imperial College of Tropical Agriculture, Trinidad, W.I, 1959.
     No. of sheets: 3
     Coverage: Complete coverage.

Guatemala

130. *Mapa de Capacidad Productiva de la Tierra.*
     Scale: 1:500,000
     Authors: Consejo Nacional de Planificacion Economica (Guatemala),
                     Area de Planificacion Intersectorial
     Publisher: Instituto Georgrafico Nacional de Guatemala, Guatemala, 1980.
     No. of sheets: 4
     Size: 111 × 109 cms   Coordinates: W92 30—W87 30/N18 30—N13
                     30

Guinea

131. *Carte de Reconnaissance Pédologique de la République de Guinée.*
     Scale: 1:250,000

Authors: Kawalec, A.; Balde, Y.
Publisher: FAO/PNUD GUI/72/004 Service National des Sols, Bulletin
                Senasol 2, 1978.
No. of sheets: 4
Type of map: Black & White
Coverage: Partial coverage.

Guinea Bissau

132. *Os Solos da Guine Portuguesa.*
      Scale: 1:1,000,000
      Authors: Silva Teixeira, A.J. da.
      Publisher: Junta de Investigacoes do Ultramar, Lisbon, 1962.
      No. of sheets: 3
      Coverage: Complete coverage.

Guyana

133. *General Soil Map of British Guiana.*
      Scale: 1:1,000,000
      Authors: Braun, E.U.G.; Derting, J.
      Publisher: British Guiana Soil Survey/FAO-SF:19/BRG Report on the Soil
                      Survey Project, Volume VII, FAO, 1964.
      Coverage: Complete coverage.

Haiti

134. *Erosion.*
      Scale: 1:250,000
      Authors: Leo, O.
      Publisher: Direction de l'Amenagement du Territoire et Protection de
                      l'Environnement (DATPE), Secretairerie d'Etat du Plan/BDPA,
                      Paris, DATPE, 1982.
      Coverage: Complete coverage.
135. Sols et Vocation de la Terre, in *Mission d'Assistance Technique Intégrée.*
      Scale: 1:50,000
      Publisher: Bureau de Développement Régional, Secrétariat Général,
                      Organisation des Etats Américains, Washington, D.C., 1972.
      Coverage: Complete coverage.

Honduras

136. Coverage: 1:1,000,000 by Direccion Ejecutiva de Catastro.

Hungary

137. *Genetikai Talajterkep (Soil Map). VII. Talajok (Soils).*
    Scale: 1:1,000,000
    Authors: Varallyay, G.; Zilahy, P.
    Publisher: Research Institute for Soil Sciences & Agricultural Chemistry,
                  Hungarian Academy of Sciences; Agricultural University of
                  Godollo, Plant Protecting and Agrochemical Centre,
                  MINAGRI.
    Coverage: Complete coverage. Also covered at 1:2,000,000 scale in *Atlas
                  of the Danubian Countries*. (See Map #8)
138. *Magyarorszag Genetikus Talajteakepe.*
    Scale: 1:200,000
    Authors: Kovacs, I.
    Publisher: Budapest, MEM Novenyvedelmies Agrokemiai Kozpontja
                  Meliosacip es Talgitani Foosztalyan, 1984.
    No. of sheets: 9

Iceland

139. *General Soil Map of Iceland* (2nd draft).
    Scale: 1:2,500,000
    Authors: Johanneson, B.
    Publisher: FAO/WSRO, Land and Water Development Division, 1963.
    Type of map: Hand-coloured
    Coverage: Complete coverage.
140. *Jardvegskort af Islandi = Soil Map of Iceland.*
    Scale: 1:750,000
    Authors: USDA, Cartographic Division
    Publisher: Washington, D.C., 1959
    Size: 53 × 74 cm
    Notes: Soil map adapted by University Research Institute, Reykjavik,
                  Iceland, from original manuscript by Dr. Iver J. Nygard.

India

141. *India, Vertisols and Associated Soils.*
    Scale: 1:4,500,000
    Authors: Sehgal, J.L.
    Publisher: National Bureau of Soil Survey & Land Use Planning, Indian
                  Council of Agricultural Research, Nagpur (NBSS Publ. 19,
                  Soils Bulletin), 1988.
    Coverage: Complete coverage.
142. *Punjab, Soils.*
    Scale: 1:1,000,000
    Publisher: Survey of India, Government of India, 1988.
    Coverage: Partial coverage.

143. *Soils of India (Suborder Associations).*
    Scale: 1:7,000,000
    Authors: Shankaranarayana, H.S.; Murthy, R.S.
    Publisher: National Bureau of Soil Survey and Land Use Planning, Indian
                      Council of Agricultural Research, Nagpur, 1987.
    Classification: FAO/UNESCO; Soil Taxonomy.
    Coverage: Complete coverage.

Indonesia

144. *Indonesia Generalized Soil Map = Indonesia, Peta Tanah Bahan.*
    Scale: 1:7,500,000
    Publisher: Badan Koordinasi Survey dan Pemetaan Masional, Gadjah
                      Mada, Yogyakarta, 1980.
    Size: 37 × 79 cms
    Notes: National Resource Atlas of Indonesia in English and
                      Indonesian.

145. *The Land Resources of Indonesia—A National Overview—Atlas, Regional*
    Physical Planning Programme for Transmigration (RepPProT), Soils (Map 8).
    Scale: 1:2,500,000
    Publisher: Land Resources Department NRI; Overseas Development
                      Administration, U.K.; Ministry of Transmigration, Directorate
                      General of Settlement Preparation, Indonesia, 1990.
    Classification: Soil Taxonomy.
    Coverage: Complete coverage.

Iran

146. The Soils of Iran.
    Scale: 1:2,500,000
    Authors: Dewan, M.L.; Famouri, J.
    Publisher: Soils Institute, Ministry of Agriculture, Iran/FAO, 1964.
    Coverage: Complete coverage.

Iraq

147. *Exploratory Soil Map of Iraq.*
    Scale: 1:1,000,000
    Authors: Buringh, P.
    Publisher: Division of Soils and Agricultural Chemistry, Directorate
                      General of Agricultural Research and Projects, Baghdad, 1957.
    Coverage: Complete coverage.

Ireland

148. Scale: 1:1,000,000
      Classification: FAO/UNESCO.
      Coverage: Covered by *Soil Map of the European Communities*. (See Map
                        #11) Also covered at 1:575,000 scale (N.J.Gardiner & T.
                        Radford, 1980).
149. *Ireland: General Soil Map.*
      Scale: 1:575,000
      Authors: Gardiner, M.
      Publisher: Dublin, National Soil Survey, 1969
      Size: 98 × 66 cm

Israel

150. *Israel Soil Associations.*
      Scale: 1:1,000,000
      Authors: Dan, I.; Koyumdjisky, H.
      Publisher: Volcani Institute for Agricultural Research/Hebrew University
                        of Jerusalem, 1971.
      Coverage: Complete coverage. Also 1:500,000 coverage (1975).

Italy

151. *Carta dei Suoli d'Italia.*
      Scale: 1:1,000,000
      Authors: Mancini, F.
      Publisher: Comitato per la Carta dei Suoli, Firenze, 1966.
      Classification: FAO/UNESCO.
      Coverage: Complete coverage. Also covered by *Soil Map of the European
                        Communities*. (See Map #11). Also local coverage at larger
                        scales.
152. *Soil Map of Sicily; Carta dei Suoli della Sicilia.*
      Scale: 1:250,000
      Authors: Fierotti, G.
      Publisher: Regione Siciliana, Assessorato Territorio ed Ambiente/
                        Universita degli Studi di Palermo, Istituto di Agronomia
                        Generale, 1988.
      Classification: FAO/UNESCO; Soil Taxonomy; CPCS French.
      Coverage: Partial coverage.

Ivory Coast

153. *Carte Pédologique de la République de Côte d'Ivoire.*
      Scale: 1:2,000,000
      Authors: Dabin, B.; Leneuf, N.

Publisher: ORSTOM, Service des Sols de la Côte d'Ivoire, 1960.
Coverage: Complete coverage.

154. *Etude Pédologique et des Ressources en Sols de la Région du Nord du 10é Parallěle en Côte d'Ivoire: Cartes des unités morphopédologiques et des paysages morpho-pédologiques, partie ivoirienne des feuilles de Nielle, de Tingrela et de Tienko.*
 Authors: Levêque, A., Trinh, S.
 Publisher: Office de la Recherche Scientifique et Technique Outre-Mer, Paris, France, 1983.
 Size: 6 maps, 46 × 64 cms.

155. *Etude Pédologique de la Région de Boundiali-Korhogo (Côte d'Ivoire): Cartographie et typologie sommaire des sols, feuille Boundiali, feuille Korhogo.*
 Scale: 1:200,000
 Authors: Beaudou, A.G., Sayol, R.
 Publisher: Office de la Recherche Scientifique et Technique Outre-Mer, Paris, France, 1980.
 Size: 4 maps, 55 × 55 cms

156. *Esquisse Pédologique de la Côte d'Ivoire à l'Echelle de 1:500,000.*
 Scale: 1:500,000
 Authors: Perraud, A.; Souchère, P. de la
 Publisher: Paris, Service Cartographique de l'O.R.S.T.O.M., 1969.
 No. of sheets: 4 (77 × 70 cm)
 Size: 1 map, 154 × 140 cm.

Jamaica

157. *Soil Maps.*
 Scale: 1:50,000
 Publisher: Regional Research Centre of the British Caribbean. Imperial College of Tropical Agriculture, Trinidad. Soil and Land Use Surveys.
 No. of sheets: 7
 Coverage: Complete coverage.

Japan

158. *Soil Map of Japan.*
 Scale: 1:500,000
 Authors: Oyama, M.
 Publisher: Yoken-do Co. Ltd, 1975.
 No. of sheets: 6
 Classification: FAO/UNESCO.
 Coverage: Complete coverage. Also coverage at 1:2,000,000 scale.

Jordan

159. *Provisional Soil Map.*
     Scale: 1:1,000,000
     Authors: Moormann, F.R.
     Publisher: FAO, 1955.
     Type of map: Draft, hand-coloured.
     Coverage: Complete coverage.

Kazakhstan

160. *Soil Map of the Kazakh, SSR.*
     Scale: 1:2,500,000
     Authors: Uspanov, U.U.; Evstileev, Ju. G.
     Publisher: Glavnoe Upravlenie Geodeenii i Kartografii pri Sovete
                     Ministrov SSSR, Moscow, 1976.
     Coverage: Complete coverage.

Kenya

161. Exploratory Soil Map of Kenya, in: *Exploratory Soil Map and Agro- Climatic Zone Map of Kenya.*
     Scale: 1:1,000,000
     Authors: Sombroek, W.G.
     Publisher: Kenya Soil Survey, Nairobi, 1980.
     Classification: FAO/UNESCO.
     Coverage: Complete coverage.
162. *Soil Map of IGADD Countries, 1:1,000,000.*
     Scale: 1:1,000,000
     Publisher: FAO-ISRIC, Wageningen, FAO-GIS, Rome, 1992.
     Classification: FAO/UNESCO, Rev. legend.
     Type of map: Digitized
     Coverage: Complete coverage.

Korea, Republic of

163. *Reconnaissance Soil Map of Korea.*
     Scale: 1:250,000
     Publisher: Korea Soil Survey, Institute of Plant Environment/UNDP/FAO
                     AGL:SF/KOR 13, Soil Survey, Technical Report 2, 1971.
     No. of sheets: 4
     Coverage: Complete coverage.
164. *Soil Map of Korea.*
     Scale: 1:1,000,000
     Publisher: Soil Survey & Physics Division, Institute of Agricultural
                     Sciences, Office of Rural Development, Suweon, 1983.
     Coverage: Complete coverage.

165. *Soils of Korea: With Generalized Soil Map of Patterns of Soil Orders and Suborders of Korea.*
     Scale: 1:1,000,000
     Authors: Um, Ki Tae
     Publisher: Agricultural Sciences Institute, Rural Development Administration, Suweon, Korea, 1985..
     Size: 27 × 48 cms
     Notes: 66p. of text. Includes bibliography.

Kuwait

166. *Reconnaissance Soil Map of Kuwait.*
     Scale: 1:500,000
     Authors: Nayeem Sayghi, A.; Al-Shawwa, F.
     Publisher: The Arab Centre for the Studies of Arid Zones and Dry Lands, Damascus/Ministry of Public Works, Kuwait, 1983.
     Coverage: Complete coverage.

Laos

167. *Plaine de Vientiane, Carte Pédologique.*
     Scale: 1:1,000,000
     Publisher: FAO/UNDP LAO/85/007, FAO, 1985.
     No. of sheets: 2   Classification: FAO/UNESCO; Russian.
     Type of map: Transparent + Black & White
     Coverage: Partial coverage.

Latvia

168. *Soil Map of the European Part of USSR.*
     Scale: 1:2,500,000
     Authors: Lobova, E.V.; Rozov, N.N.
     Publisher: USSR Academy of Sciences, Dokuchaev Institute of Soils, 1947.
     No. of sheets: 4
     Coverage: Partial coverage of ex-U.S.S.R.
     Notes: In Cyrillic.

Lebanon

169. *Soil Map of Arab Countries: Syria and Lebanon.*
     Scale: 1:1,000,000
     Authors: Ilaiwi, M.
     Publisher: The Arab Centre for the Studies of Arid Zones and Dry Lands, Soil Science Division, Damascus, 1985.
     Coverage: Complete coverage.

Lesotho

170. Soils, West and East Sheets, in: *The Land Resources of Lesotho.*
     Scale: 1:250,000
     Authors: Bawden, M.G.; Carroll, D.M.
     Publisher: Land Resources Division, Directorate of Overseas Surveys,
                     Land Resource Study 3, U.R, 1968.
     No. of sheets: 2
     Coverage: Complete coverage

Liberia

171. Soils Associations, in: *Mano River Union, Project Area, Liberia.*
     Scale: 1:1,000,000
     Authors: Van Mourik, D.; Sao, D.R.
     Publisher: Mano River Union Secretariat, Freetown, 1979.
     Coverage: Partial coverage.

Libya

172. *Soil Map of the Central Zone.*
     Scale: 1:200,000
     Publisher: Soil-Ecological Expedition of the USSR V/O
                     "Selkhozpromexport.," 1980.
     Coverage: Partial coverage.

Lithuania

173. *Soil Map of the European Part of USSR.*
     Scale: 1:2,500,000
     Authors: Lobova, E.V.; Rozov, N.N.
     Publisher: USSR Academy of Sciences, Dokuchaev Institute of Soils,
                     1947.
     No. of sheets: 4
     Coverage: Partial coverage of ex-U.S.S.R.
     Notes: In Cyrillic.

Luxembourg

174. *Carte des Sols du Grand-Duche de Luxembourg.*
     Scale: 1:100,000
     Authors: Wagener, J.
     Publisher: Ministere de l'Agriculture et de la Viticulture, Administration
                     des Services, Service de Pedologie, 1969.
     Coverage: Complete coverage. Also covered at 1:1,000,000 scale by *Soil
                     Map of the European Communities.* (See Map #11)

Madagascar

175. *Carte Pédologique de Madagascar.*
      Scale: 1:1,000,000
      Authors: Riquier, J.
      Publisher: Centre ORSTOM de Tananarive, 1968.
      No. of sheets: 3
      Coverage: Complete coverage.

Malawi

176. Soils, in: *The National Atlas of Malawi.*
      Scale: 1:1,000,000
      Publisher: Atlas Coordination Committee, 1983.
      Coverage: Complete coverage. Includes other physical resources maps.
177. *Soils/Physiography, Map 1.*
      Scale: 1:250,000
      Publisher: Land Resources Evaluation Project MLW/85/011, Ministry of
                        Agriculture UNDP-FAO, 1991.
      No. of sheets: 7
      Coverage: Complete coverage.

Malaysia

178. *Generalized Soil Map, Peninsular Malaysia.*
      Scale: 1:800,000
      Publisher: Dilukis di Bilik Lukisan Nilatan Soils, Kuala Lumpur, 1970.
      Coverage: Partial coverage.
179. *Soil Map of Sarawak.*
      Scale: 1:500,000
      Publisher: Soil Survey Division, Research Branch, Department of
                        Agriculture, Sarawak/Directorate of National Mapping,
                        Malaysia, 1968.
      No. of sheets: 2
      Classification: FAO/UNESCO.
      Coverage: Partial country coverage.
180. *The Soils of Sabah.*
      Scale: 1:250,000
      Authors: Acres, B.D.; Bower. R.P.
      Publisher: Land Resources Division, Land Resource Study 20, 1975.
      No. of sheets: 10
      Coverage: Partial coverage.

Mali

181. *Les Ressources Terrestres au Mali; Mali Land and Water Resources (Volume
      I).*
      Scale: 1:500,000

Publisher: USAID/TAMS, New York, N.Y., 1983.
Coverage: Partial coverage. Complete coverage in: *Atlas Jeune Afrique*
                  (1980).

Malta

182. *Soils Map, Malta & Gozo* (DOS-Misc. 258, 1st ed.).
      Scale: 1:30,000
      Publisher: Directorate of Overseas Surveys, U.K, 1960.
      Coverage: Complete coverage.

Mauritania

183. *Carte Pédologique du Guidimaka.*
      Scale: 1:200,000
      Authors: Audry, P.; Pereira Barreto, S.
      Publisher: ORSTOM, Centre de Dakar-Hann, 1961.
      Coverage: Partial coverage. Complete coverage in: *Atlas Jeune Afrique*
                        (1977).

Mauritius

184. *Carte Pédologique.*
      Scale: 1:50,000
      Authors: Willaime, P.
      Publisher: ORSTOM/Mauritius Sugar Industry Research Institute, Reduit,
                        1983.
      No. of sheets: 3
      Coverage: Complete coverage.

Mexico

185. *Carta Edafologica 1:1,000,000.*
      Scale: 1:1,000,000
      Publisher: Secretaria de Programacion y Presupuesto, Direccion General
                        de Geografia del Territorio Nacional, Estados Unidos
                        Mexicanos, 1981.
      No. of sheets: 8
      Classification: FAO/UNESCO (1970 modificada por DGGTENAL).
      Coverage: Complete coverage.
      Notes: Carta Edafologica, 1:250,000, 122 sheets planned (65
                        published), DGG, 1982–.

Mongolia

186. *Soil Map of Mongolia.*
      Scale: 1:2,500,000

Publisher: Bnmau-vin Barilga-Akhilga, ClaanoaatarHot, 1983.
Type of map: Photographic reproduction
Coverage: Complete coverage.
Notes: In Cyrillic.

Morocco

187. *Carte Pédologique de la Région de Settat Utilisant les Données du Satellite Landsat Thematic Mapper.*
Scale: 1:100,000
Authors: Marzouk, A. and Badraoui, M.
Publisher: Université de Sherbrooke, Centre d'Application et de
                 Recherches en Télédétection, Canada, Institut Agronomique et
                 Vétérinaire Hassan II, Maroc, 1989.
Coverage: Partial coverage.
188. *Atlas du Bassin du Sébou. Sols (Rharb), Lithologie, Types de Végétation,*
Pentes, Érosion Actuelle, Perte de Sols, Érosion Potentille Maximum.
Scale: 1:500,000
Publisher: Ministère de l'Agriculture et de la Réforme Agraire, Maroc, FAO-
PNUD, 1970.
Coverage: Partial coverage. Atlas includes other thematic maps.
189. *Carte Pédologique, Région du Souss: Agadir-Ait Baha.*
Scale: 1:100,000
Authors: Staimesse, J. P. and P. Billaux
Publisher: ORSTOM, Paris, 1978.
Coverage: Partial coverage.
190. *Cartes Pédologiques du Tadla, de la Région de Meknes-Fes et de la Bordure*
Méridionale du Rharb, de la Mamora Septentrionale et de leur Borduree
Orientale.
Scale: 1:500,000
Publisher: Ministère de l'Agriculture et de la Réforme Agraire, Maroc,
                 1966.
Coverage: Partial coverage.

Mozambique

191. *Land Resources Inventorv, Mozambique.*
Scale: 1:2,000,000
Authors: Kassam, A.H.; Van Velthuizen, H.T.
Publisher: FAO/UNDP MOZ/75/011 Land and Water Use Planning
                 Project, Field Document 35, 1982.
Classification: FAO/UNESCO (1974).
Coverage: Complete coverage.

Myanmar

192. *Soil Map of Burma.*
Scale: 1:2,600,000

Authors: Hla Aye
Publisher: Ministry of Agriculture and Forest, Land Use Division, 1963.
Type of map: Hand-coloured
Coverage: Complete coverage.

Namibia

193. *Land Types, Sheets 1 & 2.*
     Scale: 1:1,000,000
     Publisher: FAO/UNDP AG:DP/NAM/78/004 Assessment of Potential
                     Land Suitability, Technical Report 2 - Land Regions and Land
                     Use Potential, 1983.
     No. of sheets: 2
     Coverage: Complete coverage.

Nepal

194. *Generalized Soil Map of Nepal* (provisional).
     Scale: 1:1,000,000
     Publisher: Division of Soil Science & Agricultural Chemistry, Department
                     of Agriculture, Nepal, 1978.
     Type of map: Black & White
     Coverage: Complete coverage.

Netherlands

195. *Globale Bodemkaart, Generalized Soil Map.*
     Scale: 1:600,000
     Publisher: Stichting voor Bodemkartering, Soil Survey Institute,
                     Wageningen, Netherlands.
     Coverage: Complete coverage. Also covered by *Soil Map of the European
                     Communities.* (See Map #11). Coverage also at 1:250,000 and
                     1:50,000 scale by STIBOKA.
196. *Bodemkaart van Nederland.*
     Scale: 1:250,000
     Publisher: Wageningen, Stiboka, 1985.
     No. of sheets: 4

New Caledonia

197. *Carte Pédologique de Nouvelle Calédonie.*
     Scale: 1:1,000,000
     Authors: Latham, M.
     Publisher: ORSTOM, France, 1978.
     No. of sheets: 2
     Coverage: Complete coverage.

New Zealand

198. *Soil Map of the North and South Islands New Zealand.*
     Scale: 1:1,000,000
     Publisher: Department of Scientific and Industrial Research, Wellington, 1963.
     No. of sheets: 2
     Coverage: Complete coverage.
199. *Soil Maps and Extended Legend of Whakatane Borough and Environs, Bay of Plenty, New Zealand.*
     Authors: Pullar, W.A.
     Publisher: Department of Scientific and Industrial Research, Wellington, N.Z, 1972.
     Size: 2 col. maps, 45 × 69 cms. and 33 × 37 cms.

Nicaragua

200. State of Ground (L8–T8), Soils-Agricultural (L5–Ts), Soils-Engineering (L13–T13), in: *Inventario Nacional de Recursos Fisicos; National Inventory of Physical Resources.*
     Scale: 1:1,000,000
     Publisher: AID Resources Inventory Center/Corps of Engineers, Washington, D.C., USA, 1966.
     No. of sheets: 3
     Coverage: Complete coverage.

Niger

201. *Carte Pédologique de Reconnaissance de la République du Niger, Feuilles Maradi, Zinder et Niamey.*
     Scale: 1:500,000
     Publisher: ORSTOM, Centre de Recherches Pédologiques de Hann-Dakar, 1964.
     No. of sheets: 3
     Coverage: Partial coverage. Complete coverage in: *Atlas Jeune Afrique* (1980).

Nigeria

202. Soil Associations, Map 1 in: *Nigeria Land Resources Management Study.*
     Scale: 1:1,500,000
     Publisher: FAO/World Bank Cooperative Programme, Investment Centre/ODA Natural Resources Institute, 1990.
     Type of map: Photocopy + transparent
     Coverage: Complete coverage.
203. *Soils Map of Nigeria.*
     Scale: 1:500,000
     Publisher: Soil Survey Division, Federal Department of Forestry and Agricultural Land Resources, Kaduna.

    No. of sheets: 8
    Type of map: Black & White
    Coverage: Complete coverage.
204.   *Soil Map of Eastern Nigeria.*
    Scale: 1:1,000,000
    Authors: Jungerius, P.D.
    Publisher: Luxembourg, Pub. du Service Geol. du Lux., Vol. XIV, 1964.

Norway

205.   *Jordsmonnkart Over Norge.*
    Scale: 1:2,000,000
    Authors: Lâg, J.
    Publisher: Norges Landbrukshogskole, 1979.
    Coverage: Complete coverage. Covered also at 1:1,000,000 scale (M. Thoresen).
206.   *Jordartkart Over Syd- Norge.*
    Scale: 1:1,000,000
    Authors: Holmsen, G.
    Publisher: Oslo, N. Geol. Under., 1971.

Oman

207.   *General Soil Map, Sultanate of Oman.*
    Scale: 1:250,000
    Publisher: Ministry of Agriculture and Fisheries/FAO OMA/87/011 Soil Survey and Land Classification Project, 1991.
    Coverage: Complete coverage.

Pakistan

208.   *Generalized Soil Map.*
    Scale: 1:5,000,000
    Publisher: Soil Survey Project: Soil Resources in West Pakistan and Their Development Possibilities, 1970.
    Type of map: Black & White, inserted in report.
    Coverage: Complete coverage.
209.   *Landforms and Soils, Indus Plain, West Pakistan.*
    Scale: 1:250,000
    Publisher: Government of Canada for the Government of Pakistan (Colombo Plan Cooperative Project), 1956.
    No. of sheets: 25
    Coverage: Partial coverage.
210.   *Soil Map of Sind.*
    Scale: 1:1,000,000
    Authors: Choudhri, B.
    Publisher: Soil Survey of Pakistan, Lahore, 1978.
    Coverage: Partial coverage.

Panama

211. *Mapa General de Suelos de Panama.*
  Scale: 1:500,000
  Publisher: Seccion de Suelos, Departamento de Forestales y Suelos,
                  Panama, 1964.
  Type of map: Black & White
  Coverage: Complete coverage.

Papua New Guinea

212. *Soils of Papua New Guinea.*
  Scale: 1:1,000,000
  Authors: Bleeker, P.
  Publisher: CSIRO Division of Water and Land Resources, Natural
                  Resources Series No. 10, Australia, 1988.
  No. of sheets: 4
  Classification: Soil Taxonomy.
  Coverage: Complete coverage.

Paraguay

213. *Soil Map of Paraguay.*
  Scale: 1:2,000,000
  Authors: Tirado Sulsona, P.
  Publisher: U.S. Department of the Interior, Geological Survey,
                  Professional Paper 327 Plate 3, 1952.
  Coverage: Complete coverage.
214. *Suelos; Region Nororiental del Paraguay, Estudio para su Planificacion y
  Desarrollo, Cuenca del Plata.*
  Scale: 1:500,000
  Publisher: Secretaria General de la Organizacion de los Estados
                  Americanos; Republica del Paraguay, 1975.
  Coverage: Partial coverage.
215. *Mapa de Suelos. Clasificación Preliminar. República del Paraguay.*
  Scale: 1:1,000,000
  Authors: Ministerio de Agricultura y Ganaderia (Paraguay)
  Publisher: Asunción, Dpto. de Ingeniería, 1952.
  No. of sheets: 2
  Size: 9 × 23 inches
  Type of map: Lithographed copies
  Notes: Sheet 1, Región oriental; Sheet 2, Región occidental (Chaco).

Peru

216. *Inventario, Evaluacion e Integracion de los Recursos Naturales, Mapas de
  Grandes Grupos de Suelos.*

Publisher: Oficina Nacional de Evaluacion de Recursos Naturales, Lima,
                  1971–82.
No. of sheets: 16
Classification: FAO/UNESCO; National; Soil Taxonomy.
Coverage: Partial country coverage (various scales) (Pucallpa, Alto Mayo,
                  Conchucos, Chicama, Fortaleza, Casma, Villa Rica,
                  Cajamarca, Moche, Viru, Rio Grande, Ica, Canete, Chilca,
                  Piura).

217. *Mapa de suelos. Zona Cenepa—Alto Maranon.*
      Scale: 1:100,000
      Publisher: Oficina Nacional de Evaluacion de Recursos Naturales, Lima,
                      1976.
      Size: 52 × 67 cms   Coordinates: 5.00S 78.07W 4.23S 77.38W
      Notes: Accompanies the volume: "Inventario y Evaluacion de los
                      Recursos de Suelos y Forestales de la Zona Cenepa—Alto
                      Marañon" Mapa No. 1.

218. *Mapa de Suelos del Peru* (Segunda Aproximacion).
      Scale: 1:3,000,000
      Authors: Zavaleta Garcia, A.
      Publisher: Universidad Agraria "La Molina," Departamento de Suelos,
                      1964.
      Coverage: Complete coverage.

219. *Mapa de Suelos del Peru.*
      Scale: 1:1,000,000
      Publisher: Oficina Nacional de Evaluacion de Recurso Naturales, 1969.

Philippines

220. *Soil Map of the Philippines.*
      Scale: 1:1,600,000
      Authors: Mariano, J.A.; Valmidiano, A.T, 1972.
      Type of map: Hand-coloured draft + Black & White
      Coverage: Complete coverage.

221. Soil Map of the Philippines in: *ASPAC Food & Fertilizer Technology Centre,
      Technical Bulletin No. 12.*
      Scale: 1:4,800,000
      Authors: Mariano, J.A.; Valmidiano, A.T.
      Publisher: ASPAC Food & Fertilizer Technology Centre, 1972.
      Type of map: Black & White
      Coverage: Complete coverage.

222. *Soil Map, Palawan Province.*
      Scale: 1:500,000
      Authors: Barrera, A.
      Publisher: Department of Agriculture and Natural Resources, Bureau of
                      Soils, Manila, 1960.
      Coverage: Partial coverage.

Poland

223. *Mapa Gleb Polski; Soil Map of Poland.*
      Scale: 1:1,000,000
      Authors: Dobrzanski, B.; Kuznicki, F.
      Publisher: Polska Akademia Nauk, Instytut Uprawy Nawozenia i
                      Gleboznawstwa, 1974.
      Coverage: Complete coverage.
224. Soil Map of Poland with FAO Classification (Mapa gleb Polski wedlua
      Nomenklatury FAO). In: *Distinguishing Criteria and Spatial Approach to
      Polish Soils According to the FAO Classification.*
      Scale: 1:2,000,000
      Authors: Dobrzanski, B.; Kuznicki, F.; Bialousz, S.
      Publisher: OPGK Bialystok, Warsaw, 1984.
      Classification: FAO/UNESCO.
      Coverage: Complete coverage.

Portugal

225. *Carta dos Solos de Portugal.*
      Scale: 1:1,000,000
      Authors: Carvalho Cardoso, J.; Teixeira Bessa, M.
      Publisher: Servico de Reconhecimento e de Ordenamento Agrario, 1971.
      Classification: FAO/UNESCO.
      Coverage: Complete coverage. Also covered at 1:1,000,000 scale by *Soil
                      Map of the European Communities.* (See Map #11)

Qatar

226. *Reconnaissance Soil Survey and Land Classification.*
      Publisher: FAO-UNDP AGL:QAT/71/501 Hydro-Agricultural Resources
                      Survey. Technical Report 1, FAO, 1973.
      No. of sheets: 3
      Type of map: Black & White
      Coverage: Complete coverage.

Romania

227. *Harta Solurilor.*
      Scale: 1:1,000,000
      Publisher: Institutul de Cercetari pentru Pedologie si Agrochimie, 1978.
      Coverage: Complete coverage. Covered also at 1:2,000,000 in *Atlas of the
                      Danubian Countries.* (See Map #8)
228. *Marta Pedologiea a Republicii Socialiste Romania.*
      Scale: 1:500,000
      Authors: Florea, N., et al.
      Publisher: Bucuresti, Institutual Geologie si Geofizica, 1970.
      No. of sheets: 4

Russian Federation

229. *Pochvennaia Karta SSSR.*
     Scale: 1:10,000,000
     Publisher: Glavnoe Upravlenie Geodeenie i Kartografii Mvd SSSR, 1960.
     Coverage: Complete coverage of ex-U.S.S.R.
230. *Soil Map of Moldavia SSR.*
     Scale: 1:750,000
     Date: 1971
     Coverage: Partial coverage.
     Notes: In Cyrillic.
231. *Soil Map of the European Part of USSR.*
     Scale: 1:2,500,000
     Authors: Lobova, E.V.; Rozov, N.N.
     Publisher: USSR Academy of Sciences, Dokuchaev Institute of Soils,
                       1947.
     No. of sheets: 4
     Coverage: Partial coverage of ex-U.S.S.R.
     Notes: In Cyrillic.
232. *Pochvennaia Karta Sredneaziatskikh Respublik.*
     Scale: 1:2,500,000
     Authors: Kerzum, P. A., et al.
     Publisher: Moscow, GUGK, 1971.
     Size: 1 map, 44 × 97 cm
     Coverage: Turkmen, Uzbek, Tadzhik and Kirghiz republics.
     Notes: Sostavlena, oformlena i podgotovlena k pechati Nauchno-
                       redaktsionnoi kartosostavietel'skoi chast'iu GUGK v 1970 g.

Rwanda

233. *Carte des Associations des Sols du Rwanda et du Burundi.*
     Scale: 1:1,000,000
     Authors: Van Wambeke, A.
     Publisher: INEAC (Institut National pour l'Etude Agronomique du
                       Congo), Institut Geographique Militaire, Bruxelles, 1962.
     Coverage: Complete coverage.
234. *Carte des Sols du Congo Belge et du Ruanda- Urundi.*
     Scale: 1:5,000,000
     Authors: Sys, C.
     Publisher: Institut National pour l'Etude Agronomique du Congo Belge,
                       Bruxelles, 1958–59.
     Coverage: Complete coverage.
235. *Soil Map of IGADD Countries, 1:1,000,000.*
     Scale: 1:1,000,000
     Publisher: FAO-ISRIC, Wageningen, FAO-GIS, Rome, 1992.
     Classification: FAO/UNESCO, Rev. legend.
     Type of map: Digitized
     Coverage: Complete coverage.

Saint Christopher and Nevis

236. *Soil Map, St. Kitts.*
      Scale: 1:25,000
      Authors: Lang, D.M.; Carroll, D.M.
      Publisher: The Regional Research Centre, Imperial College of Tropical
                        Agriculture, University of the West Indies, Soil & Land Use
                        Surveys No. 16, 1966.
      No. of sheets: 3
      Coverage: Complete coverage.

Saint Lucia

237. *Soil Map of St. Lucia.*
      Scale: 1:25,000
      Authors: Stark, J.; Green, A.J.
      Publisher: The Regional Research Centre, Imperial College of Tropical
                        Agriculture, University of the West Indies, Soil & Land Use
                        Surveys No. 20, 1966.
      No. of sheets: 3
      Coverage: Complete coverage.

Saint Vincent and the Grenadines

238. *Soil Map of St. Vincent.*
      Scale: 1:20,000
      Authors: Watson, J.P.; Jones, T.A.
      Publisher: The Regional Research Centre of the British Caribbean at the
                        Imperial College of Tropical Agriculture, Trinidad, Soil &
                        Land Use Surveys No. 3, 1958.
      No. of sheets: 4
      Coverage: Partial coverage.

Samoa

239. *Soils and Land Use of Western Samoa.*
      Authors: Wright, A.C.S.
      Publisher: 1963.
      No. of sheets: 8
      Notes: NZ Soil Bureau Bulletin no. 22. Has: Soil Map of Savai,
                        1:95,040; Soil Map of Upolu, 1:95,040; Provisional Soil Map
                        of Upolu, 1:40,000.

Sao Tome and Principe

240. *Carta dos Solos do Principe.*
      Scale: 1:50,000

Authors: Carvalho Cardoso, J.
Publisher: Junta de Investigacoes do Ultramar, Missao de Estudos
                Agronomicos do Ultramar, 1960.
Coverage: Partial coverage.
241. *Carta dos Salos de Sao Tome.*
Scale: 1:50,000
Publisher: 1960.

Saudi Arabia

242. *General Soil Map, Kingdom of Saudi Arabia.*
Scale: 1:1,000,000
Publisher: Land Management Department, Ministry of Agriculture and
                Water, Riyadh, National Soil Survey and Land Classification
                Project (UTFN/SAU/015/SAU), 1988.
No. of sheets: 12
Classification: 7th Approximation.
Coverage: Complete coverage.
243. *General Soil Map of the Kingdom of Saudi Arabia.*
Scale:·   1:125,000
Publisher: Ministry of Agriculture and Water/U.S.-Saudi Arabian Joint
                Commission on Economic Cooperation, 1985.
Type of map: Landsat imagery.
Coverage: Complete coverage.

Senegal

244. *Carte Pédologique du Sénégal.*
Scale: 1:1,000,000
Authors: Maignien, R.
Publisher: Centre ORSTOM de Dakar-Hann, 1965.
Coverage: Complete coverage.

Seychelles

245. *Seychelles Soils.*
Scale: 1:50,000
Authors: Piggott, C.J.
Publisher: Directorate of Overseas Surveys, U.K., 1966.
No. of sheets: 4
Coverage: Complete coverage.

Sierra Leone

246. *Land Systems of Sierra Leone.*
Scale: 1:500,000

Authors: Gordon, O.L.A., Schwaar, D.C.
Publisher: FAO-UNDP Land Resources Survey Project SIL/73/002,
                 Technical Report 1, 1980.
Classification: FAO/UNESCO.
Coverage: Complete coverage.

Singapore

247. *Soil Map of Republic of Singapore.*
     Scale: 1:63,360
     Authors: Ives, D.W.
     Publisher: 1977.
     Notes: New Zealand Soil Bureau Map 169.

Solomon Islands

248. *Land Resources of the Solomon Islands: Outer Islands.*
     Scale: 1:250,000
     Authors: Wall, J.R.D.; Hansell, J.R.F.
     Publisher: Land Resources Division, Land Resource Study 18, Volume
                      18, Land Resources Development Centre, U.K, 1976.
     No. of sheets: 2
     Coverage: Partial country coverage. Other islands covered at 1:250,000
                      scale by LRDC.
249. *The Soils of the Solomon Islands.*
     Scale: 1:1,000,000
     Authors: Wall, J.R.D.; Hansell, J.R.F.
     Publisher: Land Resources Development Centre, U.K., Technical Bulletin
                      4 (Maps), 1979.
     No. of sheets: 2
     Coverage: Complete coverage.

Somalia

250. *Soil Map of IGADD Countries, 1:1,000,000.*
     Scale: 1:1,000,000
     Publisher: FAO-ISRIC, Wageningen, FAO-GIS, Rome, 1992.
     Classification: FAO/UNESCO, Rev. legend.
     Type of map: Digitized
     Coverage: Complete coverage.

South Africa

251. *Soil Map of South Africa.*
     Scale: 1:5,000,000
     Publisher: Soils Research Institute, Department Agricultural Tech.
                   Services, Pretoria, 1965.

Type of map: Black & White + hand-coloured
Coverage: Complete coverage.

Spain

252. Scale: 1:1,000,000
      Classification: FAO/UNESCO.
      Coverage: Covered by *Soil Map of the European Communities*. (See Map
                        #11) 1:100,000 series underway by ICONA/LUCDEME
                        Project.

Sri Lanka

253. *Agro-Ecological Map of Sri Lanka.*
      Scale: 1:1,000,000
      Authors: Wijesinghe, T.M.K.
      Publisher: Land and Water Use Division, Department of Agriculture,
                        Peradeniya/ITC, Enschede, 1981.
      Coverage: Complete coverage.
254. *Soil Map of Ceylon.*
      Scale: 1:500,000
      Authors: Panabokke, C.R.; De Alwis, K.A.
      Publisher: Land Use Division, Irrigation Department, Colombo, 1971.
      Classification: Soil Taxonomy.
      Coverage: Complete coverage.

Sudan

255. *Soil Map of IGADD Countries.*
      Scale: 1:1,000,000
      Publisher: FAO-ISRIC, Wageningen, FAO-GIS, Rome, 1992.
      Classification: FAO/UNESCO, Rev. legend.
      Type of map: Digitized
      Coverage: Complete coverage.
256. *Soil Resources Regions of the Blue Nile, White Nile, Gezira and Khartoum
      Provinces of the Sudan.*
      Scale: 1:1,000,000
      Authors: Purnell, M.F.; De Pauw, E.F.
      Publisher: Soil Survey Administration, Wad Medani, SUD/71/553, FAO,
                        1976.
      Type of map: Black & White, inserted in report.
      Coverage: Partial coverage.

Suriname

257. Geomorphology and Soils, in: *Suriname Planatlas.*
      Scale: 1:1,000,000

Publisher: National Planning Office of Suriname (SPS), Regional
              Development and Physical Planning Department (HARPRO);
              Organization of American States (OAS), Washington, D.C.,
              1988.
Coverage: Complete coverage.

Swaziland

258. *Soil Map of Swaziland (Southern & Northern Sheets).*
      Scale: 1:125,000
      Authors: Murdoch, G.
      Publisher: Ministry of Agriculture, Mbabane, 1968.
      No. of sheets: 2
      Coverage: Complete coverage.

Sweden

259. *Prognoskarta Över Åkerjorden i Sverige = Prognostic Map of the Arable Land
      in Sweden.*
      Scale: 1:500,000
      Publisher: Stockholm, 1967.
      No. of sheets: 4
      Size: 90 × 124 cm.
      Notes: Kungl. Lantbruksstyrelsen, Meddelanden Ser. A nr 6.

Switzerland

260. *Bodenkarte der Schweiz = Carte des Sols de la Suisse.*
      Scale: 1:1,000,000
      Authors: Frei, E.; Jugasz, P.
      Publisher: Station Federale d'Essais Agricoles, Zurich, 1964.
      Coverage: Complete coverage. Also covered by *Soil Map of Middle
                      Europe* at 1:1,000,000 (1986).

Syria

261. *Soil Map of Arab Countries: Syria and Lebanon.*
      Scale: 1:1,000,000
      Authors: Ilaiwi, M.
      Publisher: The Arab Centre for the Studies of Arid Zones and Dry Lands,
                      Soil Science Division, Damascus, 1985.
      Coverage: Complete coverage.
262. Soil Map of Syria, in: *Contribution to the Knowledge of the Soils of Syria*
      (Thesis).
      Scale: 1:1,000,000
      Authors: Ilaiwi, M.

Publisher: State University of Ghent, Belgium, 1983.
Type of map: Black & White
Coverage: Complete coverage.

Tanzania

263. *Soils and Physiography.*
      Scale: 1:2,000,000
      Authors: De Pauw, E.F.
      Publisher: FAO Project Crop Monitoring and Early Warning Systems,
                    GCP/URT/047/NET, Dar es Salaam, 1983.
      Coverage: Complete coverage.
264. *Tabora Region: Land Evaluation and Land Use Planning. Land Units.*
      Scale: 1:500,000
      Authors: Acres, B.D.; Mitchell, A.J.B.
      Publisher: Land Resources Development Centre, U.K, 1984.
      Coverage: Partial coverage.

Thailand

265. *General Soil Map of Thailand.*
      Scale: 1:1,000,000
      Authors: Vijarnsorn, P. and C. Jongpakdee
      Publisher: Soil Survey Division, Department of Land Development, 1979.
      Classification: Soil Taxonomy
      Coverage: Complete coverage.
266. *The Soils of the Kingdom of Thailand: Explanatory Text of the General Soil Map.*
      Scale: 1:1,250,000
      Authors: Moorman, F.R.; Rojamasoonthon, S.
      Publisher: Soil Survey Dvision, Ministry of Resources, Bangkok,
                    Thailand, 1972.
      No. of sheets: 2
      Notes: 59p. of text. Illustrated with maps, 2 folded in pocket 27cm.
267. *General Soil Map of Thailand.*
      Scale: 1:1,000,000
      Authors: Vijarnsorn, P.; Jongpakee, C.
      Publisher: Bangkok, SSD, M of R., 1979.
      No. of sheets: 3

Togo

268. *Carte Pédologique du Togo a l'Echelle de 1:1,000,000.*
      Scale: 1:1,000,000
      Authors: Lamouroux, R.
      Publisher: ORSTOM, Centre de Lome, Notice explicative No. 34, 1966.
      Coverage: Complete coverage.

269. *Ressources en Sols du Togo, Carte des Unités Agronomiques Déduites de la Carte Pédologique, Socle Granito-Gneissique Limite a l'Ouest et au Nord par les Monts Togo.*
      Scale: 1:200,000
      Authors: Lévêque, A.
      Publisher: ORSTOM, Paris, 1978.
      No. of sheets: 4
      Coverage: Partial coverage.

Tonga

270. *Soil Map of Tongatapu Island, Tonga.*
      Scale: 1:100,000
      Authors: Gibbs, H.S.
      Publisher: Department of Scientific and Industrial Research, Wellington, 1972.
      Coverage: Partial coverage.
      Notes: Other islands covered at 1:25,000 from New Zealand Survey Soil Bureau.

Trinidad and Tobago

271. *Soil Map of Tobago* (Sheet 3).
      Scale: 1:25,000
      Authors: Hansell, J.; Hill, I.
      Publisher: Land Capability Survey of Trinidad & Tobago, 1964.
      Coverage: Partial coverage.
272. *Trinidad, Soils.*
      Scale: 1:150,000
      Publisher: Land Capability Survey, Ministry of Agriculture and U.W.I, 1971.
      Coverage: Partial coverage.

Tunisia

273. Carte Pédologique de la Tunisie, in: *Sols de Tunisie, No 5.*
      Scale: 1:500,000
      Publisher: Direction des Ressources en Eau et en Sol (DRES), Division des Sols, Tunis, 1973.
      Coverage: Complete coverage.

Turkey

274. *Erosion Map of Turkey.*
      Scale: 1:1,000,000
      Authors: Akyurek, I.; Kavak, S.
      Publisher: TOPRAKSU Genel Mudurlugu, 1981.

275. *General Soil Map of Turkey.*
      Scale: 1:800,000
      Authors: Oakes, H.; Arikok, Z.
      Publisher: Soil Survey Section, Ankara Soil and Fertilizer Research
                        Institute, 1954.
      No. of sheets: 4
      Coverage: Partial coverage.

Tuvalu

276. *Tuvalu Land Resources Survey, Island Reports* (Nanumea, Nanumaga, Niutao,
      Nui, Vaitupu, Nukufetau, Funafuti, Nukulaelae, Niulakita).
      Authors: McLean, R.F.; Kelly, J.
      Publisher: FAO-UNDP TUV/80/011 Tuvalu Land Resources Survey;
                        Government of Tuvalu; University of Auckland, 1991.
      No. of sheets: 7
      Type of map: Black & White, inserted in report.
      Coverage: Complete coverage (various scales).

Uganda

277. *Soil Map of IGADD Countries, 1:1,000,000.*
      Scale: 1:1,000,000
      Publisher: FAO-ISRIC, Wageningen, FAO-GIS, Rome, 1992.
      Type of map: Digitized
      Classification: FAO/UNESCO, Rev. legend.
      Coverage: Complete coverage.
278. *Uganda, Soils.*
      Scale: 1:1,500,000
      Publisher: Uganda Government, 1967.
      Coverage: Complete coverage.

Ukraine

279. *Soils Atlas of Ukraine.*
      Scale: 1:2,000,000
      Authors: Sokolovski, A.N.; Voronina, A.V.
      Publisher: GUGK (Direction General of Geodesy and Cartography); Soils
                        and Agrochemistry Institute of Ukraine, 1977.
      Coverage: Complete coverage.
      Notes: In Cyrillic.

United Kingdom

280. *Soil Map of England and Wales.*
      Scale: 1:1,000,000
      Authors: Avery, B.W.; Findlay, D.C.

Publisher: Ordnance Survey, Southampton, 1975.
Coverage: Partial coverage. Complete coverage at 1:1,000,000 scale by
                Soil Map of the European Communities. (See Map #11)
                Coverage also at larger scales (1:250,000), by SSEW.

Uruguay

281. *General Soil Map of Uruguay; Carta General de Suelos del Uruguay.*
      Scale: 1:1,000,000
      Authors: Duran, A.; Kaplan, A., 1967.
      Type of map: Black & White
      Coverage: Complete coverage.

Uzbekistan

282. *Soil Map of Uzbekistan.*
      Scale: 1:1,500,000
      Publisher: Uzbekistan Academy of Agricultural Science, Institute of Soil
                     Science, 1960.
      Coverage: Complete coverage.
      Notes: In Cyrillic.

Vanuatu

283. *Pédologie, in: Nouvelles Hebrides: Atlas des Sols et de Quelques Données du
      Milieu Naturel.*
      Scale: 1:100,000
      Authors: Quantin, P. Publisher: ORSTOM, Paris, 1973–78.
      No. of sheets: 7
      Coverage: Complete coverage.

Venezuela

284. *Map of the Valley of Lake Valencia.*
      Scale: 1:20,000
      Publisher: Ministerio de Obras Publicas, Direccion General de Recursos
                     Hidraulicos, 1976.
      Coverage: States of Aragua and Carbobo. Size: 35×51 cms.
      Notes: Accompanies the volume: "Estudios de Suelos Semi-detallado,
                     Depresion del Largo de Valencia Estado de Aragua y
                     Carabobo."
285. *Suelos, Mapa de Orderes v Subordenes* (7a Aproximacion).
      Scale: 1:1,500,000
      Authors: Comerma, J.
      Publisher: Centro de Investigaciones Agronomicas, Seccion de Suelos,
                     1970.

Classification: Soil Taxonomy.
Type of map: Hand-coloured
Coverage: Complete coverage.

Vietnam

286. *Ban do Sinh thai hong nghiep dong Bang Song Cuu Long = Agro-ecological*
      map of the Mekong Delta, Vietnam = Carte Agro-ecologique du Delta du
      Mekong, Vietnam.
      Authors: Tran An Phong
      Publisher: National Institute of Agricultural Planning and Projection, Ho
                      Chi Minh City, Vietnam, 1985.
      No. of sheets: 4
      Size: 76 × 102 cms
287. *General Soil Map; Carte Générale des Sols.*
      Scale: 1:1,000,000
      Authors: Moormann, F.R.
      Publisher: Directorate for Studies and Research in Agronomy, Forestry
                      and Animal Husbandry; U.S. International Cooperation
                      Administration/FAO, 1961.
      Coverage: Complete coverage.
288. *Soil Map of Mekong Delta.*
      Scale: 1:250,000
      Authors: Ton That Chieu; Tran An Phong
      Publisher: National Institute of Agricultural Planning and Projection
                      (NIAPP) in the Investigational State Program in Mekong Delta,
                      2nd period, 1990.
      No. of sheets: 6
      Classification: FAO/UNESCO; Soil Taxonomy.
      Coverage: Partial coverage.

Yemen

289. *Yemen General Soil Map.*
      Scale: 1:500,000
      Publisher: Department of Agronomy, New York State College of
                      Agricultural and Life Sciences at Cornell University; Yemen
                      Arab Republic Ministry of Agriculture, 1983.
      Coverage: Complete coverage.
      Size: 110 × 83 cms.
      Notes: Inset: Administrative boundaries.

Yugoslavia

290. *Pedoloska Karta, Soil Map.*
      Scale: 1:1,000,000
      Authors: Nejgebauer, V.; Filipovski, G.

Publisher: Yugoslav Society of Soil Science, Geokarta, Beograd, 1959.
Coverage: Complete coverage. Covered also at 1:2,000,000 scale in *Atlas of the Danubian Countries*. (See Map #8)

Zaire

291. *Carte del Sols du Congo Belge et du Ruanda- Urundi.*
      Scale: 1:5,000,000
      Authors: Sys, C.
      Publisher: Institut National pour l'Etude Agronomique du Congo Belge, Bruxelles, 1958–59.
      Coverage: Complete coverage.
292. *Dorsale de Kivu, Cartes de Reconnaissance, Vegetation et Sols.*
      Scale: 1:500,000
      Publisher: Institut National pour l'Etude Agronomique du Congo Belge, Bruxelles, 1957–58.
      No. of sheets: 4
      Coverage: Partial coverage. Other areas covered at larger scale by INEAC.
293. *Kwango, Cartes de Reconnaissance: Sols et Vegetation.*
      Scale: 1:1,000,000
      Authors: Devred, R.; Sys, C.
      Publisher: Institut National pour l'Etude Agronomique du Congo Belge, Bruxelles, 1955.
      No. of sheets: 2
      Coverage: Partial coverage. Other areas covered at larger scale by INEAC (1950–60).

Zambia

294. Distribution of Soils of the Kaombe, Muluwe and Lunzua Series (Map 3–7). Occurrence of Deep Sandy Soils (Map 3–9), in: *Land Resources of the Northern and Luapula Provinces.*
      Scale: 1:1,000,000
      Authors: Land, D.M.; Mansfield, J.E.
      Publisher: Land Resources Division, Ministry of Overseas Development, U.K. (Land Resource Study 19, Vol. 3), 1975.
      No. of sheets: 2
      Coverage: Partial coverage.
295. *Soils Map of the Republic of Zambia.*
      Scale: 1:1,500,000
      Authors: Zambia. Survey Dept.
      Publisher: Published by the Surveyor General, Ministry of Lands and Mines, Lusaka, Zambia, 1967.
      Size: 74 × 87 cms.
      Notes: Text on verso. Scale 23.674 miles to 1 inch.

296. *Soil Map of Zambia*, 1983 edition.
    Scale: 1:2,500,000
    Authors: Veldkamp, W.J.
    Publisher: Soil Survey Mt.Makulu, Land Use Branch, Lusaka, 1983.
    Classification: FAO/UNESCO.
    Type of map: Black & White
    Coverage: Complete coverage.

Zimbabwe

297. *Provisional Soil Map of Zimbabwe, Rhodesia*, 2d ed.
    Scale: 1:1,000,000
    Publisher: Chemistry and Soil Research Institute, Department of Research
                    and Specialist Services, Causeway, 1979.
    Coverage: Complete coverage.

13. Updated Reference Resources

PETER MCDONALD

Cornell University

The following list of print resources is selective and evaluative. It covers materials published since 1980 and serves as a supplement to *Guide to Sources for Agricultural and Biological Research*, edited by J. Richard Blanchard and Lois Farrell.[1] The section on soil and fertilizers [E083—E185] in Blanchard and Farrell, with its 105 entries, most nearly matches the reference update here. *Guide to Sources* constitutes the most extensive recent listing of information sources in soil science. This is less true for fertilizers and the fertilizer industry, which have a number of more recent handbooks, some of which are listed in the update. Soil science does not lack for bibliographies; however, few are evaluative or comprehensive, so only nine were deemed of sufficient breadth for listing here.

CAB International's abstracting journal *Soils and Fertilizers*, its online counterpart *CAB Abstracts*, and the *AGRICOLA* database of the National Agricultural Library remain the primary search tools for the literature of soil science. Both of these databases are available on compact disk and online.

The subject of soil science touches on many disciplines. Some of the references listed below are not directly soils related, but are included for their usefulness as resource guides to such allied disciplines as biology, agricultural engineering, earth sciences, geology, agroclimatology, the fertilizer industry, sustainable agriculture, conservation, and farmland protection.

The alphanumeric codes in brackets at the end of citations refer to the numbers used for the same or preceding titles in Blanchard and Farrell. These are usually new editions. Maps are not included in this update, but a comprehensive list at the national or continent level appears in Chapter 12.

---

1. J. Richard Blanchard and Lois Farrell, eds., *Guide to Sources for Agricultural and Biological Research*, sponsored by the U.S. National Agricultural Library, USDA (Berkeley: University of California Press, 1981).

Reference Sources Update in Soil Science

American Society of Photogrammetry. *Multilingual Dictionary of Remote Sensing and Photogrammetry:*
  English Glossary and Dictionary Equivalent Terms. Falls Church, Va.: American Society of Photogrammetry, 1984. 343p. [French, German, Italian, Portuguese, Spanish and Russian.]
  Following a list of abbreviations and acronyms, the almost 2,000 terms are defined in English, with numbers that correspond to translations in the other languages. The English listing is cross-referenced to the definitions in the six other languages.
Bailey, G. D. *Bibliography of Soil Taxonomy, 1960–1979.* Wallingford, U.K.: CAB International, 1987. 194p.
  Lists papers, journal articles, monographs, technical reports and conference proceedings on all aspects of soil taxonomy. The soils of Spain are classified according to the U.S. Soil Conservation Service *Soil Taxonomy* (1975). Glossary of terms included.
Bertsch, Floria. *Bibliografia de Suelos de Costa Rica.* San Jose, C.R.: Universidad de Costa Rica, Facultad de Agronomia, 1987. [In Spanish].
  A comprehensive bibliography with 2,275 entries covering all aspects of soils. Divided into five subject categories: Mineral Nutrition, Fertilizers, Soil Fertility, Soil Classification and Soil Physics. The Introduction contains a brief bibliometric study of the soils literature of Costa Rica.
Boulaine, Jean. *Histoire des Pédologues et de la Science des Sols.* Paris: Institut National de la Recherche Agronomique, 1989. 290p. [In French]
  The first comprehensive overview of the history of soil science from pre-history to the present. Major milestones in the evolution of pedology and soil sciences primarily in Europe and the United States are discussed. France predominates. Third World developments have minimal treatment. Includes tables and graphs and a quick reference chronology: 7000 B.C. to 1986 A.D.; extensive bibliography pp. 259–266.
British Sulphur Corporation. *World Directory of Fertilizer Manufacturers.* 6th ed. London, U.K.: Longman, 1988. 274p.
  Describes the activities of the major world producers of fertilizers. Company information includes location of manufacturing plants, products, company structure and financial data.   [E123]
British Sulphur Corporation. *World Directory of Fertilizer Products.* London, U.K.: Longman, 1985. 142p.
  Arranged in five parts: (1) Fertilizer companies and their products. (2) Major international marketing organizations. (3) Major producers of raw materials. (4) Major national purchasing organizations. (5) A guide to the international trade of fertilizer materials.   [E124]
Choudhari, J. S. and Aminullah. *Soils Fertilizer Responses and Irrigation Waters in Arid and Semi-Arid Regions: A Bibliography 1833–1979.* Jodhpur, India: Latesh Prakashan, 1980. 140p.
  1,276 references are presented on the subject of soil fertilizer responses, mainly

from India. References were culled from all sources: monographs, journals, reports, symposia and government documents. The citations are grouped by subjects but are not evaluative. Includes an Author Index and a listing of journal abbreviations.

Dindal, Daniel L., ed. *Soil Biology Guide*. New York: J. Wiley, 1990. 1349p.

A comprehensive guide to North American microbial and invertebrate soil inhabitants. It represents the most current research results of the biology, taxonomy and ecology for each soil biotic group. The forty-eight chapters of varying length cover the main soil biota groups, including such major taxonomic groups as Nematoda, Bacteria, Fungi, Algae, Arthropoda and Protozoa.

Duchaufour, Philippe. *Ecological Atlas of the Soils of the World*, translated from the French by G. R. Mehuys, C. R. De Kimpe, and Y. A. Martel. New York: Masson, 1978. 178p.

Classification of soils on the basis of their ecology. Ten major sections: (1) The Basis of an Ecological System of Soil Classification. (2) Immature Soils. (3) Calcimagnesium Soils. (4) Isohumic Soils and Vertisols. (5) Brunified and Eluviated Soils. (6) Podzolized Soils. (7) Hydromorphic Soils. (8) Fersialitic Soils. (9) Ferruginous and Ferralitic Soils. (10) Sodic Microstructure in Various Profile Studies. Includes bibliography 169–174p.

Eisenbeis, Gerhard, and Wilfried Wichard. *Atlas on the Biology of Soil Arthropods = Atlas zur Biologie de Bodenarthropden*. Berlin and New York: Springer-Verlag, 1987. 437p. [In English and German]

A comprehensive guide with 1,133 scanning electron microscope photographs, 192 plates and 219 figures. Each Order of biota is divided into its distribution patterns, characteristics, behaviour, taxonomy, breeding habits, life cycles and food preferences. Twenty pages of references. Indexed by subject and scientific names.

English-Chinese; Chinese-English Dictionary of Soil Science. Prepared by J.C. Xie, K.S. Cao, and J. X. Chen. Beijing: Science Press, 1988. 820 p. Contains 30,000 English-Chinese terms and 10,000 Chinese-English on soil geography, soil chemistry and physics, soil-plant nutrition, soil biochemistry and microbiology, soil use, ecology and environmental protection. An appendix on soil taxonomy is included.

Alphabetical listing of soil science terms and soil types in English with Chinese equivalent. Common English abbreviations defined in appendix. The book is pocket size.

Farm Chemicals Handbook '90. Willoughby, Ohio: Meister Publishing Co., 1990. 700p.

The seventy-sixth edition. This biennial handbook is divided into six major sections: Master Index; Fertilizer Dictionary; Regulatory File; Storage, Handling and Application Equipment; a Listing of Chemical Fertilizer Manufacturers and Distributors. Notable for more environmentally-related information which has been incorporated.   [E050]

Food and Agriculture Organization. *Agroclimatological Data for Latin America and the Caribbean*. Rome: FAO, 1984. 406p. [*FAO Plant & Protection Series*, no. 24].

This volume presents agroclimatological data for 800 agricultural stations through Latin America. Opening chapters explain the agroclimatological tables and the analytical tables on rainfall variability. Includes listings of the agroclimatological stations, tables of agroclimatological parameters and growing seasons as well as rainfall variability tables.

Food and Agriculture Organization. *Torrent Control Terminology*. Rome: Food and Agricultural Organization, 1981. 156p. [*FAO Conservation Guide* no. 6]

Torrent control terms in French, German, English, Spanish and Italian in an alphabetical listing cross-indexing and illustrations of important aspects of torrential process and control.

Food and Agriculture Organization. *Watershed Management Field Manual: Slope Treatment Measures and Practices*. Rome, Italy: FAO, 1988. 148p. [*FAO Conservation Guide* no. 13/3]

Provides detailed assistance to professionals concerned with planning and implementation of watershed management activities. In two parts. Part I: Land Preparation Methods for Afforestation. Part II: Terraces and Ditches. Brief bibliography at the end.

Glossary of Soil Science Terms. 3d ed. Madison, Wis.: Soil Science Society of America, 1987. 44p.

Four sections follow the alphabetical listing of "acceptable terms." (1) Soil structure grades. (2) Soil texture terms. (3) Soil water terms. (4) Tillage terminology. Four appendices cover: (1) Obsolete terms. (2) An outline of the U.S. soil classification system. (3) Tillage terminology procedural guide. (4) New designations for soil horizons and layers. A table of conversion factors for SI units and non-SI units is provided.   [E122]

Goldman, Steven J., Katherine Jackson, and Taras Bursztynsky. *Erosion and Sediment Control Handbook*. New York: McGraw-Hill Book Co., 1986. 1 vol.

A comprehensive how-to-do-it guide organized into ten chapters covering the major subject areas of erosion control. Provides detailed information on evaluating erosion potential, developing a regulatory program for erosion/sediment control, planning for erosion/sediment control and the design and construction of measures to prevent erosion and sediment loss.

Gupta, I. C., and P. Gupta. *Twentieth Century Soil Salinity Research in India: An Annotated Bibliography. Supplement*. Bikaner: Alfa Publishers and Distributors, 1990. 1 vol.

References to about 2,400 monographs, journal articles, technical reports and government documents on all aspects of soil salinity research, primarily in India. Includes appendices: (I) Lists important nineteenth century publications on soil salinity; (II) Covers the botanical and Hindi names of crops. Includes cross-referenced subject index, author index and source index. Supplement follows same format through 1989 publications.

Gupta, I. C., K. N. Pahwa, and J. Yadav. *Twentieth Century Soil Salinity Research in India: An Annotated Bibliography, 1901–1983*. New Delhi: Concept Publishing Co., 1984. 792p.

Handbook of Ground Water Development. Compiled by the Roscoe Moss Co. New York: J. Wiley & Sons, 1990. 493p.

A multidisciplinary guide to the latest theories and techniques for the exploration, extraction, use and management of groundwater. In three sections: (1) The physical geology and hydrodynamics of water in the ground. (2) Design and construction with a chapter on corrosion and incrustation. (3) Management of wells and well field operations.

Harrison, A. F. *Soil Organic Phosphorus: A Review of World Literature*. Wallingford, U.K.: CABI, 1987. 257p.

This review covers: (1) The methods used to determine organic phosphorus content of soils. (2) The physicochemical nature of organic phosphorus. (3) The influences of land-use and management practices. (4) The changes in organic phosphorus during pedogenesis. (5) The factors affecting its rate of mineralization. Contains about 900 references, mostly in English and primarily for developing countries.

Hignet, Travis P., ed. *Fertilizer Manual*. Boston: M. Nijhoff Publishers, 1985. 363p.

Prepared by the International Fertilizer Development Center in collaboration with the United Nations Industrial Development Organization as a replacement of the 1967 edition. Serves as a reference source on fertilizer production technology, with emphasis on fertilizer industry planning for developing countries. In five parts. Part I deals with the history of fertilizers and the world outlook. Part II covers the production and transportation of all important nitrogen fertilizers. Part III evaluates the characteristics of phosphate fertilizers. Part IV deals with potash fertilizers. Part V covers planning, pollution control, and the economics of the major fertilizer products world-wide. Selected references follow the five parts.

Index to Soil Surveys in California. Compiled by the Department of Conservation, Division of Land Resources Protection. Sacramento, Calif.: Division of Land Resources Protection, 1982. 84p. [With an Appendix: *An Index to the SCS Soil Surveys in California*. 86p.]

This guide is a list of the published surveys of California covering the subjects of soils, forests, water resources and land reclamation. Includes surveys issued by the SCS, the U.S. Forest Service, U.S. Bureau of Land Management, U.S. Bureau of Reclamation, the California Dept. of Water Resources, the California Dept. of Forestry and surveys published by the University of California. The appendix lists SCS Soil Surveys by region and county.

International Fertilizer Industry Association. *Glossary of Fertilizer Terms*. English, German, Spanish and French. Paris: International Fertilizer Industry, Ltd., 1986. 71p.

A comprehensive list of terms associated with fertilizers and their application. There is a multilingual table of main fertilizer types. Cross-referenced in index.

International Peat Society. *Russian, English, German, Finnish, Swedish Peat Dictionary*. Helsinki, Finland: International Peat Society, 1984. 595p.

Produced by the IPS National Committee of the USSR, the first section of this work contains an alphabetical listing of Russian terms which are numbered with the corresponding word equivalent in the other languages arranged beneath. The other sections follow the same pattern with indexes of alphabetically arranged German, Finnish, Swedish and English terms numbered respectively, by means of which the corresponding Russian and other-language words can be located in the first section.

International Society of Soil Science. *List of Members. English, German and French.* Amsterdam: Drukkerij Systema, 1991. (1st ed. 1954–)

Jongerius, A., and G. K. Rutherford. *Glossary of Soil Micromorphology: English, French, German, Spanish, and Russian.* Wageningen: Centre for Agricultural Publishing and Documentation, 1979. 138p.

Compiled by the Sub-Commission on Soil Micromorphology of the International Society of Soil Science (ISSS), the glossary contains terms in use through 1974. Origins of all terms is given with various quotations about use by prominent specialists. Cross-referenced by means of the relevant entry and term numbers. A subject index for all five languages is provided with gender(s) of each term listed by language.

Kachelman, D. L. *Fluid Fertilizer Reference Manual.* Muscle Shoals, Ala.: National Fertilizer Development Center, 1989. 28 refs., 48 figures, 99 tables [*TVA Bulletin* Y-210].

Fluid fertilizer technology information presented under the subjects: Nitrogen solutions; ammonium phosphate solutions; potassium chloride solutions; NP, NK and NPK solutions; suspensions; acid fluid fertilizers; general and miscellaneous topics. Common solution grades are listed in tables.

Kilmer, Victor J. *Handbook of Soils and Climate in Agriculture.* Boca Raton, Fla.: CRC Press, 1989. 438p.

An attempt to bring together pertinent information on all aspects of soil science and agroclimatology for nonspecialists. Provides a quick reference guide to subjects such as: climate of the United States, soil classification, soil physics, soil chemistry, soil microbiology and soil organic matter. There are several chapters on soil fertility and fertilizers. Soil and water management are covered in the last two chapters. Includes a glossary of thirty-two pages. References follow each chapter.

Landon, J. R., ed. *Booker Tropical Soils Manual: A Handbook for Soil Survey and Agricultural Land Evaluation in the Tropics and Subtropics.* London: Booker Agricultural International Ltd.; and New York: Longman, 1984. 450p. (1991 paperback edition available.)

Provides detailed information on land resource field studies, survey organization and practice, classification and mapping of soils, land evaluation, soil physics, soil chemistry, salinity and sodicity of soils, and soil and land suitability maps. The fifteen Annexes cover all aspects of practical implementation of tropical soil land evaluation. Twenty-four pages of references. Indexed by author and subject.

Lozet, Jean, and Clement Mathieu. *Dictionary of Soil Science.* 2d ed., revised and enlarged. Paris and Rotterdam: A. A. Balkema, 1991. 282p.

Over 2,800 terms used in soil science and related disciplines, relating to science, mineralogy, petrology and micromorphology. Appendices outline the major classification systems. Brief bibliography included.

McGowan, V., and V. Skerman. *World Directory of Collections of Cultures of Microorganisms*. Brisbane, Australia: World Data Centre on Microorganisms, 1982. 641p.

A directory containing lists of collections and lists of species of algae, bacteria, fungi, yeasts, lichens, protozoa, viruses and samples of animal tissues in the various collections. Many of the collections are strong in soils fauna.

Nanda, Meera, ed. *Planting the Future: A Resource Guide to Sustainable Agriculture in the Third World*. Minneapolis, Minn.: University of Minnesota, 1990. 176p.

A systematic resources guide to Third World sustainable agriculture, much of it to do with soils, soil management and conservation. Two-hundred twenty-five groups and 246 resources from forty-six countries are listed. Section one is "Reports On Groups and Farming Practices" divided alphabetically by country. Section two is "Bibliography of Published Works by Resource Guide Groups." Section three is a "List of Audio-Visual Resources." Section four deals with "Trends, Conclusions and Recommendations." Section five is the Appendices, which includes a Country Index, Regional Index, Alphabetical Index, Subject Index, further resources and a brief glossary.

National Directory of Farmland Protection Organizations. By Nancy Bushwick and Hal Hiemstra. Washington, D.C.: National Association of State Departments of Agriculture Research Foundation, 1983. 87p.

A reference document on farmland protection activities in the United States. While not intended as a comprehensive overview, much of the information deals with sources for soil conservation assistance. Organized geographically by state, with a National section at the beginning, and two appendices: (1) Summary of State Agricultural Departments Farmland Protection Activities and Addresses. (2) A selected listing of Educational Materials on Farmland Preservation issued by various Cooperative Extension Services.

Nelson, L. B. *History of the U.S. Fertilizer Industry*. Muscle Shoals, Ala.: Tennessee Valley Authority, 1990. 537p.

This history begins with the earliest uses of organic fertilizers in pre-history, with a primary focus on the transformation of the United States fertilizer industry from its inception to its worldwide impact today. It offers a reference source for teachers and students of agricultural history.

Nielsson, Francis T., ed. *Manual of Fertilizer Processing*. New York: M. Dekker Inc., 1987. 526p.

An update of the 1960 edition of *Chemistry and Technology of Fertilizers* by Vincent Sauchelli. Provides a comprehensive guide to the fertilizer industry with emphasis on administrative decision-making in solving production problems in the Third World. Nineteen chapters cover the main fertilizer groups as well as environmental regulations. References follow each chapter.   [E159]

Nieves-Bernabe, M., R. Bienes-Alla, and V. Gomez-y-Miguel. *Key to Spanish Soils = Clave de los Suelos Espanoles*. Madrid, Spain: Mundi-Prensa Libros, S. A., 1988. 142p. [In Spanish]

Odet, J. *Fertilization Guide for Vegetable Crops = Memento Fertilization des Cultures Legumieres*. Paris: Centre Technique Interprofessionel des Fruits et Legumes, 1989. 398p. [In French].

An update of the 1982 edition giving general principles behind fertilizer use and practical guidance on: soil maintenance and improvement; types of fertilizers and their effects with methods for application; soil and plant analysis and its interpretation; fertilization and solution preparation; nutrient deficiencies and toxicities. Includes a fertilizer requirements guide for fifty-two selected crops. Appendices list analytical laboratories and fertilizer consultants with names and addresses.

Oliver, John, and Rhodes Fairbridge. *Encyclopedia of Climatology*. New York: Van Nostrand Co., 1987. 986p. [Encyclopedia of Earth Sciences, Vol. XI]

Each of the approximately 250 major entries has been written by an authority in the field. References follow each entry. The encyclopedia is cross-referenced by a set of three indexes: the Author Citation Index, a Subject Index and the Geographical Index.

Olson, Gerald. *Soils and the Environment: A Guide to Soil Surveys and Their Application*. New York: Chapman and Hall, 1984. 219p.

This guide discusses the problems of land abuse and the results, concentrating principally on the scientific description of soils, delineation, interpretation and application of pertinent survey information in a subject format. The chapters titles are: Soil Profile Descriptions; Laboratory Analyses; Soil Classification; Computerized Data Processing; Engineering Applications; Agricultural Land Classification; Erosion Control; Yield Correlations; Archeological Considerations; Planning for the Future. A glossary appears at the end.

Orvedal, A. C. *Bibliography of Soils of the Tropics*. Washington, D.C.: Office of International Cooperation and Development and the Soil Conservation Service, USDA, 1975–1983. 5 vols. [U.S. *AID Technical Series Bulletin* no. 17]

Vol 1. *Tropics in General and Africa*. Vol 2. *Tropics in General and South America*. Vol 3. *Tropics in General, Middle America and West Indies*. Vol 4. *Tropics in General and Islands of the Pacific and Indian Ocean*. Vol 5. *Tropics in General and Tropical Mainland Asia, Pakistan, Nepal and Bhutan*. An alphabetical listing by region citing references and maps relating to soils collected since 1945 by the Soil Geography Unit of the U.S. Soil Conservation Service. Each volume is partially an update of previous volumes. Over 20,000 citations listed in all languages with English translations.   [E106]

Pahwa, K. N., and I. C. Gupta. *World Literature on Reclamation and Management of Salt Affected Soils (1950–1981)*. New Delhi: Associated Publishing Co., 1982. 352p.

Contains 811 abstracts, arranged alphabetically by author. All titles are given in English and those translated from other languages are given in abbreviated form at the end of the reference. Cross-referenced with an Author Index, Subject Index and Geographical Index.

Plaisance, Georges. *Dictionary of Soils = Dictionnaire des Sols*. Translated from the French by Margaret Saidi. New Delhi: Amerind Publishing Co., 1981. 1109p. (Available from U.S. National Technical Information Service.)

An English translation of the 1958 edition. Roughly 11,000 French terms are defined in English, including derivation and source. Covers all aspects of agronomy and soil science.   [E118]

Resource Conservation Glossary. Ankeny, Iowa: Soil Conservation Society of America, 1982. 193p.

Contains 4,000 terms used in resource management. The Subject scope includes agronomy, air resources, anthropology, aquaculture, cartography, conservation (all aspects), computer science, ecology, engineering, erosion and sedimentation, fish and wildlife biology, fertilizers, forestry, geography and geology, horticulture, hydrology and irrigation, landuse planning, outdoor recreation, range management, remote sensing, salinity control, soils and tillage, water resources, waste management and the weather.

Soil Mechanics Glossary: Terms Used in Latin America = Vocabulario de Mecanica des Suelos: Terminos Usados en Latinoamerica. English, Spanish, Portuguese and French. (Mexico: Sociedad Mexicana de Mecanica de Suelos, 1977. 108p.

Two-thousand two-hundred and thirty-one (2,231) terms in English followed by their translations in Spanish, Portuguese and French. Covers all aspects of soil mechanics with local usages identified by the name of country.

Soil Science Dictionary: English and Malay = Istilah sains tanah, bahasa Inggeris-bahasa Malaysia, bahasa Malaysia-bahasa Inggeris. Kuala Lumpur: Kementerian Pendidikan Malaysia, 1989. 258p.

Soils of the British Isles. Wallingford, U.K.: CABI, 1990. 463p.

Collates the accumulated information on soil variation in the British Isles stemming from the extensive soil surveys and laboratory studies since 1950. Classification into lithomorphic soils, podzols, man-made soils, brown soils, gley and peat soils. One-hundred eighty-seven (187) profiles are described.

Somani, L. L. *Dictionary of Soils and Fertilizers.* New Delhi: Mittal Publications, 1989. 5 parts.

Twelve-thousand (12,000) entries covering all aspects of soil science, fertilizers and related subjects.

Somani, L. L. *Soil Physical Research in India 1925–1985; An Annotated Bibliography.* New Delhi: Associated Publishing Co., 1987.

Contains 1,291 abstracts from Indian and non-Indian journals. Arranged alphabetically by author. Indexed by author, source and subject.

Swindale, Leslie, ed. *Soil-Resource Data for Agricultural Development.* Honolulu, Hawaii: Hawaii Agricultural Experiment Station, 1978. 306p.

A state-of-the-art compilation of soil resource data for the tropics including information on the classification and collection of soil surveys, soil survey interpretation, soil-resource data for land-use planning, case studies of the use of soil-resource data for agricultural development, agrotechnology transfer and soils management in the humid tropics. This is an edited and revised compilation of papers presented at conference on the planning and implementing of soil-resource data for agricultural development (Hyderabad, India, January 18–23, 1976). Appendices contain the proceedings of the seminar, keynote speeches, a list of participants and the organizations represented.

Trzyna, Thaddeus, ed. *Preserving Agricultural Lands: An International Annotated Bibliography.* Claremont, Calif.: California Institute of Public Affairs, 1984. 82p.

A guide to the literature of land conservation worldwide published between 1970–1983 with emphasis on the United States. Includes references to monographs, articles, reports, government documents and other bibliographies which were selected by an advisory board at the California Institute of Public Affairs. While not dealing exclusively with soil conservation, much of the focus deals with this and related subjects. Includes a two page list of other bibliographies and an Author Index.

U.S. Soil Conservation Service, Soil Survey Staff. *Keys to Soil Taxonomy.* 4th ed. Blacksburg, Va.: Virginia Polytechnic Institute and State University, 1990. 422p.

A pocket guide to soil taxonomy, providing an update to taxonomic keys required for the classification of soil types and the revisions in the key that have been approved by the USDA Soil Conservation Service. Incorporated are all amendments approved by date of publication c1986. Early chapters are introductory with explanatory notes on categories, family differentiae, and an identification guide. Later chapters cover the main soil types. Indexed, with SI conversion table at the end.

U.S. Soil Conservation Service. *Engineering Field Manual for Conservation Practices.* Lincoln, Neb.: U.S. Dept. of Agriculture, Soil Conservation Service, 1984. 2 vols.

Provides field guidance for soil and water conservation practices, using basic engineering principles, techniques and procedures for planning, design, installation and maintenance. Contains seventeen chapters on such subjects as Engineering Surveys, Hydraulics, Preparation of Engineering Plans, Terraces, Drainage, Irrigation, and Construction and Construction Material.

U.S. Soil Conservation Service. *List of Published Soil Surveys: 1991.* Washington, D.C.: SCS, 1992, 23p.

Published yearly, this full listing of USDA Soil Conservation Service soil surveys covers the years 1899–1991 by state and county. A list of State Conservationists and their addresses are given. Indicates 4,187 were published of which 2,405 are in print.   [E184]

Vosick, Diana. *The Economics of Soil Erosion: A Handbook for Calculating the Cost of Off-Site Damage.* [Prepared by the American Farmland Trust]. Washington, D.C.: American Farmland Trust, 1987. 136p.

Provides detailed information about the various categories of soil erosion damage and the methods to calculate them. Two case studies are described. Contains extensive bibliography.

Water Resources Management in Asia: A Selective Bibliography with Introductory Essays. Honolulu, Hawaii: East-West Environment and Policy Institute, 1984. 497p.

This bibliography was produced by the Asian Water Resources Management Project. In seven parts: Part I presents a brief historical overview of Asian water resources. Part II is a reference guide to the bibliography, discussing how the material was organized and the methods of cross-reference employed. Parts III to VII present the 861 alphabetical author listings, supplemented by subject, country and author indexes.

Weiss, M., Mea Whittington, and L. Tiegen. *Weather in U.S. Agriculture: Monthly Temperature and Precipitation by State and Farm Production Region, 1950–1984.* Washington, D.C.: USDA Economic Research Service, 1985. 244p. [*USDA Statistical Bulletin* no. 737].

Two major tables are included: State Tables and Farm Production Region Tables. Both are indexed and cover monthly tabulations of temperature and precipitation rates for the years 1950–1984. A weighted-average process is used to aggregate across the climate divisions within the contiguous United States. Two sets of aggregation weights are used: geographic areas and areas of harvested cropland. References are given at the beginning.

Western Fertilizer Handbook. 7th ed. U.S. California Fertilizer Association. Danville, Ill.: Interstate Printers & Publishers, Inc., 1985. 288p.

Covers agricultural regions of the western United States. Contains basic information on the properties of soil, water and plants, fertilizer products, their application and management, soil amendments and irrigation water quality. A brief summary of western laws governing fertilizer use is included as well as a glossary, reference tables for the inclusion of conversion factors.   [E176]

World Meteorological Organization. *Glossary of Terms Used in Agroclimatology.* Enlarged Edition. World Meteorological Organization, 1984. 244p. [*Agricultural Meteorology Report* no. 20.]

Contains agrometeorological terms as well as relevant hydrological, pedological and general agricultural terms, names of crops and fauna and terms pertaining to agrometeorological aspects of crop and animal pathology.

14. Historical Soil Science Literature of the United States

ROY W. SIMONSON
College Park, Md.

PETER MCDONALD
Cornell University

The evolution of the study of soils in the United States started late and progressed slowly, making limited headway through most of the nineteenth century. During the first decades of independence, the background sciences such as chemistry and geology were in early stages of development, not yet capable of coping with systems as complex as soils. Appreciable growth in knowledge in both fields was still needed. Furthermore, although the great bulk of the population was dependent on farming for its livelihood, few people had the time or the money to look beyond their daily needs. Methods of cultivation and planting were largely handed down from one generation to the next. The earliest investigations were thus left to wealthy planters, among whom some were interested enough to conduct trials of various kinds to improve production from their fields.

The first trials and all subsequent studies of soils have been made against a background of one or more of three conceptions of soils. One of those originated at least as early as Roman times and the other two since the colonial period of American history.[1] The first conception held soil to be a medium for plant growth, the loose mantle on the land surface that provided a foothold and mechanical support for plants. That conception has not disappeared but has been modified to cover uptake of nutrient elements from soil as well as provision of mechanical support. A second conception, following the birth and development of the science of geology, held soil to be the weathered mantle of rock on the land surface. That conception brought with it a theory of soil formation, namely, that soils are formed by rock weathering or deposition of weathered rock materials. The second concep-

---

1. R. W. Simonson, "Concept of Soil," *Advances in Agronomy* 20 (1968): 1–47.

tion was widely held during the last half of the nineteenth century and the first half of the twentieth century. The third conception which originated in the latter part of the nineteenth century held that soil is "a natural body having a definite genesis and distinct nature of its own," to quote from one of the early Russian papers.[2] Soils are considered to be open systems to which substances can be added and from which they can be lost. Moreover, the direction and intensity of these processes are controlled by a combination of climate, living organisms, parent materials, and topography. This conception has been spreading among American soil scientists for about seventy years, especially during the last forty.

Most early publications reflect the first conception, namely of soil as a medium for plant growth. Later works elaborated the conception of soil as weathered rock. The second conception has largely disappeared from the literature of soil science, replaced by the third. Only within the present century have publications been prepared against the background of soil as a natural historical body. Some blending of the first and third conceptions has occurred during the last thirty years or so.

An effort is made in this report to identify conceptions of soil reflected in publications cited. The purpose of the approach is to bring out changes in the American literature of soil science over time. A better knowledge than that of the authors would be needed for comparable coverage of world literature.

A. Colonial Period

Among people in the United States, including leaders such as George Washington and Thomas Jefferson, the conception of soil was an early version of a medium for plant growth, something that provided mechanical support. Many wealthy planters were acquainted during the colonial period with new developments in agriculture in England.[3] Copies of some of these books were available in the colonies. A good illustration is the *Horse Shoeing Husbandry*.[4] Jethro Tull demonstrated on his farm near Oxford, England, that cultivation with horse-drawn implements increased crop yields. His colorful explanation of the results reflected the prevailing conception of soil—well pulverized soil material was taken in by the "lacteal" mouths of plant roots. Other English books of the period and more from the continent

    2. Peter Fireman, *Russian Soil Investigations* (1901 [*Experiment Station Record*, vol. 12]). Part 1, pp. 704–712; Part 2, pp. 807–818.
    3. L. C. Gray, *History of Agriculture in the Southern United States to 1860* (Washington, D.C.: Carnegie Institution of Washington, 1933).
    4. Jethro Tull, *Horse Shoeing Husbandry* (London: Jethro Tull, 1730).

have been identified in the many editions of *Soil Conditions and Plant Growth*.[5]

As early as 1760, George Washington applied manure, marl, and river mud to his fields at Mount Vernon.[6] In the postwar period, he also applied gypsum, known as land plaster, to improve growth of clover. He planted buckwheat to be plowed under as green manure.[7] Careful records were kept and the results were shared with friends and acquaintances but findings were not published.

Like Washington, Jefferson took a lively interest in his plantation, with as many experiments at Monticello as were conducted at Mount Vernon.[8] The persistence of Jefferson's interest in soil improvement is intriguing; in a letter to a friend he wrote, "We can buy an acre of new land cheaper than we can manure an old acre."[9] Yet he never did drop the interest. It extended from soil amendments and crop rotations to farm implements. Jefferson invented a moldboard plow in 1790 and a hillside plow in 1808.[10] Despite his great breadth of knowledge, his understanding of soils was elementary, as was normal at the time. Gray cites a remark by Jefferson that leaving soil bare was harmful because the sun would "absorb the nutritious juices of the earth."[11] Notebooks kept by Jefferson on farm operations from 1773 to 1816 plus some related writings were published in the current century, but they carry limited information about experiments at Monticello.[12]

In the early nineteenth century, occasional tracts and small books on agriculture were published in this country although most still came from Europe. Gray (1933) cites a *Treatise on Practical Farming*, published by John A. Binnes of Frederick-town, Maryland, in 1803. Directed to farmers rather than scientists, the book reflects the early conception of soil as a medium for plant growth.

B. The Nineteenth Century

The first appearance of the conception of soil as weathered rock was in Albany County, New York, in a geological survey for the "improvement of

5. E. W. Russell, *Soil Conditions and Plant Growth*, 10th ed. (London: Longman, 1973).

6. A. O. Craven, "Soil Exhaustion as a Factor in the Agricultural History of Virginia and Maryland, 1606–1860," *University of Illinois Studies in Sociology* 13 (1926): 1–179.

7. Gray, *History of Agriculture in the Southern United States* . . .

8. Ibid.

9. E. M. Betts, *Thomas Jefferson's Farm Book* (with commentary and relevant extracts from other writings) (Princeton, N.J.: American Philosophical Society and Princeton University Press, 1953).

10. W. H. Gardner, "Early Soil Physics into the Mid-20th Century," *Advances in Soil Science* 4 (1986): 1–101.

11. Gray, *History of Agriculture in the Southern* . . .

12. Betts, *Thomas Jefferson's Farm Book*.

agriculture."[13] The text of their report indicates that Eaton and Beck were using two conceptions of soil. When its formation was under discussion it was weathered rock. But when recommendations were being made for management, it was considered a medium for plant growth. This work was the first "soil survey" made in the United States and also the first to make recommendations by "kinds" of soils.

One year after the geological survey in New York was made, chemistry was applied to the study of soils by Edmund Ruffin.[14] He began a crusade for liming soils by writing an article on their composition and their improvement by calcareous manures. In publishing the article, the *American Farmer* referred to Ruffin as "the first American soil scientist."[15] He had applied marl to fifteen acres in 1818. The results made liming the central interest for the remainder of his life.[16]

The first edition of a book, *An Essay on Calcareous Manures* was published by Ruffin eleven years after his first article promoted the liming of soils.[17] The major part of the book reflected a conception of soil as a medium for plant growth but in the few remarks about soil formation, Ruffin turned to the theory of rock weathering. The book contributed to the belief that chemistry would soon solve all problems in the utilization of soils even though the application of chemistry by Ruffin had been limited in scope.

Ruffin had studied *Elements of Agricultural Chemistry* by Sir Humphry Davy.[18] An important part of the book records the current understanding of soils chemistry, on the one hand, and the existing uncertainties about the growth of plants, on the other. The debate about a "vital principle" in plant growth was in full swing at the time; much remained to be learned about plant nutrition. Davy reported that the quantities of free acids (sulfuric, phosphoric, nitric, acetic, etc.) were so small as to be negligible. Ruffin accepted that statement but argued that there was something more "which for want of a better name I shall call soil acidity."[19] Realizing that he was challenging the best-known English chemist of the time, Ruffin added, "When a person disputes learned authority it behooves him to lay out his

13. A. Eaton and T. R. Beck, *A Geological Survey of the County of Albany* (Albany, N.Y.: The Agricultural Society of Albany County, N.Y., 1820).

14. Edmund Ruffin, "On the Composition of Soils, and Their Improvement by Calcareous Manures," *The American Farmer* 3 (40) (1821): 313–320.

15. Ibid., p. 313.

16. Gray, *History of Agriculture in the Southern* . . .

17. Edmund Ruffin, *An Essay on Calcareous Manures* (Petersburg, Va.: J. W. Campbell, 1832).

18. Humphry Davy, *Elements of Agricultural Chemistry: A Course of Lectures for the Board of Agriculture* (London: Longmans, Hurst, Rees, Orme, and Brown, 1813).

19. Ruffin, "On the Composition of Soils, and Their Improvement by Calcareous Manures," p. 314.

evidence."[20] Ruffin did just that in the longest chapter of his book, using the method of converging lines of evidence where no one line is conclusive. Ruffin did not claim to be the first to apply calcareous manures, but he demonstrated the benefits more clearly than before and vigorously promoted the practice.

Most agricultural books published in the United States during the nineteenth century were reprints of European editions. Among them, especially prominent and influential, was the first edition of *Organic Chemistry in Its Application to Agriculture and Physiology* by Justus Liebig.[21] It had been translated from German for prior publication in England. Other editions of the book were published later in this country. Liebig's text reflected a conception of soil as a medium for plant growth, one that provided mechanical support.

Printed in Cambridge, Mass., the first American edition created a stir in agricultural circles in New England because of the appendices. One was an excerpt from the writing of Samuel A. Dana which was in direct conflict with the ideas expressed by Liebig. He had made the first successful attack upon the theory that plants absorbed humus as their food,[22] one held long in Europe. Liebig also dismissed summarily the idea of any "vital principle" as responsible for the growth of plants. Dana, on the other hand, subscribed to both ideas at the time.

Dana published his own book, *Muck Manual for Farmers—A Treatise on the Physical and Chemical Properties of Soils*, the year after Liebig's book was first printed in the United States.[23] Although it also reflected a conception of soil as a medium for plant growth, it disagreed in many ways with the book by Liebig. In that contest of ideas, Liebig won. His book became widely accepted in the United States, one consequence being that his ideas affected the soil science literature of the country for the remainder of the nineteenth and the first third of the twentieth centuries. Rossiter has suggested that the first edition by Liebig was one of the most important scientific books ever published.[24]

Added to the effects of American editions of European books were publications by a few men who went to Europe for advanced training. Their

20. Ibid., p. 316.

21. Justus Liebig, *Organic Chemistry in Its Application to Agriculture and Physiology* (Cambridge, Mass.: J. Owen, 1841).

22. Richard Bradfield, "Liebig and the Chemistry of the Soil," in *Liebig and after Liebig: A Century of Progress in Agricultural Chemistry*, edited by F. R. Moulton (1942 [American Association for the Advancement of Science Publication no. 16]), pp. 48–55.

23. Samuel A. Dana, *Muck Manual for Farmers—A Treatise on the Physical and Chemical Properties of Soils* (Lowell, Mass.: Daniel Bixby, 1842).

24. M. W. Rossiter, *The Emergence of Agricultural Science: Justus Liebig and the Americans, 1840–1880* (New Haven, Conn.: Yale University Press, 1975).

influence was far out of proportion to their numbers. Examples are Samuel W. Johnson, a prime mover in establishing the Connecticut Agricultural Experiment Station, and E. W. Hilgard, who was first the State Geologist and Professor of Chemistry in Mississippi and later Director of the California Agricultural Experiment Station. Each of these men contributed freely to soil science literature.

C. Early Periodicals

Information about soils intended for farmers rather than scientists was published in agricultural journals during much of the nineteenth century. One year after the appearance of Ruffin's book, Jesse Buell, a gentleman farmer near Albany, New York, started *The Cultivator*.[25] It became the first successful agricultural journal. Later, the title was changed to *Country Gentleman*. Under the two titles the journal lasted for seventy-eight years. The motto on the masthead was "to improve the soil and the mind." The journal covered all agricultural topics. Articles about soils focused on their management—plowing, methods of cultivation, amendments, drainage, and the like, all reflecting the conception of soil as a medium for plant growth. Prepared by American and European authors, the articles were consistently addressed to farmers.

A second example of a successful journal directed to farmers was *The Southern Cultivator*, started in Georgia in 1835. It lasted for ninety-three years.[26] Daniel Lee, an agricultural chemist, was an early publisher to whom much of the success of the journal can be attributed. An early editorial announced that issues of the journal would be priced to reach the "most humble tiller of the soil" and that the "primary objective was to restore the exhausted lands of the South."[27] That objective is an eloquent statement of the prevailing understanding of the soils of the region. People did not yet realize that hte fertility of the soils had been low when they were first cleared in the southeastern Untied States. Some improvement in the low levels of fertility were possible during the life of *The Southern Cultivator* but the big changes were to come later.

In contrast to *The Cultivator, Country Gentleman,* and *The Southern Cultivator*, most agricultural journals failed after a few years of press. Even some journals intended for agricultural scientists failed, e.g., *Agricultural*

---

25. Albert Demaree, *The American Agricultural Press 1819–1860* (New York: Columbia University Press, 1941).
26. Ibid.
27. *The Southern Cultivator* 1 (1835): 1.

Science. A rough estimate of the failures of agricultural journals puts the number in the thousands.[28]

Articles about soils and their properties that appeared in farm journals were meant to permit application of existing knowledge to farm practice. They did not report new findings for soil scientists. The general background for them was the conception of soil as a medium for plant growth. Any long-time value of the articles would be as examples of efforts to disseminate useful information rather than as part of the core literature recording the development of the science itself.

D. The Beginnings of Soil Science

A monograph by E. W. Hilgard on the geology and agriculture of Mississippi, published twenty-eight years after the book by Ruffin, provides another milestone in the literature. Both geology and chemistry were applied to the study of soils of the state over a period of five years to provide the substance of the monograph.[29] The 391 pages of the monograph are divided about equally between the sections on geology and agriculture. The latter section opens with a long discussion of soils and their properties, decidedly advanced for its day. Comparable discussions were not available in other books or journals at that time. The monograph was thus a harbinger of the many future contributions to soil science and its literature.[30]

Another and later exception to the conception of soil as the mantle of weathered rock was to be found in the annual report of the U.S. Geological Survey for 1891. The report included what is in effect a monograph on "The Origin and Nature of Soils" by N. S. Shaler.[31] The long article was reprinted separately in 1892. Shaler's essay was ahead of its time and got little attention from practitioners of soil science. Although much of his discussion of soil origin reflects a conception of soil as the weathered rock mantle, Shaler recognizes that living organisms must take part if a soil is to be formed. Weathering alone is not enough. He also takes a step toward the conception of soil as a natural historical body in the following sentence: ". . . he should clearly see that this mass of debris, which at first sight

28. Stephen Stuntz, *List of Agricultural Periodicals of the United States and Canada Published During the Century: July 1810 to July 1910* (1941 [USDA Miscellaneous Publication no. 398]).

29. E. W. Hilgard, *Report on the Geology and Agriculture of the State of Mississippi* (Jackson, Miss.: E. Barksdale, State Printer, 1860).

30. M. G. Cline, "Historical Highlights in Soil Genesis, Morphology, and Classification," *Soil Science Society of America Journal* 41 (2) (1977): 250–254.

31. N. S. Shaler, "The Origin and Nature of Soils," Extracted from *U.S. Geological Survey 12th Annual Report (1890–91)*, Part I (1892), pp. 213–345.

seems a mere rude mingling of unrelated materials, is in truth a well organized part of nature, which has beautifully varied and adjusted its functions with forces which operate it . . ." Shaler noted the harmful effects of soil erosion and added: ". . . This slight superficial and inconstant covering of the earth should receive a measure of care which is rarely given to it . . ."[32]

The founding of agricultural institutions in the United States began in earnest during the 1860s. The bill establishing the U.S. Department of Agriculture was signed by Lincoln in 1862. The Morrill Land Grant Act became law the next year, providing for the establishment of a college of agriculture and mechanic arts in every state. Twenty-four years later the Hatch Act went far toward putting agricultural experiment stations on a solid footing.

During the formative years of the land-grant colleges and agricultural experiment stations, their contributions to the literature of soil science were limited. Such contributions began to appear more frequently, however, in the last decade of the nineteenth century and the flow gathered momentum in the first part of the twentieth century. Soil scientists in these agricultural institutions initially published findings in bulletins and circulars. Few monographs and books were published during the last part of the nineteenth century. Some well-known scientists such as T. B. Osborne of Connecticut produced no monographs. Others, including E. W. Hilgard, for whom the journal *Hilgardia* was named in California, prepared no monographs after moving to that state until he had retired. Besides the bulletins and circulars published by the state experiment stations, some were also published by the U.S. Department of Agriculture. Generally, these publications were directed to farmers rather than to scientists. Two examples of bulletins intended for scientists were by Hilgard, *A Report on the Relations of Soil to Climate*,[33] and Milton Whitney, *Some Physical Properties of Soils and Their Relation to Moisture an Crop Distribution*.[34] Both were published in 1892 by the U.S. Department of Agriculture. The markedly different viewpoints of the two men that led to bitter controversy later are evident in those bulletins.[35] Unfortunately very little of the information originally presented in the bulletins and circulars of the state experiment stations and the U.S. Department of Agriculture has been reprinted. As a consequence, occasional gems lie buried in old bulletin publications.

---

32. Ibid., p. 223.
33. E. W. Hilgard, *A Report on the Relations of Soil to Climate* (Washington, D.C.: USDA, 1892 [USDA Weather Bureau Bulletin no. 3]).
34. Milton Whitney, *Some Physical Properties of Soils in Their Relation to Moisture and Crop Distribution* (Washington, D.C.: USDA, 1892 [USDA Weather Bureau Bulletin no. 4]).
35. Hans Jenny, *E. W. Hilgard and the Birth of Modern Soil Science* (Italy: Collana Della Rivista "Agrochimica," 1961).

Books about soils written primarily for other scientists or for the teaching of soil science at the college level began to appear in the late 1890s and early 1900s. One of the first was *Soil: Its Nature, Relations, and Fundamental Principles of Management* by Franklin H. King.[36] Ten years later, *Soils—Their Formation, Properties, and Composition* appeared, written by E. W. Hilgard.[37] It covered the waterfront for its day and was meant to serve primarily as a reference book for scientists. In 1909, T. L. Lyon and E. O. Fippin published *The Principles of Soil Management*.[38] The following year Cyril G. Hopkins brought out his well-known *Soil Fertility and Permanent Agriculture*, intended as both a college text and a reference book.[39] The conception of soil as a medium for plant growth forms the entire background for all of these early books, except that of Hilgard, whose book reflects two conceptions of soil—soil as the weathered rock mantle and soil as a medium for plant growth.

A new line of publications in soil science began in the United States when the U.S. Department of Agriculture issued the first volume of soil surveys in 1900.[40] Maps had been made to show the distribution of soils of four small areas in different parts of the United States. The maps were then published together with descriptions of the kinds of soils plus information about their use and management in agriculture. The Division of Soils, later the Bureau of Soils, with Milton Whitney as its chief, made and published the soil surveys. After its first few years, the Bureau published massive volumes of field operations in soil surveys every year. Activities of the Bureau of Soils had a substantial impact on the soil science literature of the United States. History of the soil survey program from its beginnings in 1899 through 1970 has been sketched recently.[41]

In addition to its soil survey reports, the Bureau of Soils published bulletins and circulars on a number of topics. These pamphlets affected the soil science literature of the country more than their numbers would suggest. One bulletin, *The Chemistry of the Soil as Related to Crop Production*,[42]

36. F. H. King., *Soil: Its Nature, Relations, and Fundamental Principles of Management* (New York: Macmillan, 1895).

37. E. W. Hilgard, *Soils—Their Formation, Properties, and Composition* (New York: Macmillan, 1906).

38. T. L. Lyon and E. O. Fippin, *The Principles of Soil Management* (New York: Macmillan, 1909).

39. C. G. Hopkins, *Soil Fertility and Permanent Agriculture* (New York: Ginn & Co., 1910).

40. Milton Whitney, *Field Operations of the Division of Soils, 1899* (1900 [USDA Report no. 64]).

41. R. W. Simonson, *Historical Highlights of Soil Survey and Soil Classification with Emphasis on the United States, 1899–1970* (Wageningen, Netherlands: International Soil Reference and Information Centre, 1989 [Technical Paper no. 18]).

42. Milton Whitney and F. K. Cameron, *The Chemistry of the Soil as Related to Crop Production* (1903 [USDA Bureau of Soils Bulletin no. 22]).

precipitated bitter controversy within the country[43] and ridicule from overseas.[44] Other bulletins have better withstood the test of time. Four examples will be cited. Buckingham published *Studies on Movement of Soil Moisture* in 1907; his ideas lay fallow for about twenty years but have since been widely adopted.[45] Three bulletins published several years later have also been important contributions. These are *The Movement of Soil Material by the Wind*,[46] *Soil Erosion*,[47] and *A Study of the Soils of the United States*.[48] Free reported quantities of sediments moved by wind in the western parts of the country. The bulletin also included a seventy-five-page bibliography covering a half-dozen languages. McGee reported erosion conditions, including photographs of spectacular gullies that have received more publicity later, and outlined methods of erosion control. Coffey reviewed past proposals for the classification of soils, outlined an ideal basis, and proposed recognition of four broad classes on the basis of soil surveys completed to date. Three of the four broad classes are recognized as orders in the current American system (1975).

E. The Twentieth Century

One more source of literature came into existence with the founding in 1907 of the American Society of Agronomy. It inaugurated publication of the *Journal of the American Society of Agronomy*, the name of which was later changed to *Agronomy Journal*. Primary emphasis of the journal has been on crops and on crop-soil relations but in early years articles were included on soils apart from plant growth. One early example is a report by Coffey on soil classification.[49] Another more recent article by Kellogg con-

43. (a) E. W. Hilgard, *Chemistry of Soils as Related to Crop Production—Bureau of Soils Bulletin no. 22*, Proceedings of the 17th Annual Convention Association American Agriculture Colleges and Experiment Stations, Nov. 1903 (1904 [USDA Office of Experiment Stations Bulletin no. 142]), pp. 117–121. (b) C. G. Hopkins, *The Present Status of Soil Investigation*, Proceedings of the 17th Annual Convention Association American Agriculture Colleges and Experiment Stations, Nov. 1903 (1904 [USDA Office of Experiment Stations Bulletin no. 142]), pp. 95–104.
44. E. J. Russell, "The Recent Work of the American Soil Bureau," *Journal of Agricultural Science* 1 (1905): 327–346.
45. Edgar Buckingham, *Studies on Movement of Soil Moisture* (1907 [USDA Bureau of Soils Bulletin no. 39]).
46. E. E. Free, *The Movement of Soil Material by the Wind* (1911 [USDA Bureau of Soils Bulletin no. 68]).
47. W. J. McGee, *Soil Erosion* (1911 [USDA Bureau of Soils Bulletin no. 71]).
48. G. N. Coffey, *A Study of the Soils of the United States* (1912 [USDA Bureau of Soils Bulletin no. 85]).
49. G. N. Coffey, "The Present Status and Future Development of Soil Classification," *Journal of the American Society of Agronomy* 8 (1916): 239–243.

sisted of suggestions for reading by soil scientists.[50] With the passing of time the proportion of articles focused entirely on soils has diminished. The background for a great majority of soil-related articles has been the conception of a medium for plant growth. The *Agronomy Journal* continues to be a world leader in publishing information on crop and soil improvement.

The first American periodical devoted entirely to the subject of soils was founded by J. G. Lipman of Rutgers University, New Jersey, as *Soil Science* in 1916. That journal covered the full spectrum. It overlapped the *Journal of the American Society of Agronomy* to some extent in articles on soil-crop relations. The proportion of articles focused exclusively on soils diminished somewhat after the first twenty years while that on soil-crop relations increased. Most articles in *Soil Science* during its first twenty years reflected one or both of two conceptions of soil—as a medium for plant growth or as the mantle of weathered rock. Subsequently, the second of these gradually faded as the one of soil as a natural historical body replaced it. Articles continue to reflect two conceptions.

The total volume of publications in agricultural sciences increased sharply after World War I with a consequent increase in soil science literature. A selected list of soil-related periodicals and serials being published in the United States prior to 1930 is given in Table 14.1. The list is intended to show the diversity of publications.

Table 14.1. Selected American soil-related serials and periodicals pre-1930

Journal	First issued
Physical Review	1893–
The American Fertilizer	1894–1950
Forestry and Irrigation	1902–1908
Soil Culture	1903–1905
Fertilizer Review and Buyer's Guide	1910–
Journal of the American Society of Agronomy	1913–1948
Journal of Agricultural Research	1913–1949
Soil Science	1916–
Journal of Soil Improvement	1917–1920
Junior Soldiers of the Soil	1919–1923?
Agricultural Lime News Bulletin	1920–1927
American Soil Survey Association Bulletin	1921–1936
Baynes' Soil Improver	1923–1925?
Mississippi Soil Improvement Journal	1923–1930?
Hilgardia	1925–
National Fertilizer Review	1926–1955

50. Charles E. Kellogg, "Reading for Soil Scientists, Together with a Library," *Journal of the American Society of Agronomy* 32 (1) (1940): 867–876.

Among the journals and serials listed in Table 14.1 two carried most of the articles about soils. Those were *Soil Science* and the *Journal of the American Society of Agronomy*. Occasional papers about soils appeared in each of the *Journal of Agricultural Research* and *Hilgardia*. The *American Soil Survey Association Bulletin* was restricted to articles focused on soils, but the volumes, published annually, were small and had very limited circulation. Only the two journals were thus important in the United States for the first two decades after World War I.

A new periodical was added when the Soil Science Society of America was organized in 1936. That society absorbed the American Soil Survey Association and the Soil Section of the American Society of Agronomy. At first the new society published annual volumes of *Proceedings*, consisting of papers presented at its meetings. Later, publication was shifted to a quarterly and finally to a bimonthly journal. Organized in the same pattern as the International Society of Soil Science, the American society grouped papers according to subject matter, e.g., soil physics, soil chemistry, soil biology, and so on. Conceptions of soils reflected in the articles in the early years were like those in the papers in *Soil Science*, one or both of a medium for plant growth or the mantle of weathered rock. Later the society *Journal* included a higher proportion of articles reflecting the conception of soil as natural historical body than did other periodicals. Many papers were still focused on soil as a medium for plant growth or reflected a combination of two conceptions. The *Soil Science Society of America Journal* has become the largest carrier in the United States of articles focused on soils. The journal also has a major place among those of the world.

The literature of soil science expanded after World War I. Books included college texts, summaries of the current status in a field, and descriptions of soils of a region. One widely used textbook was *The Nature and Properties of Soils* by T. L. Lyon and H. O. Buckman.[51] Two other texts were *Farm Soils, Their Management and Fertilization* by E. L. Worthen[52] and *The Soil and the Microbe* by S. A. Waksman and R. L. Starkey.[53] A major work covering the current status in its field was *Principles of Soil Microbiology* by S. A. Waksman.[54] It also served as a primary reference. Additional examples were *Theory and Practice in the Use of Fertilizers* by F. E. Bear,[55] *Soil Physics* by L. D. Baver,[56] and *Factors of Soil Formation*

---

51. T. L. Lyon and H. O. Buckman, *The Nature and Properties of Soils* (New York: Macmillan, 1922).

52. E. L. Worthen, *Farm Soils, Their Management and Fertilization* (New York: Wiley, 1927).

53. S. A. Waksman and R. A. Starkey, *The Soil and the Microbe* (New York: Wiley, 1931).

54. S. A. Waksman, *Principles of Soil Microbiology* (Baltimore, Md.: Williams and Wilkins Co., 1927).

55. F. E. Bear, *Theory and Practice in the Use of Fertilizers* (New York: Wiley, 1929).

56. L. D. Baver, *Soil Physics* (New York: Wiley, 1940).

by H. Jenny.[57] These last two were also used as advanced texts. Descriptions of soils of an area were given in *The Soils and Agriculture of the Southern States* by H. H. Bennett[58] and in the massive monograph on *Soils of the United States* by C. F. Marbut.[59] More information on soils of the country was provided by *Soils and Men* which was intended for the public at large but instead became a reference book for soil scientists for the next twenty years.[60]

For most of these books, the background conception of soil was as a medium for plant growth with little concern for origin. Exceptions were the monograph by Marbut and the book by Jenny in which the background conception is of soil as a natural historic body. That concept also runs through some articles in *Soils and Men*, whereas others in the same book reflect the conception of a medium for plant growth. A combination of two conceptions—a medium for plant growth and the weathered rock mantle—serve as the background for the book by Lyon and Buckman.

Concern about soil erosion has been responsible for appreciable additions to the literature of soils science during the present century. This concern prompted action during the late 1920s and early 1930s although damage done by erosion had been recognized much earlier.[61] The first plots for the study of soil erosion were laid out in Missouri in 1914.[62] More vigorous statements about the evils began to appear in the late 1920s, as for example a bulletin by Bennett and Chapline.[63] Congress appropriated funds in 1930 for studies in the control of erosion, and a total of ten stations were established in different parts of the country.[64] The imagination of the public was captured by the large dust storms and the attendant publicity in the first half of the 1930s. The concern prompted the establishment of the Soil Erosion Service in the U.S. Department of Interior, later transferred to the U.S. Department of Agriculture and renamed the Soil Conservation Service, in the early years of the New Deal. H. H. Bennett, who had earlier been in charge of soil erosion investigations in the Bureau of Chemistry and Soils,

57. Hans Jenny, *Factors of Soil Formation* (New York: McGraw-Hill, 1941).

58. H. H. Bennett, *The Soils and Agriculture of the Southern States* (New York: Macmillan, 1921).

59. C. F. Marbut, *Soils of the United States* (Washington, D.C.: USDA Atlas of American Agriculture, 1935).

60. U.S. Department of Agriculture, *Soils and Men: Yearbook of Agriculture 1938* (Washington, D.C.: U.S. Govt. Print. Off., 1938).

61. (a) Shaler, "The Origin and Nature . . ." (b) W. J. Spillman, *Soil Conservation* (1910 [USDA Farmer's Bulletin no. 406]). (c) McGee, *Soil Erosion*.

62. M. F. Miller, "Early Investigations Dealing with Water Runoff and Soil Erosion," *Journal of the American Society of Agronomy* 38 (7) (1946): 657–660.

63. H. H. Bennett and W. R. Chapline, *Soil Erosion, a National Menace* (1928 [USDA Circular no. 33]).

64. G. M. Browning, "History and Challenge in Soil and Water Conservation and Management," *Soil Science Society of America Journal* 41 (2) (1977): 254–259.

became chief of the new organization. Through its activities more than 3,000 local soil conservation districts have been formed to combat erosion. Also established to promote soil conservation was the Soil Conservation Society of America, now named the Soil and Water Conservation Society of America, which puts out a journal. The program for control of soil erosion has generated a large array of publications of various dimensions, including a book, *Soil Conservation* by H. H. Bennett.[65]

F. Publishers

No investigation on the evolution of the literature of soil science would be complete without mention of the important commercial publishers in the field. During the nineteenth century, the great American publisher in agriculture was Orange Judd of New York, himself an agriculturalist of note. His firm published both popular and scholarly works on a wide range of subjects, including many titles on soils. Richard Allen's *American Farm Book: Being a Practical Treatise on Soils, Manures, Draining, Irrigation . . .* published by Judd in 1849 is a good example of a work aimed at the general reader.[66] But many of the early scientists who wrote extensively about soils were also published by Judd. Among them was Samuel Johnson, a co-founder of the Agricultural Experiment Station at Connecticut, who as early as 1859 presented a paper before the Smithsonian Institution outlining his theories on the uses of tillage, drainage and soil amendments. Judd published Johnson's *Peat and its Uses, as Fertilizer and Fuel* in 1866, and his *How Crops Feed* in 1870. Orange Judd was not alone in this arena however. Commercial firms such as Saxton and Macmillan, also of New York, and Bixby in Boston, followed on the heels of Judd's success and were influential in the arena of agricultural publishing until the close of the nineteenth century, and in the case of Macmillan, are still in business today.

In the twentieth century, new commercial houses gained ascendancy, notably McGraw-Hill and later John Wiley and Sons, both of New York. Also of New York, the up-and-coming firm of Van Nostrand Reinhold published Milton Whitney's *Soil and Civilization* in 1925 among other important titles. In Baltimore, Williams and Wilkins began publishing in agriculture, notably Waksman's *Principles of Soil Microbiology* in 1927. These East Coast publishers dominated the industry until World War II. One of the first big publishers not established in the East was William Brown of Dubuque,

65. H. H. Bennett, *Soil Conservation* (New York: McGraw-Hill, 1939).

66. Richard Allen, *American Farm Book: Being a Practical Treatise on Soils, Manures, Draining, Irrigation . . .* (Judd, 1849).

Iowa, who published, in 1949, Louis Thompson's classic *Soil Fertility*. Much of the agricultural literature read in the United States, however, still came from the United Kingdom. Important agricultural publishers from the United Kingdom included Clarendon Press, Longman-Green, Chapman and Hall, and Edward Arnold.

G. Early Indexes and Bibliographies

As with any growing body of literature, abstracts, bibliographies and indexes proliferated in the early twentieth century to make readily accessible the sheer volume of soils research literature. Perhaps the earliest attempt at a systematic index of this kind in the United States was the *Experiment Station Record* put out by the USDA in collaboration with the land grant experiment stations. Started in 1890, the *Record* listed the publications of the sixty-four land-grant experiment stations and gave brief abstracts for each entry. In 1903, this was followed by a *General Index to the Experiment Station Record*. Commercial ventures followed. In 1922, the H.W. Wilson Company began its *Agricultural Index*, which covered all of agriculture and some forestry. Both the USDA publication and Wilson's had subject access and soils, in both, were prominently represented.

But it was not until the 1930s that soil bibliographies came into their own. In England, the first comprehensive serial bibliography on soils was compiled by the Commonwealth Bureau of Soils, which began publishing its *Bibliography of Soil Science, Fertilizers and General Agronomy* in 1931 at the start of the Great Depression. It was published triennially until 1962. In 1938 the landmark *Soils and Fertilizers* was started. In the United States a flood of stand-alone bibliographies were published at this time on diverse aspects of soils. *A Selected Bibliography on Erosion and Silt Movement*, by Gordon Williams[67] and *Bibliography of Organic and Forest Soils, 1926 to 1934*, by the Central States Forest Experiment Station in Columbus, Ohio, are two early examples.[68]

H. Popular Literature

Any investigation into the early literature of soil science must grapple with the problem of its place in the related discipline of agronomy. As was discussed in Chapter 3, throughout most of the nineteenth century, soil

67. (a) Samuel Johnson, *Peat and its Uses, as Fertilizer and Fuel* (Judd, 1866). (b) Samuel Johnson, *How Crops Feed* (Judd, 1870). Gordon Williams, *A Selected Bibliography on Erosion and Silt Movement* (Washington, D.C.: U.S. Govt. Print Off., 1937).

68. *Bibliography of Organic and Forest Soils, 1926 to 1934* (Columbus, Ohio: Central States Forest Experiment Station).

science was inextricably bound to the broad study of agronomy, which has been defined as "that branch of agriculture that deals with the theory and practice of field crop production and soil management." (Webster's 2d ed. 1934). For the purposes of this investigation, works classed as agronomy will be dealt with in the volume on Crop Science; only those works with a preponderance of soil research have been included here. There are still problems which need clarification. Soil science is one of the few agricultural disciplines with almost no popular periodical literature to speak of. By comparison, between 1850 and 1950 there were over ninety English-language popular bee-keeping journals alone, and several hundred on poultry.[69] This is not to say that the subject of soils was not covered in the popular press. Rather it speaks to the fact that almost all successful farm journals, in covering many subjects at once, also had a soils column where aspects of soil management were discussed. Since the majority of these non-specific magazines fall logically in the broader category of agronomy and not in soils per se, they will be dealt with elsewhere.

As has been noted, citation analysis will not necessarily get at the popular literature of a period. Because of this shortcoming, there is really no quantitative method with which to define the extent of this type of literature and it remains a vexing problem. In soils, this is partly solved in any case by the lack of a large popular literary heritage. Nevertheless, careful searching in the shelf lists of Mann Library, coupled with extensive examinations of subject entries in the *Dictionary Catalog of the National Agricultural Library*, as well as perusal of listings of new titles in old sales catalogs and publication advertisements of various popular publishing houses, offers a rough means by which to measure the extent of the material.

What one discovers is that popular soil monographs were indeed published but the number was never extensive. Some were published by companies selling farm machinery. A prime example of this type of work was the popular *Soil Culture and Modern Farm Methods* written by W. E. Taylor, Director of the Soil Culture Department at the John Deere Company in Moline, Illinois.[70] It was in its 5th printing by 1924. Another soil genre popular through the years, yet scholarly in nature, took up broad historical themes equating cultivation of the soil with the rise of western civilizations. Whitney's *Soil and Civilization*[71] and *Soil: Its Influence on the History of the United States* by Archer Hulbert are representative.[72]

69. Henry Murphy et al., "Primary Historical Literature: 1850–1950," in *The Literature of Animal Science and Health* (Ithaca, N.Y.: Cornell University Press, 1992).

70. W. E. Taylor, *Soil Culture and Modern Farm Methods* (Moline, Ill.: J. Deere & Co., 1924).

71. Whitney, *Soil and Civilization* (New York: Van Nostrand Reinhold, 1925).

72. Archer Hulbert, *Soil: Its Influence on the History of the United States* (New Haven, Conn.: Yale University Press, 1930).

The fertilizer industry, perhaps more than anyone, published the bulk of the popular material, distributing literally hundreds of pamphlets to small farmers through regional branch offices of their associations and, to a lesser extent, through the local grange. But the Grange, or Patrons of Husbandry, founded in 1867 chiefly to curb the power of railroad companies and the grain elevators was often at odds with big business. Wisely, in its early years, the fertilizer industry championed itself as a *friend* of the farmer, offering amendments to improve his soil. Indeed, many of their pamphlets had folksy titles: *Folks and Fields Need Lime* (1921) was distributed by the International Harvester Company of Chicago; *Good Fertilizers Make Good Farms* (1917) reads another by the National Fertilizer Association and *Larger Crops With the Guess Work Left Out* (1912) was published by the Bowker Company of Boston. One improbable title was printed by the Myers' Company of New York whose unabashed motto was "Chilean Nitrate of Soda Propaganda" (selling, among other fertilizers, top grade imported guano) with the title *Food for Tomatoes Will Soon Feed You.* Far from scholarly, these tracts are nevertheless of great interest to historians as a pictorial reflection of farm life in rural America. These simple pamphlets have a "word-to-the-wise" charm to them, based on time-honored agricultural knowledge.

More substantial monographs of a popular kind were aimed primarily at garden growers and their soils and to a lesser extent to farmers. Important publishers of this genre included the A. T. De La Mare Company of New York; The Rural Publishing Company, also of New York and the above mentioned Orange Judd. On the Delaware, Chynoweth and Co., and his competitor, the improbable Atlee Burpee and Co., both of Philadelphia carried many titles on soils. The popular *Lockwood's Manuals* series, with several titles on soils, was published by Lockwood, Crosby and Sons, of London and the famous *Cassell's Agricultural Text-Books,* were published by Cassell and Co., also of London with offices in Paris and Melbourne. Finally, no historical list of soil publishers would be complete without a brief mention of a small clique of influential publishers whose works form the historical nucleus of the alternative agricultural movement. Faber and Faber in London, Devin-Adair in New York, and the Rodale Press of Emmaus, Pennsylvania are the three stalwarts of this genre. These publishers will increasingly be seen as a historical link with the low-input, labor intensive farming practices of the past. For this reason, several titles pertaining to "organic" or, more correctly, sustainable treatments of soil improvement have been included in the core historical lists.

I. Importance of the Historical Literature

Soil science literature of the past serves several functions and thereby assumes importance to the present. First, it is the best available history of the development of soil science itself. The pathways followed, including digressions and dead ends, in that development are reflected in the publications of the past. Ruffin (1832) applied some early chemistry to soils before promoting his crusade for lime applications. Later, Hilgard (1860) used more advanced chemistry plus geology in his studies and went beyond both in his understanding of soils and their origin (Cline, 1977).[73] Coffey synthesized the results of the soil surveys completed to date with prior proposals for systems of classification and then recommended that four broad groups of soils be recognized.[74] He also spelled out requirements for an ideal system of soil classification, which still look good.

These three examples mark steps in the past development of soil science in the United States. A second and more important function served by soil science literature of the past, whether recognized or not, is that the ideas of the present have been derived from those of the past. The classification of soils has gone through a number of stages; systems have been built on various bases, each of which has been derived directly or indirectly from the past.[75] Soil properties identified as important early on still get consideration in current schemes. The place and significance of such properties and combinations of properties can be better understood if the history recorded in the literature is known. The need for adjusting liming rates to the nature of soils rests on much better information now that it did for Ruffin because of the knowledge of cation exchange and clay minerals goes far beyond "something more than stiffness or sandiness."[76] Present understanding of the movement of moisture in soils draws on findings made by Buckingham (1907) early in the present century.

A third function of the literature of the past follows from the direct incorporation of earlier ideas into the understandings of the present. In soil classification, for example, a group of soils originally called Chernozems in Russia, also represented in the Great Plains of North America, has largely

---

73. Cline, "Historical Highlights in Soil Genesis, Morphology and Classification."

74. Coffey, *A Study of the Soils* . . .

75. R. W. Simonson, "Soil Classification in the Past—Roots and Philosophies," International Soil Reference and Information Centre, Wageningen, Netherlands, *Annual Report*, 1984 (1985), pp. 61–68. (Reprinted as chapter 18, in D. S. Fanning and M. C. B. Fanning, *Soil Morphology, Genesis and Classification* [New York: Wiley, 1989], pp. 135–152.)

76. Ruffin, "On the Composition of Soils, and Their Improvement by Calcareous Manures," p. 317.

been retained in the current American system as Borolls.[77] Ideas propounded by Buckingham (1907) about movement of moisture in soils are part of those held today. Use of the old in such ways is reminiscent of the construction of early Christian churches in Rome. Side by side in those churches are columns of different styles and stones. A granite column stands next to one of fluted marble. A little inquiry brings out that the columns came from Roman temples that had been torn down. The columns had also been parts of still earlier structures.

Useful today as well are elements of past literature as examples of careful reasoning, sometimes even of pungent or poetic expression. Ruffin (1832) restricted his generalizations about occurrence of acid soils to that part of Virginia "below the falls of our rivers," i.e., the Coastal Plain.[78] He believed that acid soils could be found elsewhere but lacked confirming data. When taking issue with Sir Humphry Davy about soil acidity, Ruffin used converging lines evidence that could be studied to good advantage today by budding soil scientists.

The pungent language of the past is lacking today, which seems a pity. Modern science demands precise dispassionate language. Before the modern era, writing was decidedly more colorful, as for example, the following paragraph from the preface in Hilgard's report on Mississippi (1860):

> It is commonly and cheaply charged upon professional men generally, and upon those cultivating the exact sciences in particular, that they have a perverse disposition to wrap up everything known in an unintelligible jargon or "big words," as they are currently termed. It is expected of them that they should develop new ideas (such as result from the special study of any subject), but that they should use no *new terms* in expressing or communicating them; which is simply impossible. No one can expect to be *taught* without *learning*; let him catch the *idea*, and it will matter little to him whether the word expressing the same be Greek, Latin, or Chinese; English words already possessing definite meanings cannot be used to express new ideas. If he cannot take the time, or the trouble to learn the *idea*, he ought not complain that he cannot understand the *term*.[79]

The language in soil science literature of the past also includes poetic passages. Perhaps the milieu was more romantic. One good example comes

77. Soil Survey Staff, *Soil Taxonomy: A Basic System of Soil Classification for Making and Interpreting Soil Surveys* (1975 [USDA Handbook no. 436]).

78. Ruffin, *An Essay on the Calcareous Manures*, p. 3.

79. Hilgard, *Report on the Geology and Agriculture of the State of Mississippi* (Mississippi: E. Barksdale Jackson, state printer, 1860), p. xiv.

from the report by Shaler (1892) on the nature and origin of soils: "The primitive men, at least in their savage state, had very little influence on the soil—much less, indeed, than many species of lower animals. As long as man trusted to the chase, to fishing, or to the resources afforded by wild fruits and grains for their subsistence, and to chance stones picked up along streams for their weapons, they were practically without influence upon the soil. When, however, our kind took the first long step upward in the arts and began to till the earth, a new and momentous influence was introduced into the assemblage of soil conditions."[80]

J. Core Historical Literature for Preservation: Rationale and Methodologies

The purpose of this historical component is to identify the soil science literature worthy of long-term preservation. The need is real. Upwards of 45% of all nineteenth century printed matter is in ruinous condition due to the acidity of the paper upon which it was printed. The identification and preservation of this literature will provide a permanent record for future scholarly and historical use. The same methods used to arrive at the current core literature were applied on the historical literature. This identification concentrated on the literature of soils published between 1850 to 1950. Prior to the earlier date, it was felt all printed matter was worthy of primary preservation. To arrive at the core after 1850, historical source documents were subjected to citation analysis. The titles were then reviewed by scholars in the field. While a variety of data was gathered, a conscious decision was made for the primary emphasis to be on works published in English, with greatest emphasis on United States or Canadian imprints. Excluded from consideration were government documents, such as USDA handbooks and bulletins, as well as the experiment station bulletins of the land grant universities, most of which have already been preserved on microfilm. Three distinct types of publication important to soil science were identified and analyzed: (A) scientific monographs, (B) scholarly journals, and (C) popular publications, both books and magazines.

Source Documents

The historical literature was analyzed by using citations provided in seminal works on soils for the period 1850–1950. These source documents were

80. Shaler, "The Origin and Nature of Soils," p. 215.

recommended by scholars in the field with 30 years of professional experience in teaching and research as well as by the members of the Soil Science Steering Committee for the Core Agricultural Literature Project. Selecting source documents to represent an even balance of subjects was part of the initial screening process.

Table 14.2 lists the source documents the Core Project used to analyze the historical literature of soils appropriate for college-level education and advanced research of the time. Earlier editions of some of the works listed below were devoid of citations, notably Russell's book which is now in its 11th edition. It was necessary to use the eighth edition to find sufficient citations for analysis. Furthermore, classic texts such as Charles Darwin's *The Formation of Vegetable Mould, through the Action of Worms . . .*[81] and Liebig's *Organic Chemistry . . .*[82] were rarely cited in the source documents, even though they were often mentioned in the texts.

The few journal articles analyzed were broad, historical overviews published to commemorate some significant date, either the founding of the parent publication, or in the case of *Soil Science Society of America Journal's* Volume 41, to honor the 1976 U.S. Bicentennial. These overview articles are an important repository of historical information on the evolution of soil science.

For each monographic citation careful data were recorded, including date, publisher (e.g. government, experiment station, or commercial house) and country of publication. The analysis which follows focuses primarily on English language documents and serials; foreign language publications and publishers, unless exceptionally represented in the counts, were excluded from comparative analysis.

K. Historical Soil Science Journals 1850–1950

In Table 14.1, selected non-evaluative periodicals were listed. All types were represented to show the sorts of publications commonly associated with soils management in the early years of its development. Analysis of the historical source documents provided the data to quantitatively rank the scientific periodicals and these are shown in Table 14.3. Since the primary aim of the historical component of the Core Project analysis is to set preservation priorities for English language material only, no foreign language journals are listed. Of foreign titles which ranked highly, far and away, the

81. Charles Darwin, *The Formation of Vegetable Mould, Through the Action of Worms . . .*, 4th ed. (New York: Saxton, 1856).
82. Liebig, *Die Organische Chemistry in ihrer Anwendung auf Agricultur und Physiologie.*

Table 14.2. Source documents for historical soil science

Source monographs for historical analysis

Ayres, Quincy C. *Soil Erosion and its Control.* 1st ed. New York and London: McGraw-Hill, 1939. 365p.

\# Baver, L. D. *Soil Physics.* 3d ed. New York: J. Wiley and Sons, 1956. 489p.

Bear, Firman E. *Soils and Fertilizers.* New York: J. Wiley and Sons, 1942. 374p.

Black, C. A. *Soil-Plant Relationships.* New York: Wiley and Sons, 1957. 336p.

\# Collings, Gilbeart H. *Commercial Fertilizers: Their Source and Use.* Philadelphia: Blakiston and Sons, 1934. 356p.

Emerson, Paul. *Principles of Soil Technology.* New York: Macmillan, 1930. 402p.

\# Gardner, Frank D. *Soils and Soil Cultivation.* Philadelphia: John Winston, Co., 1918. 223p.

Jenny, Hans. *Factors of Soil Formation: A System of Quantitative Pedology.* New York: McGraw-Hill, 1941. 281p.

\# Joffe, Jacob S. *Pedology.* 2d ed. New Brunswick, NJ: Pedology Publications, 1949. 662p. (1st ed. 1936)

\# Mohr, Edward C. *Tropical Soils: A Critical Study of Soil Genesis.* New York: Interscience Publishers, 1954. 497p.

\# Mosier, J. G. and A. F. Gustafson. *Soil Physics and Mangement.* Philadelphia: J. B. Lippincott, 1917. 442p.

Robinson, Gilbert W. *Soils: Their Origin, Constitution and Classification* . . . London and New York: T. Murby and
Co., 1932. 390p.

Roe, Harry B. *Moisture Requirements in Agriculture: Farm Irrigation.* New York: McGraw-Hill, 1950. 413p.

\# Russell, Edward J. *Soil Conditions and Plant Growth.* 8th ed. New York: J. Wiley, 1949. 793p.

\# Smith, J. Warren. *Agricultural Meteorology.* New York: Macmillan Co., 1920. 304p.

\# Thompson, Louis, M. *Soils and Soil Fertility.* New York: McGraw-Hill, 1957. 451p.

\# Tisdale, Samuel L. *Soil Fertility and Fertilizers.* New York: Macmillan Co., 1956. 430p.

Waksman, Selman A. *Humus: Origin, Chemical Composition* . . . 2d ed. Baltimore: Williams and Wilkins, 1938.
526p.

Source journal articles for historical analysis

Childs, E. C. and E. G. Young. "Soil Physics: Twenty-Five Years On," *Journal of Soil Science* vol. 25 no. 4,
December 1974.

\# "Fifty Years of Progress in Soil Science," *Geoderma* vol. 12, no. 4., 1974. (50th Anniversary of the International
Society of Soil Science.)

\# Gardner, Walter. "Early Soil Physics into the Mid-20th Century," *Advances in Soil Science, Vol. 4.* New York:
Springer Verlag, 1986. pp. 1–75.

\# Simonson, Roy W. "Historical Aspects of Soil Survey and Soil Classification," *Soil Survey Horizons* vol. 27 no. 1–3. Parts I,II,III. Spring, Summer, Fall 1986.

\# "Soil Classification Special Issue," *Soil Science* vol. 62, no. 2, 1949.

Soil Science Society of America: Journal Vol. 41, no. 2. "Bicentennial Papers" March-April 1977. pp. 221–265.

Gardner, Walter H. "Historical Highlights in American Soil Physics."

Thomas, G. W. "Historical Developments in Soil Chemistry."

Clark, Francis E. "Soil Microbiology—It's a Small World."

Viets, Frank G. "A Perspective on Two Centuries of Progress in Soil Fertility and Plant Nutrition."

Cline, M. G. "Historical Highlights in Soil Genesis, Morphology and Classification."

Browning, George, M. "History and Challenges in Soil and Water Conservation Management."

Russel, Darrell A. and Gerald Williams. "History of Chemical Fertilizer Development."

= Sources from which monographs only have been added to the list.

Table 14.3. Core historical English language scholarly journals in soil science

Journals	Citation counts	Filmed
Soil Science (1916–):	420	Vol. 1–25, 52–113
Proceedings of the SSSA (1937–):	182	Vol. 11–20
Journal of the American Society of Agronomy (1907–):	163	Vol. 5–38
Journal of Agricultural Science (1905–):	154	Vol. 1–92
Journal of the American Chemical Society (1878–):	50	Vol. 1– (to current issue)
Journal of Agricultural Research (1913–):	49	Vol. 1–78
Agricultural Engineering (1920–):	38	Vol. 1–44
Journal of the Society of Chemical Industry (1882–):	32	Vol. 1–68
Plant Physiology (1926–):	29	Vol. 1– (to current issue)
Transactions of the American Geophysical Union (1920–):	24	Vol. 1–25 (also reprinted)
Proceedings of the Royal Society, London (1847–):	21	Vol. 1–75
Biochemical Journal (1906–):	19	Vol. 1–130
Hilgardia (1925–):	19	Vol. 1–41
Science (1883–):	19	Vol. 1– (to current issue)
Botanical Gazette (1875–):	18	Vol. 1–3
Transactions of the Faraday Society (1905–):	17	Vol. 1–67
Empire Journal of Experimental Agriculture (1933–):	16	Vol. 1–25
Botanical Review (1935–):	14	Vol. 1–34
Journal of Physical Chemistry (1896–):	14	Vol. 1–20
Nature (1869–):	14	Vol. 1– (to current issue)
Ecology (1920–):	10	Vol. 1–36
Journal of Forestry (1917–):	9	Vol. 1– (to current issue)
Journal of the Royal Society of Agriculture (1840–):	9	*No Film available in U.S.*
Plant and Soil (1948–):	8	Vol. 1–8
Soils and Fertilizers (1938–):	8	Vol. 1–17
Soil Conservation (1935–):	7	Vol. 1–30
Journal of the American Ceramic Society (1918–):	5	Vol. 1– (to current issue)

most commonly cited of these was the German, *Zeitschrift fur Pflanzener-nahrung Dungung Bodenkunde* which was first published in 1922. This was followed by the French *Comptes Rendus des Seances de l'Academie d' Agriculture de France*, published semimonthly since 1915, but never in August/September, of course, when the French take their holidays. It should be noted that this periodical supersedes one of the oldest agricultural journals in Europe, the *Compte Rendu Mensuel*, published by the firm of P. Renouard since 1837, the founding date of the Academie d'Agriculture de France. In 1915, this publication merged with the newer title.

Of the 4,430 citations analyzed, 3,188 or 71.9% were to serials. Five of these twenty-seven journals deal broadly with agriculture as a whole, notably the *Journal of the Agronomy Society of America* (later *Agronomy Journal*) and the British *Journal of Agricultural Science* published at Cambridge. Indeed, seven of the nine articles appearing in the first issue of this latter journal (Vol. 1, no. 1, 1905) dealt with soils. Eight journals deal with the subjects of chemistry and engineering, with the *Journal of the American Chemical Society* ranking highest. Three journals are botanical in scope, the still popular *Plant Physiology* being the most heavily cited. *Science, Nature, Ecology* and the *Proceedings of the Royal Society of London* are all general in scope; their inclusion here is somewhat problematic for the high counts their articles received do not always mean they published articles on soils. Rather it speaks to their wide-spread readership, and their editorial policy to publish insightful articles on a wide range of subjects, such as microbiology and meteorology, which bear on soils research peripherally. The remaining seven titles, beginning with *Soil Science* in 1916, can properly be said to be soils specific and represent the core English language historical journals.

Eight of the journals began publication in the 19th century, the oldest being the *Journal of the Royal Society of Agriculture* a society founded in 1838 with the expressed objective to "provide information contained in agricultural publications and in other scientific works as has been proved by practical experience to be useful to the cultivators of the soil."[83] Their *Journal*, which was inaugurated in 1840, was the vehicle by which this information was to be disseminated. Surprisingly, the majority of the titles listed in Table 14.3 have already been microfilmed which is not entirely surprising since the universe of periodical literature in soils is small.

L. Core Historical Monographs

Of the 4,430 citations analyzed, 1,242 or 28.1% were to monographs. Since the stated purpose of this work was to find the core historical soils

83. H. M. Jenkins, *The Royal Agricultural Society of England* (London: Clowes & Sons, 1878).

literature in the United States and Canada and, secondarily, of England and the rest of the English-speaking Commonwealth, it is not surprising that of the 1,014 monograph citations for which country of publication data were kept, 841 or 82.9% were published in the United States. No country of publication data were kept for conferences. This was due, in part, to the fact that the data were often problematic; extreme cases have been recorded of a conference that was held in one country, carried a convening society's imprint from another, and was finally printed in an entirely different country. While this is an extreme example of the problem, it does show the inherent problem of designating a country of publication for conferences; to circumvent these discrepancies, the 170 citations to conferences in this analysis carry no country designation.

During the early period in the development of soil science in the United States, beginning in the 1890s, the major form of scholarly communication was overwhelmingly the bulletins of the land-grant agricultural experiment stations and to a lesser extent those of the USDA. Although these publications were issued in numbered series, they can properly be considered monographs, because they all have distinct titles, and in some cases are of considerable length.

The agricultural experiment stations of five land-grant universities stand out in soils. Highest ranked was the University of Missouri at Columbia, with forty-three citations. Second was California with thirty-five. This was followed by Cornell/New York State Agricultural Experiment Stations with twenty-nine. Utah was next with nineteen, and fifth was Iowa with eighteen citations. In all, forty-three states were represented in the experiment station counts. For government bulletins, no single title stands out as exceptionally important, although in aggregate over forty titles were commonly cited 32.2% of the time. Compared to current publications, we find the figures in Table 14.4 almost entirely reversed, with organizations and commercial publishing houses coming far out on top (see Chapter 9), and the importance of land-grant AES bulletins almost non-existent. Several factors account for this change. For one, there were few soil journals of quality pub-

Table 14.4. United States monographs by publisher type

Type	Percentage
Commercial	24.1%
Government	32.2%
Organizations	2.6%
Land-grant experiment stations	38.4%
University presses	2.7%

lished in English, between 1890 and the 1940s, which would publish the research of most university or government scientists. Secondly, the emphasis today on acquiring tenure and its attenuating need for a strong track record of established research publication in reputable journals, was absent from the congenial atmosphere of scientific investigation prior to 1950. Finally, the pamphlet-sized publications of the experiment stations and the USDA were ideally suited for the dissemination of research in the field prior to 1950. These publications were widely distributed, were well indexed by the U.S. Department of Agriculture, and most research stations were eager to publish the work of their own scientists' findings. The high counts for USDA bulletin publications, and those of the experiment stations during this period, is an interesting historical anomaly. The authors have heard several emeritus professors lament that these publications now lack for readers and are no longer considered of importance.

Commercial publishing houses accounted for 24.1% of the United States total. McGraw-Hill was the highest ranked, with J. Wiley and Sons and Macmillan close seconds. World-wide, the highest ranked publishing houses were from Germany, which is not surprising since many prominent soil scientists of the era who published in English were from Europe and Russia, where German was the *lingua franca* of science; among these were important scientists such as Jenny, Ramann, Wollny, and Hilgard all of whom wrote books which are of seminal historical importance.

What was surprising was the low count organizations and societies received in the historical counts. It should be noted, however, that whereas organizations such as the American Society of Agronomy began publishing their *Journal* as early as 1907, it was not until the 1940s that they started publishing monographs. Their first was the proceedings of a symposium titled, *Hunger Signs in Crops* edited by Gove Hambidge in 1941.[84] The Soil Science Society of America did not publish a monograph until the 1950s. Rather, the most prominent organizational publishers for this period were the American Potash Institute, the National Fertilizer Association and the Carnegie Institute of Washington. Table 14.5 gives the place of publication data for monographs cited.

Of the 138 conference citations, most were to the series of international soil science congresses, which began with the gathering in Budapest, Hungary in April, 1909, under the title "Reunion of the First International Conference of Agro-Geologists," hosted by the Royal Hungarian Institute of Science. It was followed by another "reunion" in Stockholm in 1910. No meetings were held through the war years 1914–1919. When the "agro-

84. G. Hambidge, ed., *Hunger Signs in Crops* (Washington, D.C.: American Society of Agronomy, National Fertilizer Association, 1941).

Table 14.5. Place of publication of monographs cited in source documents

Country	Number	Percent
United States	841	82.9%
Germany	79	7.7%
United Kingdom	45	4.4%
France	16	1.5%
Russia	11	1.0%
Other[a]	22	2.5%

[a]Australia, Austria, Belgium, Canada, Denmark, Finland, Japan, Kenya, Netherlands, Philippines, Spain, Sweden.

geologists" met again, this time in Prague in 1922, their name and focus were changed to "Pedology" and by the Rome meeting in 1924, moves were afoot to found the International Society of Soil Science as we know it today. The first official meeting of the ISSS was held in Washington, D.C. in 1927, presided over by Jacob Lipman of Rutgers University. Table 14.6 lists the first four international meetings; differences in citation counts for these four pre-1950 conferences were statistically negligible.

The top twenty core historical monographs are listed below. Two of these monographs are part of the above-mentioned proceedings series, namely the first and third International Congresses of Soil Science. It is interesting to note that the second congress, held in Russia in 1930, is not listed and may represent a bias of exclusion on the part of the source document authors in part because most of the papers presented at the second congress are in Russian, German and French with only a few in English. Four of the top monographs are government publications, the important *Soil Survey Manual* of 1937 by Charles Kellogg,[85] Richards' volume on saline soils,[86] Marbut's *Soils of the United States*[87] and surprisingly, the narrative overview *USDA Yearbook: Soils and Men,* published in 1938, which is obviously a land-mark in the field.[88] Faulkner's *Plowman's Folly* deserves mention for it was polemical in its day, and like the works of Aldo Leopold and others, this work in particular explored environmental themes which today are much in

85. Charles E. Kellogg, *Soil Survey Manual* (Washington, D.C.: U.S. Govt. Print. Off., 1937).

86. L. A. Richards, ed., *Diagnosis and Improvement of Saline and Alkali Soils* (Washington, D.C.: Soil and Water Conservation Research Branch, Agricultural Research Service, U.S. Department of Agriculture, 1969). (USDA Agriculture Handbook no. 60).

87. Curtis F. Marbut, Hugh H. Bennett, J. E. Lapham, and M. H. Lapham, *Soils of the United States* (Washington, D.C.: Govt. Print. Off., 1913).

88. United States Department of Agriculture, *Soils and Men: Yearbook of Agriculture 1938* (Washington, D.C.: Govt. Print. Off., 1938).

Table 14.6. International congresses of the International Society of Soil Science

	Date	Place	President
1st	1927	Washington, D.C.	Jacob G. Lipman
2nd	1930	Leningrad, USSR	K. K. Gedroits
3rd	1935	Oxford, U.K.	Sir John Russell
4th	1950	Amsterdam, Neth.	C. H. Edelman

vogue.[89] Another work which caused great debate among early soil scientists for its outright criticisms of established scientific doctrines, is Cyril Hopkins *Soil Fertility and Permanent Agriculture*, which was one of the first works to look at the socio-economic aspects of soil and soil loss and proposed radically new theories to address these.[90] Two of the volumes, namely Darwin's treatise on worms and Liebig's on chemistry were published in the nineteenth century, while the rest are distributed evenly from 1900 to 1950.

The United Kingdom published five of the top titles, the United States fourteen. The title by Mohr, *Soils of Equatorial Regions,* is the only tropical soils monograph listed, originally published in Peiping, China.

Top Twenty Historical Monographs by Citation Counts

1. Russell, Edward J. Russell's Soil Conditions and Plant Growth. 11th ed., edited by Alan Wild. Burnt Mill, U.K. and New York: Longmans and J. Wiley, 1988. 991p. (Rev. ed. of: Soil Conditions and Plant Growth, 10th ed., 1973. 1st ed., Longmans, 1912. 8th ed., 1950. 635p.)
2. International Congress of Soil Science, 1st, Washington, D.C., 1927. Proceedings . . . , edited by S. A. Waksman and R. B. Deemer. Washington, D.C., 1928. 178p.
3. Richards, L. A., ed. Diagnosis and Improvement of Saline and Alkali Soils. Washington, D.C.: Soil and Water Conservative Research Branch, Agricultural Research Service, U.S. Dept. of Agriculture, 1969. 160p. (USDA Agriculture Handbook no. 60) (1st issued 1947. 2d issued 1954.)
4. Kellogg, Charles E. Soil Survey Manual. Washington, D.C.: U.S. Govt. Print. Off, 1937. 136p. (Rev. ed., U.S. Dept. of Agriculture, 1951. 503p.)
5. Hilgard, Eugene W. Soils Their Formation, Properties, Composition, and Relations to Climate and Plant Growth in the Humid and Arid Regions. New York and London: Macmillan, 1906. 593p.
6. Jenny, Hans. Factors of Soil Formation: A System of Quantitative Pedology. New York and London: McGraw-Hill, 1941. 28p.

89. Edward H. Faulkner, *Plowman's Folly* (Norman, Okla.: University of Oklahoma Press, 1974).

90. Cyril G. Hopkins, *Soil Fertility and Permanent Agriculture* (Boston and New York: Ginn and Co., 1910).

7. Darwin, Charles. The Formation of Vegetable Mould, Through the Action of Worms: With Observations on Their Habits. London: J. Murray, 1881. 326p.

8. Lyon, Thomas L., and Harry O. Buckman. The Nature and Properties of Soils. New York: Macmillan, 1922. 588p. (5th ed., 1952.)

9. Glinka, Konstantin D. The Great Soil Groups of the World and Their Development. Ann Arbor, Mich.: Edwards Brothers, 1927. 235p.

10. Marbut, Curtis F., Hugh H. Bennett, J. E. Lapham, and M. H. Lapham. Soils of the United States. Washington, D.C.: Govt. Print. Off, 1913. 791p.

11. Faulkner, Edward H. Plowman's Folly. Norman, Okla.: University of Oklahoma Press, 1943. 156p. (10th printing, 1974, with new foreword.)

12. Hall, A. Daniel. The Book of the Rothamsted Experiments. 2d ed. New York: E.P. Dutton, 1917. 332p.

13. Bear, Firman E. et al. Diagnostic Techniques for Soils and Crops, Their Value and Use in Estimating the Fertility Status of Soils and Nutritional Requirements of Crops. Washington, D.C.: American Potash Institute, 1948. 308p.

14. United States. Dept. of Agriculture. Soils and Men: Yearbook of Agriculture 1938. Washington, D.C.: Govt. Print. Off., 1938. 1232p.

15. Mohr, Edward C. J. The Soils of Equatorial Regions, with Special Reference to the Netherlands East Indies. Ann Arbor, Mich.: J. W. Edwards, 1944. 766p. (Translated from the Nederlandsch by R. L. Pendleton.)

16. Baver, Leonard D. Soil Physics. 2d ed. New York: J. Wiley, 1948. 398p.

17. Hopkins, Cyril G. Soil Fertility and Permanent Agriculture. Boston and New York: Ginn and Co., 1910. 653p.

18. International Congress of Soil Science, 3d, Oxford, 1935. Transactions . . . London: T. Murby and Co., 1935–36. 3 vols. (Articles in English, German, or French.)

19. Liebig, Justus. Organic Chemistry in its Applications to Agriculture and Physiology = Organische Chemie in ihrer Anwendung auf Agricultur und Physiologie, edited by Lyon Playfair. 1st Am. ed. Cambridge, Mass.: J. Owen, 1841. 435p.

20. Marshall, Charles E. The Colloid Chemistry of the Silicate Minerals. New York: Academic Press, 1949. 195p.

M. Identifying and Ranking the Monographs

The citations from the source documents provided the data for a quantitative assessment of the literature which is deemed important for both academic instruction and research in soil science for the period 1850–1950. These quantitatively ranked monographs were then submitted for review by scholars in the field. The reviewers, shown in Table 14.7, were asked to rank the titles identified in the analysis using the following criteria:

(1) The work has had an important influence in the subject field.

(2) The work is the first of its kind to be published or record a major advance in the field.

(3)   The work embodies the historical record of changes in the field.
(4)   The title is valuable because it is a superior work of a leader in the field of soil science.

Table 14.7. List of scholarly reviewers

W. H. Allaway	F. E. Clark	Roy W. Simonson
Emeritus Professor	Fort Collins, Colo.	College Park, Md.
Cornell University	Douglas Lathwell	G. W. Thomas
Ithaca, N.Y.	Cornell University	Emeritus Professor
Marlin G. Cline	Ithaca, N.Y.	University of Kentucky
Emeritus Professor	Peter Nye	Lexington, Ky.
Cornell University	Oxford, United Kingdom	M. Thorne
Ithaca, N.Y.		Urbana, Ill.

The quantitative listing of monographs was graded by scholars using a three category scale: Category 1 = an important historical title worthy of preservation priority: Category 2 = worth preserving, but of secondary importance: Category 3 = a title of marginal historical value. These subject specialists, all of whom have extensive research and teaching experience, ranked the list accordingly and also made suggestions of titles for inclusion. The few titles suggested were added to the list and evaluated.

The final step in analyzing these data came by combining the quantitative citation ranking with the accumulated rankings made by the reviewers. Of the 493 titles identified by citation analysis, fifty-five of them received four or more marks by the reviewers in category 3, namely "titles of marginal historical value." These were either dropped from this primary listing, or were transferred to other historical subject lists, or in the case of a few particular titles, were left on the list if extenuating reasons were found. These included, but were not limited to, titles of books that had copious woodcuts or prints: the book was a 1st edition by a reputable publisher: the book, as a physical object, was deemed important because of distinguishing characteristics such as binding, fly-leaf matter, agricultural advertisements and the like.

The final monograph ranking was calculated using a formula that combined the citation hits with the category rankings assigned by the reviewers which were weighted as follows:

Final ranking = (citation hits) + (3 × # Column 1) + (2 × # C2) + (# C3)

In order to set preservation priorities, those monographs receiving the highest scores are first rank and are deserving of immediate attention. The remainder are ranked second and should be preserved when the first tier of material has been reformatted or conserved. Of the 438 titles, 213 or 48.6% are of the first rank, and 225 or 51.4% are of the second.

Core Historical Monograph List, 483 titles

Second   Adams, Frank D. The Birth and Development of the Geological Sciences. New York; Dover, 1938. 506p. (Reprinted in 1954.)

First rank Agee, Alva. Crops and Methods for Soil Improvement. New York; Macmillan, 1912. 246p.

Second   Agee, Alva. Right Use of Lime in Soil Improvement. New York; Orange Judd Co., 1919. 89p.

Second   Aikman, Charles M. Manures and the Principles of Manuring. Edinburgh; W. Blackwood, 1894. 592p. (3d impression, Edinburgh and London; W. Blackwood and Sons, 1910. 592p.)

Second   Aiyer, A. K. Yegna N. Principles of Crop Husbandry in India. Bangalore City; Bangalore Press, 1948. 475p.

First rank Allen, Oscar N. Experiments in Soil Bacteriology. Minneapolis, Minn.; Burgess Pub. Co., 1949. 126p.

First rank Allen, Richard L. American Farm Book; or Compend of American Agriculture: Being a Practical Treatise on Soils, Manures, Draining, Irrigation, Grasses, Grain, Roots, Fruits, Cotton, Tobacco, Sugar Cane, Rice. New York; Orange Judd and Co., 1849. 325p.

First rank Allison, L. E. et al. Diagnosis and Improvement of Saline and Alkali Soils, edited by L. A. Richards. Riverside, Calif.; Regional Salinity Laboratory, 1947. 157p. (Reproduced from typewritten copy.) (Rev. ed., Washington, D.C.; Soil and Water Conservative Research Branch, Agricultural Research Service, U.S. Dept. of Agriculture, 1969. 160p. [USDA Agriculture Handbook no. 60]

Second   American Chemical Society. Potassium Symposium; Papers presented before the Division of Fertilizer Chemistry of the American Chemical Society, Buffalo, N.Y., Sept. 1942. Baltimore; Williams and Wilkins, 1943. 126p.

Second   American Plant Food Council, Inc. Our Land and Its Care. 1st ed. Washington, D.C., 1947. 64p. (2d ed., 1948.)

Second   American Society of Civil Engineers. Soil Mechanics and Foundations Division. Soil Mechanics Nomenclature. New York; Headquarters of the Society, 1941. 15p.

Second   Anderson, Frederick I. The Farmer of To-morrow. New York; Macmillan, 1913. 308p.

First rank Anderson, Thomas. Elements of Agricultural Chemistry. Edinburgh; Adam and Charles Black, 1860. 299p.

First rank Andrews, William B. The Response of Crops and Soils to Fertilizers and Manures. 1st ed. Mississippi State College, 1947. 459p. (2d ed., 1954. 463p.)

Second   Armco Inc. Handbook of Culvert and Drainage Practice, For the Solution of Surface and Subsurface Drainage Problems. Crawfordsville, Ind.; W.

Q. O'Neall Co., 1930. 349p. (2d ed., Middletown, Conn.; Armco Culvert Manufacturers Association, 1937. 472p.)

First rank   Association of Official Analytical Chemists. Official Methods of Analysis of the Association of Official Analytical Chemists. Washington, D.C.; Association of Official Analytical Chemists, (1st ed. 1919. 1st-10th ed. published by the association under an earlier name: Association of Official Agricultural Chemists.)

Second   Ayres, Arthur H. Influence of the Composition and Concentration of the Nutrient Solution on Plants Grown in Sand Cultures. 1917. 394p.

First rank   Ayres, Quincy C. Soil Erosion and Its Control. 1st ed. New York and London; McGraw-Hill, 1936. 365p.

Second   Ayres, Quincy C., and D. Scoates. Land Drainage and Reclamation. 1st ed. New York; McGraw-Hill Book Co., Inc., 1928. 419p. (2d ed., 1939. 496p.)

First rank   Bagnold, Ralph A. The Physics of Blown Sand and Desert Dunes. New York; Morrow and Co., 1942. 265p.

First rank   Bailey, Liberty H. The School-Book of Farming: A Text for the Elementary Schools, Homes and Clubs. New York; Macmillan, 1920. 388p.

First rank   Balfour, Evelyn B. The Living Soil: Evidence of the Importance to Human Health of Soil Vitality, with Special Reference to Post-War Planning. London; Faber and Faber, 1943. 246p. (New rev. ed. titled: The Living Soil and The Haughley Experiment, 1975. 383p.)

Second   Baren, Johan van. Microscopial Physical and Chemical Studies of Limestones and Limestone Soil from the East Indian Archipelago. Wagenignen, Netherlands; H. Veenman and Zonen, 1928. 195p.

Second   Barker, Percy B. A Manual of Soil Physics. Boston and New York; Ginn and Co., 1915. 101p.

Second   Barnard, Charles. Talks About the Soil in its Relation to Plants and Business. A Book of Observations and Experiments for the Use of Schools, Students, and Farmers. New York; Funk and Wagnals, 1894. 127p. (Copyright 1886.)

Second   Barrett, Thomas J. The Earthmaster System for Intensive Propagation and Use of Domesticated Earthworms in Soil-Building. Roscoe, Calif.; Earthmaster Publications, 1942. 24p. (5th rev. ed., titled: Earthworms, Their Intensive Propagation and Use in Biologial Soil-Building: The Earthmaster System, 1948. 59p.)

Second   Barrett, Thomas J. Harnessing the Earthworm: A Practical Inquiry . . . Boston; B. Humphries, 1948. 184p.

Second   Bates, G. H. Weed Control. London; Farmer and Stock-Breeder, 1948. 235p.

First rank   Baver, Leonard D. Soil Physics. New York; J. Wiley, and London; Chapman and Hall, Ltd., 1940. 370p. (4th ed. by L. D. Baver, W. H. Gardner, and R. G. Wilford, 1972. 498p.)

First rank   Bear, Firman E. Soil Management. New York; J. W. Wiley and Sons, 1924. 268p. (4th ed. titled: Soils and Fertilizers. New York; Wiley, 1953. 420p.)

First rank   Bear, Firman E. Theory and Practice in the Use of Fertilizers. New York; J. Wiley and Sons, 1929. 348p. (2d ed., New York; J. Wiley, and London; Chapman and Hall, 1938. 360p.)

First rank   Bear, Firman E. et al. Diagnostic Techniques for Soils and Crops, Their Value and Use in Estimating the Fertility Status of Soils and Nutritional Requirements of Crops. Washington, D.C.; American Potash Institute, 1948. 308p.

Second   Beaumont, Arthur B. Artificial Manures; or The Conservation and Use of Organic Matter for Soil Improvement. New York; Orange Judd Co., 1943. 155p.

Second   Beaumont, Arthur B. Garden Soils, Their Use and Conservation. New York; Orange Judd Co., 1948. 280p.

Second   Bell, James M., and Paul M. Gross. Elements of Physical Chemistry. New York and London; Longman, 1929. 446p.

First rank   Bennett, Hugh H. Elements of Soil Conservation. 1st ed. New York and London; McGraw-Hill, 1947. 406p. (2d ed., 1955. 358p.)

First rank   Bennett, Hugh H. Soil Conservation. 1st ed. New York and London; McGraw-Hill, 1939. 993p.

First rank   Bennett, Hugh H. The Soils and Agriculture of the Southern States. New York; Macmillan, 1921. 399p.

First rank   Bennett, Hugh H., and Robert V. Allison. The Soils of Cuba. Washington, D.C.; Tropical Plant Research Foundation, 1928. 410p.

Second   Bergey, David H., chmn. Bergey's Manual of Determinative Bacteriology: A Key for the Identification of Organisms of the Class Schizomycetes. Baltimore, Md.; Williams and Wilkins, Co., 1923. 461p. (New ed. titled, Bergey's Manual of Systematic Bacteriology, 1984 + . 3 vols.)

First rank   Bertramson, B. R., and J. L. White. Soil Chemistry Notes. Purdue, Ind.; Purdue University, 1948. 326p.

First rank   Black, John D., Marion Clawson, Charles R. Sayre, and Walter W. Wilcox. Farm Management. New York; Macmillan, 1947. 1073p.

First rank   Bornebusch, Carl H. The Fauna of Forest Soil. Skovbundens Dyreverden. Copenhagen; Nielsen and Lydiche, 1930. 224p. (Reprinted from Det Forstlige Forsygsvcsen i Danmark vol. XI.)

First rank   Briggs, Lyman J., and H. L. Shantz. The Wilting Coefficient for Different Plants and its Indirect Determination. Washington, D.C.; Govt. Print. Off., 1912. 83p.

Second   Brown, Victor J. Soil Stabilization after C. A. Hogentogler. A Short Course Based on Exhibit of U.S. Bureau of Public Roads displayed at the 1938 Road Show. Chicago, Ill.; Gillette Publishing Co., 1939. 141p.

Second   Browne, Charles A. A Source Book of Agricultural Chemistry. Waltham, Mass. and New York; The Chronica Botanica Co. and G. E. Stecher, 1944. 290p.

Second   Bruce, Maye E. From Vegetable Waste to Fertile Soil, Quick Return Compost. London; Faber and Faber, 1943. 52p.

First rank   Buckman, James. Science and Practice in Farm Cultivation. London, Eng.; R. Hardwicke, 1865. 358p.

Second   Buffum, Burt C. Arid Agriculture: A Hand-Book for the Western Farmer and Stockman. Worland, Wyo.; B. C. Buffum, 1909. 443p.

First rank   Bunce, Arthur C. The Economics of Soil Conservation. Ames, Iowa; Iowa State College Press, 1942. 227p.

Second   Burgess, Paul S. Soil Bacteriology Laboratory Manual. Easton, Pa.; The Chemical Publishing Co., 1914. 123p.

Second   Burkett, Charles W. Soils: Their Properties, Improvement, Management and the Problems of Crop Growing and Crop Feeding. New York; Orange Judd Co., 1907. 303p.

Second   Burt, Frederick A. Soil Mineralogy: A Discussion of Mineralogy in its Application to Soil Studies. New York; D. Van Nostrand Co., 1927. 82p.

First rank   Cameron, Farnk K. The Soil Solution, the Nutrient Medium for Plant Growth. Easton, Pa.; The Chemical Publishing Co., 1911. 136p.

First rank   Campbell, Hardy W. Campbell's 1907 Soil Culture Manual: A Complete Guide to Scientific Agriculture as Adapted to the Semi-Arid Regions . . . Lincoln, Neb.; H. W. Campbell, 1907. 320p.

First rank   Cannon, William A. Physiological Features of Roots, with Special Reference to the Relation of Roots to Aeration of the Soil. Washington, D.C.; Carnegie Institution of Washington, 1925. 168p.

Second   Carew, Norman. Ploughman's Wisdom. London; Faber and Faber, 1949. 167p.

Second   Caribbean Commission. Soil Science in the Caribbean; Report of the Soils Conference, Puerto Rico, Mar. 1950. Port-of-Spain, Trinidad; Kent House, 1950. 265p.

Second   Cave, Herbert. Fertilizers, Their Sources, Manufacture, and Uses. London and New York; Sir I. Pitman and Sons, Ltd., 1926. 116p.

Second   Chapman, Paul W. et. al., eds. Conserving Soil Resources: A Guide to Better Living. Atlanta, Ga.; T. E. Smith, 1950. 355p.

First rank   Chynoweth, James B., and William H. Bruckner. American Manures; and Farmers' and Planters' Guide: Comprising a Description of the Elements and Composition of Plants and Soils. Philadelphia; Chynoweth and Co., 1871. 260p.

First rank   Clarke, Frank W. The Data of Geochemistry. Washington, D.C.; Govt. Printing Office, 1908. 716p. (Issued also as House Doc. no. 725, 60th Cong., 1st sess.) (5th ed., 1924. 841p. [U.S. Geological Survey Bulletin no. 770])

First rank   Clarke, George R. The Study of Soil in the Field. Oxford, Eng.; Clarendon Press, 1936. 142p. (5th ed., Oxford; Clarendon Press, 1971. 145p.)

First rank   Coffey, George N. A Study of the Soils of the United States. Washington, D.C.; Govt. Print. Off, 1912. 114p.

First rank   Collings, Gilbeart H. Commercial Fertilizers: Their Sources and Use. Philadelphia, Pa.; Blakiston's Son and Co., Inc., 1934. 356p.

Second   Collingwood, Herbert W. Fertilizers and Fruit: A Trip Among Growers in the Famous Hudson River Fruit District. New York; The Rural Pub. Co., 1893. 27p.

Second   Comber, Norman M. An Introduction to the Scientific Study of the Soil. New York; Longmans, and London; Arnold and Co., 1927. 192p.

First rank   Comber, Norman M., H. Trefor Jones, and J. S. Willcox. An Introduction to Agricultural Chemistry. New York; Longmans, and London; Arnold and Co., 1947. 315p.

First rank   Commonwealth Bureau of Soil Science. Tropical Soils in Relation to Tropical Crops . . . Harpenden, U.K.; Imperial Bureau of Soil Science, 1936. 60p.

First rank   Conn, Herbert W. Agricultural Bacteriology: A Study of the Relation of Germ Life to the Farm, with Laboratory Experiments for Students. Philadelphia; P. Blakiston's Son and Co., 1901. 412p. (3d ed., rev. by Harold J. Conn, 1918. 357p.)

Second   Corbet, Alexander S. Biological Processes in Tropical Soils: With Special References to Malaysia. Cambridge, U.K.; W. Heffer and Sons Ltd., 1935. 156p.

Second   Cornell University. N.Y.S College of Agriculture, Dept. of Agronomy. The Role of Phosphorus in Crop Production. A Mimeographed Record of the Seminar. Ithaca, N.Y.; Cornell University, 1941. 393p.

First rank   Cox, Joseph F. Crop Production and Soil Management. New York; J. Wiley, 1925. 516p.

Second   Cox, Joseph F., and Lyman Jackson. Field Crops and Land Use. New York; J. Wiley, and London; Chapman and Hall, 1942. 473p.

First rank   Cox, Joseph F., and Lyman E. Jackson. Crop Management and Soil Conservation. New York; J. Wiley, and London; Chapman and Hall, Ltd., 1937. 610p. (2d ed., 1948. 572p.)

First rank   Craven, Avery O. Soil Exhaustion as a Factor in the Agricultural History of Virginia and Maryland, 1606–1860. Urbana, Ill.; The University of Illinois, 1926. 179p.

Second   Crozier, William, and Peter Henderson. How the Farm Pays. The Experience of Forty Years of Successful Farming . . . New York; P. Henderson, 1884. 400p.

Second   Cubbon, Miles H., and M. J. Markuson. Soil Management for Greenkeepers. Amherst, Mass.; 1933. 152p.

Second   Cunningham, Jules C., and W. H. Lancelot. Soils and Plant Life as Related to Agriculture. New York; Macmillan, 1916. 348p.

Second   Cutler, Donald W., and Lettice Crump. Problems in Soil Microbiology. London and New York; Longmans, 1935. 104p.

First rank   Dana, Samuel L. A Muck Manual for Farmers. Lowell; D. Bixby, 1842. 242p. (5th ed. titled, A Muck Manual for Farmers: A Treatise on the Physical and Chemical Properties of Soils . . . New York; Orange Judd, 1855. 312p.)

First rank   Darwin, Charles. The Formation of Vegetable Mould, Through the Action of Worms: With Observations on Their Habits. London; J. Murray, 1881. 326p.

Second   Davenport, Eugene. The Farm . . . New York; Macmillan, 1927. 462p.

Second   Davis, Lucius D. Improving the Farm; or Methods of Culture that Shall Afford a Profit and at the Same Time Increase the Fertility of the Soil. Newport, R.I.; Davis and Pitman, 1880. 187p.

First rank   Davy, Humphry. Elements of Agricultural Chemistry in a Course of Lectures for the Board of Agriculture. Philadelphia; B. Warner, 1821. 304p.

Second   Dies, Edward J. Titans of the Soil: Great Builders of Agriculture. Chapel Hill, N.C.; University of North Carolina Press, 1949. 213p.

First rank   Dyer, Bernard S. A Chemical Study of the Phosphoric Acid and Potash Contents of the Wheat Soils of Broadbalk Field, Rothamsted. London; Dulau and Co., 1901. 290p. (Philosophical Transactions of the Royal Society of London Series B no. 194)

Second   Dyke, William. Manures and Fertilizers. London; W. H. and L. Collingridge, 1924. 142p.

Second   Eckstein, Oskar. Potash Deficiency Symptoms = Kennzeichen des Kalimangels = Signes de Manque de Potasse. Berlin; Verlagsgesellschaft fur Ackerbaum, 1937. 234p.

First rank   Eden, Thomas. Elements of Tropical Soil Science. London; Macmillan, 1947. 136p.

Second   Emerson, Frederick V. Agricultural Geology. New York; J. Wiley, 1920. 319p.

First rank Emerson, Paul. Principles of Soil Technology. New York; Macmillan, 1930. 402p.

Second   Emerson, Paul. Soil Characteristics: A Field and Laboratory Guide. 1st ed. New York; McGraw-Hill, 1925. 222p.

First rank Emmons, Ebenezer. Agriculture of New York: Comprising an Account of the Classification, Composition and Distribution of the Soils and Rocks . . . Albany, N.Y.; C. Van Benthuysen and Co., 1846–54. 5 vols.

Second   Ernle, Rowland E. P. English Farming, Past and Present. New ed. London and New York; Longmans, Green, and Co., 1912. 504p. (6th ed., Chicago; Quadrangle Books, 1962. 559p.)

First rank Falkner, Frederic. The Farmer's Treasure: A Practical Treatise on the Nature and Value of Manures, Founded from Experiments on Various Crops . . . New York; D. Appleton, 1844. 138p.

First rank Faraday Society. Physico-Chemical Problems Relating to the Soil. London, 1922. 368p.

Second   Farm Economy: A Cyclopedia of Agriculture for the Practical Farmer and His Family . . . Special ed. Minneapolis, Minn.; H. L. Baldwin Pub. Co., 1915. 1196p.

First rank Faulkner, Edward H. Plowman's Folly. Norman; University of Oklahoma Press, 1943. 265p. (10th printing, with new foreword, 1974. 156p.)

Second   Faulkner, Edward H. A Second Look. 1st ed. Norman; University of Oklahoma Press, 1947. 193p. (The sequel to Plowman's Folly.)

First rank Fenneman, Nevin M. Physiography of Eastern United States. New York and London; McGraw-Hill, 1938. 714p.

First rank Fenneman, Nevin M. Physiography of Western United States. New York and London; McGraw-Hill, 1931. 534p.

First rank Fletcher, Stevenson W. Soils: How to Handle and Improve Them. New York; Doubleday, Page and Co., 1907. 438p.

First rank Flint, Richard F. Glacial Geology and the Pleistocene Epoch. New York; J. Wiley, and London; Chapman and Hall, 1947. 589p.

Second   Fortier, Samuel. Use of Water in Irrigation. 1st ed. New York; McGraw-Hill Book Co., Inc., 1915. 265p. (3d ed., 1926. 420p.)

First rank Fraps, George S. Principles of Agricultural Chemistry. Easton, Pa.; The Chemical Publishing Co., 1913. 493p.

First rank Fred, Edwin B. A Laboratory Manual of Soil Bacteriology. Philadelphia and London; W. B. Saunders Co., 1916. 170p.

First rank Fred, Edwin B., Ira L. Baldwin, and Elizabeth McCoy. Root Nodule Bacteria and Leguminous Plants. Madison, Wis.; University of Wisconsin, 1932. 343p.

First rank Freundlich, Herbert. Colloid and Capillary Chemistry. New York; Dutton and Co., 1922. 883p. (Translated from the 3d German ed. by H. Stafford Hatfield.)

First rank Freundlich, Herbert. New Conceptions in Collodial Chemistry. London; Methuen and Co. Ltd., 1926. 147p.

Second   Gardner, Frank D. Soils and Soil Cultivation, a Non-Technical Manual on the Management of Soil for the Production and Maintenance of Fertility. Philadelphia and Chicago; J. C. Winston Co., 1918. 223p.

Second   Gardner, Frank D. Successful Farming: A Ready Reference on all Phases of Agriculture for Farmers of the United States and Canada . . . Philadelphia, Pa.; J. C. Winston Co., 1916. 1088p.

Second   Gardner, William. Fertilisers and Soil Improvement: Description, Application and Comparative Value. London; Lockwood and Sons, 1927. 184p.

Second   Garner, Harold V. et al. Profit from Fertilizers. London; C. Lockwood, 1936. 182p. (3d and rev. ed., 1948. 160p.)

Second   Garrett, Stephen D. Root Disease Fungi: A Treatise on the Epidemiology of Soil-Borne Disease in Crop Plants . . . Waltham, Mass.; The Chronica Botanica Co., 1944. 177p.

First rank   Garrett, Stephen D. Soil-Borne Fungi and the Control of Root Disease. Harpenden, U.K.; Imperial Bureau of Soil Science, 1939. 54p. (Imperial Bureau of Soil Science. Technical Communication no. 38)

First rank   Gedroiz, K. Alkali Soils, Their Origin, Properties and Improvement; A Scientific Popular Sketch = Solontsy, ikh Proiskhozhdenie, Svoistva i Melioratsiisa; Nauchno-populisarnyi Ocherk. Nosovka; Izdanie Nosovskoi Sel'sko-khozisalistven. Opytnoistantsii, 1928. 76p.

First rank   Gilman, Joseph C. A Manual of Soil Fungi. Ames, Iowa; The Collegiate Press, Inc., 1945. 392p. (Rev. 2d ed., 1957. 450p.) (A revision of: A Summary of the Soil Fungi, 1927.)

First rank   Glinka, Konstantin D. The Great Soil Groups of the World and Their Development. Ann Arbor, Mich.; Edwards Brothers, 1927. 235p.

Second   Goessmann, Charles A. Commercial Fertilizers; Paper from the 26th Annual Report of the Secretary of the State Board of Agriculture . . . Boston, Mass.; Rand, Avery and Co., 1879. 41p.

Second   Goodrich, Charles L. The First Book of Farming. New York and Doubleday; Page and Co., 1905. 259p.

Second   Gorrie, Robert M. The Use and Misuse of Land. Oxford; Clarendon Press, 1935. 80p.

Second   Gorrie, Robert M. Soil and Water Conservation in the Punjabj. Simla, 1946. 290p.

Second   Gough, H. C. A Review of the Literature on Soil Insecticides. London; The Imperial Institute of Entomology, 1945. 161p.

Second   Graber, Laurence F., and H. L. Ahlgren. Agronomy, Principles and Practices. Dubuque, Iowa; W. C. Brown Co., 1948. 151p.

Second   Graham, Edward H. Natural Principles of Land Use. London; Oxford University Press, 1944. 274p. (2d printing, New York; Greenwood Press, 1969. 274p.)

Second   Graham, Michael. Soil and Sense. London; Faber and Faber, 1946. 274p.

First rank   Gray, A. N. Phosphates and Superphosphate. London; International Superphosphate Manufacturer's Association, 1930. 275p. (2d ed., London; E. T. Heron and Co., Ltd., and New York; Interscience Publishers, Inc., 1944 and 1945. 416p.)

First rank   Greaves, Joseph E., and Ethelyn O. Greaves. Bacteria in Relation to Soil Fertility. New York; Van Nostrand Co., 1925. 239p.

Second   Gregory, James J. H. Fertilizers: Where the Materials Come From . . . Boston, 1886. 65p.

Second   Griffith, John H. Physical Properties of Earths. Ames, Iowa; Iowa State College, 1931. 128p. (Iowa Engineering Experiment Station Bulletin no. 101)

Second   Griffiths, Arthur B. Manures and Their Uses: A Handbook for Farmers
           and Students. London; George Bell and Sons, 1889. 159p.
First rank Gustafson, Axel F. Conservation of the Soil. New York and London;
           McGraw-Hill, 1937. 312p.
Second   Gustafson, Axel F. Handbook of Fertilizers: Their Sources, Make-Up, Ef-
           fects and Use. New York; Orange Judd Co., 1928. 122p. (4th ed., 1944.
           172p.)
First rank Gustafson, Axel F. Soils and Soil Management. 1st ed. New York and
           London; McGraw-Hill, 1941. 424p.
Second   Gustafson, Axel F. Using and Managing Soils. New York; McGraw-Hill,
           1948. 420p.
Second   Gustafson, Axel F. et al. Conservation in the United States. Ithaca, N.Y.;
           Comstock Pub. Co., 1947. 477p. (3d ed., 1949. 534p.)
First rank Hall, A. Daniel. The Book of the Rothamsted Experiments. New York; E.
           P. Dutton, 1905. 294p. (2d ed., 1917. 332p.)
Second   Hall, A. Daniel. The Feeding of Crops and Stock: An Introduction to the
           Science of the Nutrition of Plants and Animals. New York; E. P. Dutton,
           1912. 298p.
First rank Hall, A. Daniel. Fertilisers and Manures. London; J. Murray, 1909. 384p.
           (5th ed. titled, Fertilizers and Manures, 1955. 333p.)
Second   Hall, A. Daniel. Reconstruction and the Land: An Approach to Farming in
           the National Interest. London; Macmillan, 1942. 286p.
First rank Hall, A. Daniel. The Soil: An Introduction to the Scientific Study of the
           Growth of Crops. London; J. Murray, 1903. 311p. 286p. (New and rev.
           ed. by G.W. Robinson, 1945. 322p.)
Second   Halligan, James E., ed. Fundamentals of Agriculture. Boston and New
           York; D. C. Heath and Co., 1911. 492p.
Second   Halligan, James E. Soil Fertility and Fertilizers. Easton, Pa.; Chemical
           Publishing Co., 1912. 397p.
First rank Hambidge, Gove, ed. Hunger Signs in Crops. A Symposium. Washington,
           D.C.; American Society of Agronomy and National Fertilizer Association,
           1941. 327p. (3d ed., Howard B. Sprague, ed. New York; McKay, 1964.
           461p.)
First rank Harris, Franklin S. Soil Alkali: Its Origin, Nature, and Treatment. New
           York; J. Wiley, 1920. 258p.
Second   Harris, Franklin S., and George Stewart. The Principles of Agronomy: A
           Text-Book of Crop Production for High-Schools and Short-Courses in Ag-
           ricultural Colleges. New York; Macmillan, 1915. 451p.
Second   Hartman, Robert J. Colloid Chemistry. Boston and New York; Houghton
           Mifflin Co., 1939. 556p.
Second   Haskell, Sidney B. Farm Fertility. New York; Harper, and London;
           Brothers, 1923. 115p.
Second   Hauser, Ernst A. Colloidal Phenomena: An Introduction to the Science of
           Colloids. New York and London; McGraw-Hill, 1939. 294p.
First rank Hilgard, Eugene W. Soils their Formation, Properties, Composition, and
           Relations to Climate and Plant Growth in the Humid and Arid Regions.
           New York and London; Macmillan, 1906. 593p.
Second   Hinkle, Samuel F. Fertility and Crop Production: A Handbook for the Stu-
           dent and Farmer. Sandusky; S. F. Hinkle, 1925. 338p.

First rank   Hoagland, Dennis R. Lectures on the Inorganic Nutrition of Plants. Waltham, Mass.; Chronica Botanica Co., 1944. 226p. (2d rev. print., 1948. 226p.)

Second   Hogentogler, Chester A. Engineering Properties of Soil. 1st ed. New York and London; McGraw-Hill, 1937. 434p.

First rank   Hopkins, Cyril G. Soil Fertility and Permanent Agriculture. Boston and New York; Ginn and Co., 1910. 653p.

First rank   Hopkins, Cyril G. The Story of the Soil, from the Basis of Absolute Science and Real Life. Boston; R.G. Badger, 1910. 350p. (6th rev. ed., 1913. 362p.)

First rank   Hopkins, Cyril G., and J. H. Pettit. Laboratory Manual for Soil Fertility. Urbana, Ill., 1905. 60p.

Second   Hopkins, Donald P. Chemicals, Humus, and the Soil: A Simple Presentation of Contemporary Knowledge and Opinions about Fertilizers, Manures, and Soil Fertility. London; Faber and Faber ltd., 1945. 278p. (Rev. ed., New York; Chemical Pub. Co., 1957. 288p.)

Second   Howard, Albert, and Yeshwant D. Wad. The Waste Products of Agriculture: Their Utilization as Humus. London and New York; H. Milford, Oxford University Press, 1931. 167p. (Continued by, An Agricultural Testament, 1940. 253p.)

Second   Howard, Albert. The Soil and Health: A Study of Organic Agriculture. New York; The Devin-Adair Co., 1947. 307p. (London ed. by Faber and Faber has title: Farming and Gardening for Health or Disease.)

Second   Howard, Louise E. The Earth's Green Carpet. 1st ed. Ammaus, Pa.; Rodale Press, 1947. 258p.

Second   Hudson, Charles T. The Rotifera: Or, Wheel-Animalcules. London; Longmans, 1886. 3 vols.

Second   Hulbert, Archer B. Soil: Its Influence on the History of the United States . . . New Haven; Yale University Press, and London; Oxford University Press, 1930. 227p.

Second   Hunt, Thomas F., and Charles W. Burkett. Soils and Crops, with Soils Treated in Reference to Crop Production. New York; Orange Judd Co., 1913. 541p.

First rank   Hutcheson, Thomas B., and T. K. Wolfe. The Production of Field Crops: A Textbook of Agronomy. 1st ed. New York; McGraw-Hill, 1924. 499p. (2d ed. by T. B. Hutcheson, T. K. Wolfe, and M. S. Kips, 1936. 445p.)

First rank   Ignatieff, Vladimir, ed. Efficient Use of Fertilizers. Washington, D.C.; Food and Agriculture Organziation, 1949. 182p. (FAO Agricultural Studies no. 9)

First rank   International Congress of Soil Science, 1st, Washington, D.C., 1927. Proceedings . . . , edited by S. A. Waksman and R. B. Deemer. Washington, D.C., 1928. 178p.

First rank   International Congress of Soil Science, 2d, Leningrad and Moscow, 1930. Proceedings . . . Moscow; State Publishing House of Agricultural, Cooperative and Collective Farm Literature (Selkolkhozgis), 1932–35. 7 vols.

First rank   International Congress of Soil Science, 3d, Oxford, 1935. Transactions . . . London; T. Murby and Co., 1935–36. 3 vols. (Articles in English, German, or French.)

First rank International Geological Congress, 11th, Stockholm, 1910. Compte Rendu de la XIe Session du Congres Geologique International. Stockholm; Kungl. Boktryckeriet, P. A. Norstedt, 1912. 2 vols.

First rank International Society of Soil Science. Pedology in USSR; Papers for the 3d International Congress of Soil Science, Oxford, 1935. Moscow; Soviet Section, 1935. 222p.

First rank International Society of Soil Science. The Problem of Soil Structure; Transactions of the 1st Commission of the International Society of Soil Science. Moscow; Soviet Section, 1933. 132p.

First rank International Society of Soil Science. Transactions of the 5th Commission of the International Society of Soil Science = Comptes Rendus de la Cinquieme Commission de l'Association Internationale de la Science du Sol = Verhandlungen der Funften Kommission der Internationalen Bodenkundlichen Gesellschaft, Helsinki VII, 1938. Helsinki; Valtioneuvoston Kirjapaino, 1938.

Second Isgur, Benjamin. An Introduction to Soil Science. Boston, Mass.; Agricultural Scientific Pub. Co., 1938. 239p.

First rank Israelsen, Orson W. Irrigation Principles and Practices. New York; J. Wiley, and London; Chapman and Hall, 1932. 422p.

Second Jack, Walter T. The Furrow and Us. Philadelphia, Pa.; Dorrance and Co., 1946. 158p.

First rank Jacks, Graham V. Land Classification for Land-Use Planning. Harpenden, U.K.; Imperial Bureau of Soil Science, 1946. 90p. (Imperial Bureau of Soil Science, Technical Communication no. 43)

Second Jacks, Graham V., and R. O. Whyte. Erosion and Soil Conservation. Aberystwyth, U.K.; Imperial Bureau of Pastures and Forage Crops, 1938. 206p. (Also published as: Imperial Bureau of Soil Science; Technical Communication no. 36.)

First rank Jacks, Graham V., and R. O. Whyte. The Rape of the Earth: A World Survey of Soil Erosion. London; Faber and Faber Ltd., 1939. 312p. (American ed. titled: Vanishing Lands: A World Survey of Soil Erosion. New York; Doubleday Doran, 1939. 332p.)

Second Jeffery, Joseph A. Text-Book of Land Drainage. New York; Macmillan, 1916. 256p.

Second Jenkins, Herbert T. Soil Mechanics Laboratory Manual: Physical Properties of Soils. Ithaca, N.Y.; Comstock Pub. Co., 1947. 109p.

First rank Jenny, Hans. Factors of Soil Formation: A System of Quantitative Pedology. New York and London; McGraw- Hill, 1941. 28p.

First rank Jenny, Hans. Properties of Colloids. Berkeley, Calif.; University of California, 1938. 136p.

Second Joffe, Jacob S. The ABC of Soils. New Brunswick, N.J.; Pedology Publications, 1949. 283p.

First rank Joffe, Jacob S. Pedology. New Brunswick, N.J.; Rutgers University Press, 1936. 575p. (2d ed., New Brunswick, N.J.; Pedology Publications, 1949. 662p.)

First rank Johnson, Samuel W. How Crops Feed. A Treatise on the Atmosphere and the Soil as Related to the Nutrition of Agricultural Plants. New York; Orange Judd, 1870. 375p. (2d ed. titled: How Crops Grow. A Treatise on the Chemical Composition, Structure and Life of the Plant, for Students of Agriculture . . . Rev. and enl. ed. 1891. 416p.)

First rank   Johnson, Samuel W. Peat and Its Uses, as Fertilizer and Fuel. New York;
              Orange Judd, 1866. 168p.
Second   Johnson, Vance. Heaven's Tableland: The Dust Bowl Story. New York;
              Farrar, Straus, 1947. 288p.
Second   Johnston, James F. W. Catechism of Agriculutral Chemistry and Geology.
              St. John, New Brunswick; Barnes, 1861. 68p.
Second   Johnston, James F. W. Elements of Agricultural Chemistry and Geology.
              New York; C. M. Saxton, 1856. 381p.
Second   Jones, Harry C. The Elements of Physical Chemistry. New York and Lon-
              don; Macmillan, Co., 1902. 565p. (4th ed., rev. and enl., 1909. 650p.)
Second   Keeble, Frederick W. Fertilizers and Food Production on Arable and Grass
              Land. Oxford, U.K.; University Press, 1932. 196p.
First rank   Keen, Bernard A. The Physical Properties of the Soil. London and New
              York; Longmans, 1931. 380p.
First rank   Kelley, Walter P. Cation Exchange in Soils. New York; Reinhold Pub.
              Corp., 1948. 144p.
First rank   Kellogg, Charles E. An Exploratory Study of Soil Groups in the Belgian
              Congo. Brussels; Institut National Pour l'Etude Agronomique de Congo
              Belge, 1949.
First rank   Kellogg, Charles E. Soil Survey Manual. Washington, D.C.; U.S. Govt.
              Print. Off, 1937. 136p. (Rev. ed., U.S. Dept. of Agriculture, 1951.
              503p.)
First rank   Kellogg, Charles E. The Soils that Support Us: An Introduction to the
              Study of Soils and Their Use by Men. New York; Macmillan, 1941. 370p.
First rank   Kendrew, Wilfrid G. Climate: A Treatise on the Principles of Weather and
              Climate. Oxford, U.K.; Clarendon Press, 1930. 329p.
First rank   Kendrew, Wilfrid G. The Climates of the Continents. Oxford, U.K.;
              Clarendon Press, 1922. 387p.
First rank   King, Franklin H. Investigations in Soil Management (In 3 parts). Wash-
              ington, D.C.; Government Printing Office, 1905. 205p.
First rank   King, Franklin H. Irrigation and Drainage. New York; Macmillan, 1906.
              502p.
First rank   King, Franklin H. Soil Management, edited by the author's widow, Mrs.
              C. B. King. New York; Orange Judd Co., 1914. 311p.
First rank   King, Franklin H. The Soil, Its Nature, Relations, and Fundamental Prin-
              ciples of Management. New York and London; Macmillan, 1895. 303p.
First rank   King, Franklin H. A Text Book of the Physics of Agriculture. 2d ed.
              Madison, Wis.; F. H. King, 1901, 1899. 604p. (5th ed., 1910. 604p.)
First rank   Kittredge, Joseph. Forest Influences: The Effects of Woody Vegetation on
              Climate . . . New York; McGraw-Hill Book Co., 1948. 394p. (Reprint,
              New York; Dover, 1973.)
First rank   Knox, Gordon D. The Spirit of the Soil. London; Constable and Co., and
              New York; Van Nostrand, 1916. 242p. (Supplementary title: An Account
              of Nitrogen Fixation in the Soil by Bacteria and of the Production of Aux-
              imones in Bacterized Peat.)
First rank   Kramer, Paul K. Plant and Soil Water Relationships. New York; McGraw-
              Hill, 1949. 347p. (2d ed., 1969. 482p.)
Second   Kruyt, Hugo R. Colloids: A Textbook, translated from the manuscript by
              H. S. van Klooster. New York; Willey, and London; Chapman and Hall,
              1927. 262p.

Second   Krynine, Dimitri P. Notes on Applied Soil Physics for Graduate Students in Civil Engineering. Ann Arbor, Mich.; Edwards Brothers, Inc., 1937. 76p.

First rank Krynine, Dimitri P. Soil Mechanics: Its Principles and Structural Applications. 1st ed. New York and London; McGraw-Hill Book Co., Inc., 1941. 451p. (2d ed., 1947. 511p.)

First rank Kubiena, Walter L. Micropedology. Ames, Iowa; Collegiate Press, Inc., 1938. 243p. (Kubina's 1937 lecture series.)

Second   Lambert, Thomas. Bone Products and Manures: An Account of the Most Recent Improvements in the Manufacture of Fat, Glue, Animal Charcoal, Size, Gelatine, and Manures. London; Scott, Greenwood, 1901. 162p.

Second   Lapham, Macy H. Crisscross Trials: Narrative of a Soil Surveyor. Berkeley, Calif.; W. E. Berg, 1949. 246p.

Second   Laurie, Alexander, and J. B. Edmond. Fertilizers for Greenhouse and Garden Crops. New York; A. T. De La Mare Co., Inc., 1929. 147p. (2d ed. titled, Soil and Fertilizers . . . and D. C. Kiplinger. 1948. 128p. (1st ed. 1930)

First rank Lawes, John B. et al. The Soil of the Farm. New York; Orange Judd Co., 1883. 107p.

First rank Lawes, John B., and J. H. Gilbert. Report of Experiments on the Growth of Wheat for the Second Period of Twenty Years in Succession on the Same Land. London; W. Clowes and Sons, 1885. 97p.

Second   Lawrence, William J. C., and J. Newell. Seed and Potting Composts: With Special Reference to Soil Sterilization. London; George Allen and Unwin Ltd., 1939. 128p.

First rank Leeper, Geoffrey W. Introduction to Soil Science. Victoria, Australia; Melbourne University Press, 1948. 222p. (4th ed., 1964. 253p.)

First rank Liebig, Justus. Organic Chemistry in its Applications to Agriculture and Physiology = Organische Chemie in ihrer Anwendung auf Agricultur und Physiologie, edited by Lyon Playfair. 1st Am. ed. Cambridge, Mass.; J. Owen, 1841. 435p.

Second   Lipman, Jacob G., and Percy E. Brown. Laboratory Guide in Soil Bacteriology. n.p., 1911. 87p.

Second   Little, James M. Erosional Topography and Erosion: A Mathematical Treatment . . . San Francisco; Lithotone printed by A. Carlisle and Co., 1940. 104p.

Second   Lloyd, Strauss L. Mining and Manufacture of Fertilizing Materials and their Relation to Soils. New York; D. Van Nostrand Co., 1918. 153p.

Second   Lobeck, Armin K. Geomophology: An Introduction to the Study of Landscapes. 1st ed. New York and London; McGraw-Hill, 1939. 731p.

First rank Lohnis, Felix and Fred. E. B. Textbook of Agricultural Bacteriology. 1st ed. New York; McGraw-Hill, 1923. 283p.

First rank Lowdermilk, Walter C. Conquest of the Land Through Seven Thousand Years. Washington, D.C., 1944. 72p.

First rank Lutz, Harold J., and Robert F. Chandler. Forest Soils. New York; J. Wiley, 1946. 514p.

First rank Lyon, Thomas L. Soils and Fertilizers. New York; Macmillan, 1918. 255p.

First rank Lyon, Thomas L., and Harry O. Buckman. The Nature and Properties of Soils. New York; Macmillan, 1922. 588p. (5th ed., 1952.)

First rank   Lyon, Thomas L., and Elmer O. Fippin. The Principles of Soil Management. New York; Macmillan, 1909. 531p.

First rank   Lyon, Thomas L., Elmer O. Fippin and Harry O. Buckman. Soils, Their
Properties and Management. New York; Macmillan, 1915. 764p.

Second   Macdonald, Duncan G. F. Hints on Farming and Estate Management . . .
5th ed. London; Longmans, 1865. 547p. (10th ed. titled, Estate Management. London; D. Steel, 1868. 726p.)

Second   Maclure, William. Observations on the Geology of the United States of
North America, with Remarks on the Probable Effects that may be Pro-
duced by the Decomposition of the Different Classes of Rocks on the Na-
ture and Fertility of Soils . . . Philadelphia; American Philosophical
Society, Transations, 1818. 91p. (Reissued, Palo Alto, Calif.; Academic
Reprints, 1954.)

Second   Macself, Albert J. Soils and Fertilizers. London; Butterworth, 1926. 224p.

First rank   Malden, Walter J. Tillage. London; G. Bell, 1891. 156p.

Second   Malherbe, Izak D. Soil Fertility. English version of the 6th Afrikaans ed.
London; Oxford University Press, 1948. 296p.

First rank   Marbut, Curtis F., Hugh H. Bennett, J. E. Lapham, and M. H. Lapham.
Soils of the United States. Washington, D.C.; Govt. Print. Off, 1913.
791p.

Second   Marr, John E. Agricultural Geology. London; Methuen, 1903. 318p.

First rank   Marshall, Charles E. The Colloid Chemistry of the Silicate Minerals. New
York; Academic Press, 1949. 195p.

First rank   Marshall, Charles E. Colloids in Agriculture. London; E. Arnold and Co.,
1935. 184p.

Second   Marson, Thomas B. Soil and Security. Edinburgh; Oliver and Boyd, 1947.
135p.

Second   Massey, Wilbur F. Practical Farming: A Plain Book on Treatment of the
Soil and Crop Production; Especially Designed for the Every-Day Use of
Farmers and Agricultural Students. New York; Outing Pub. Co., 1907.
323p.

Second   McCall, Arthur G. Field and Laboratory Studies of Soils: An Elementary
Manual for Students of Agriculture. 1st ed., 1st thousand. New York; J.
Wiley, 1915. 77p.

Second   McCall, Arthur G. The Physical Properties of Soils: A Laboratory Guide.
New York; Orange Judd Co., 1917. 102p.

Second   McConnell, Primrose. The Elements of Agricultural Geology: A Scientific
Aid to Practical Farming. London; Crosby Lockwood, 1902. 329p.

Second   McConnell, Primrose. The Elements of Farming. London; Vinton and Co.
Ltd., 1896. 149p.

Second   McConnell, Primrose. Soils: Their Nature and Management, a Practical
Handbook. London, Paris, New York, Toronto and Melbourne; Cassell
and Co., Ltd., 1908. 104p.

First rank   Merrill, George P. The First One Hundred Years of American Geology.
New Haven; Yale University Press, 1924. 773p.

First rank   Merrill, George P. A Treatise on Rocks, Rock-Weathering and Soils. New
York and London; Macmillan and Co., 1913. 400p.

Second   Meyer, Adolph F. The Elements of Hydrology. 1st ed. New York; J.
Wiley, 1917. 487p. (2d ed. rev., London; Chapman and Hall, 1928.
522p.)

Second   Michigan. State Highway Dept. Field Manual of Soil Engineering. Lansing, Mich.; State Highway Commission, 1946. 304p.

Second   Mickey, Karl B. Man and the Soil: A Brief Introduction to the Study of Soil Conservation. Chicago; International Harvester Co., 1945. 110p.

First rank Millar, Charles E. Soils and Soil Management. St. Paul, Minn.; Webb Book Pub. Co., 1929. 477p. (2d ed., 1937. 477p.)

First rank Millar, Charles E., and L. M. Turk. Fundamentals of Soil Science. New York; J. Wiley and Sons., Inc., and London; Chapman and Hall, Ltd., 1943. 462p. (8th ed., New York and Chichester; Wiley, 1990. 360p.)

Second   Miller, Merritt F. The Soil and Its Management . . . Boston; Ginn and Co., 1924. 386p.

First rank Mitchell, R. L. The Spectorgraphic Analysis of Soils, Plants and Related Materials. Harpenden, U.K.; Commonwealth Bureau of Soil Science, 1948. 183p.

First rank Mohr, Edward C. J. The Soils of Equatorial Regions with Special Reference to the Netherlands East Indies. Ann Arbor, Mich.; J. W. Edwards, 1944. 766p. (Translated from the Nederlandsch by R. L. Pendleton.)

First rank Mohr, Edward C. J. Tropical Soil Forming Processes and the Development of Tropical Soils, with special reference to Java and Sumatra. Peiping, 1933. 200p. (Translated from the Dutch: De Grond an Java en Sumatra. 2d ed. by Robert L. Pendelton, 1930.)

Second   Moore, Harry C. Dictionary of Fertilizer Materials and Terms. Philadelphia, Pa.; The American Fertilizer, 1938. (New ed. titled, Dictionary of Plant Foods, by Arnon L. Mehring, 1958. 51p.)

Second   Moore, Ransom A., and Charles P. Halligan. Plant Production. Part I. Agronomy. Part II. Horticulture. New York and Cincinnati, Ohio; American Book Co., 1919. 428p.

First rank Morrow, George E., and Thomas F. Hunt. Soils and Crops of the Farm. Chicago, Ill.; Howard and Wilson, 1892, 1891. 303p. (New ed., New York; Orange Judd Co., 1910. 303p.)

First rank Morton, John. On the Nature and Property of Soils; Their Connexion with the Geological Formation . . . London; J. Ridgway, 1838. 235p. (4th ed. enl. titled, The Nature . . ., 1843. 432p.)

Second   Mosier, Jeremiah G. Laboratory Manual for Soil Physics. Urbana, Ill.; University of Illinois, 1905. 66p.

Second   Mosier, Jeremiah G., and A. F. Gustafson. Soil Physics and Management. Philadelphia and London; J. B. Lippincott Co., 1917. 442p.

First rank Mosier, Jeremiah G., and A. F. Gustafson. Soil Physics Laboratory Manual. Boston and New York; Ginn and Co., 1912. 71p.

Second   Munro, John M. H. Soils and Manures. London; Cassell, 1892. 275p.

Second   Murray, John A. Soils and Manures. New York; D. Van Nostrand Co., 1910. 354p. (3d ed., rev. and enl., titled, The Science of Soils and Manures, 1925. 298p.)

First rank Muskat, Morris. The Flow of Homogeneous Fluids Through Porous Media. New York; McGraw-Hill, 1937. 763p. (2d ed., Ann Arbor, Mich.; J. W. Edwards, Inc., 1946. 763p.)

First rank Nash, John A. The Progressive Farmer: A Scientific Treatise on Agricultural Chemistry, the Geology of Agriculture; on Plants, Animals, Manures, and Soils. Applied to Practical Agriculture. New York; Moore, 1857. 254p.

Second   National Fertilizer Association. Soil Fertility Conference in the Northern States. 1926. Washington, D.C.; National Fertilizer Association, 1926.

Second   Oliver, George. Friend Earthworm: Practical Application of a Lifetime Study of Habits of the Most Important Animal in the World. Oceanside, Calif.; Oliver's Earthworm Farm School, 1941. 112p.

First rank   Osborn, Fairfield. Our Plundered Planet. 1st ed. Boston; Little, Brown, 1948. 217p. (2d ed., New York; Pyramid Books, 1970. 176p.)

Second   Parker, Edward C. Field Management and Crop Rotation: Planning and Organizing Farms; Crop Rotation Systems; Soil Amendment with Fertilizers; Relation of Animal Husbandry to Soil Productivity; and Other Important Features of Farm Management. St. Paul, Minn.; Webb Pub. Co., 1920. 507p.

Second   Parrish, Percy and A. Ogilvie. Calcium Superphosphate and Compound Fertilisers: Their Chemistry and Manufacture. London; Hutchinson's Scientific and Technical Publication, 1939. 322p. (A revised ed. of: Artificial Fertilisers: Their Chemistry, Manufacture and Application. 2d ed. rev., 1946. 279p.)

Second   Patrick, Austin L., and Fred G. Merkle. Soils and Soil Fertility: Prepared for Use by Winter School Classes at the Pennsylvania State College. Ann Arbor, Mich.; Edward Brothers, 1924. 114p.

Second   Pearson, Haydn S. Successful Part-time Farming. New York; Whittlesey House, and London; McGraw-Hill, 1947. 322p.

First rank   Peech, Michael. Methods of Soil Analysis for Soil-Fertility Investigations. Washington, D.C.; U. S. Govt. Print. Office., 1947. 25p. (First issued as a mimeographed report by the Bureau of Plant Industry, Soils and Agricultural Engineering in March 1945.)

Second   Pfeiffer, Ehrenfried. Bio-Dynamic Farming and Gardening, Soil Fertility Renewal and Preservation. New York and London; Anthroposophic Press and Rudolf Steiner Publishing Co., 1938. 220p. (Translated from the German by Fred Heckel.)

Second   Pfeiffer, Ehrenfried. Soil Fertility, Renewal and Preservation: Bio-Dynamic Farming and Gardening. East Grinstead, U.K.; Lanthorn Press, 1983. 184p. (New rev. ed., London; Faber and Faber, 1947. 196p.)

Second   Phillips, Alexander H. Mineralogy: An Introduction to the Theoretical and Practial Study of Minerals. New York; Macmillan, 1912. 699p.

Second   Picton, Lionel J. Nutrition and the Soil: Thoughts on Feeding. Introductory Essay on Creative Medicine by Jonathan Forman. New York; Devin-Adair, 1949. 374p.

Second   Pieters, Adrian J. Green Manuring: Principles and Practice. New York; J. Wiley; and London; Chapman and Hall, 1927. 356p.

First rank   Piper, Clarence S. Soil and Plant Analysis: A Laboratory Manual of Methods . . . Adelaide; The University of Adelaide, 1942. 368p. (New ed., New York; Interscience, 1950. 368p. [CSIRO Pamphlet no. 8]).

Second   Pirsson, Louis V., and Charles Schuchert. Introductory Geology for Use in Universities, College, Schools of Science, etc., and for the General Reader. New York; Wiley, 1924. (4th ed. titled, Outlines of Historical

Geology, by Charles Schuchert and Carl O. Dunbar. New York; J. Wiley, and London; Chapman and Hall, 1941. 291p.)

Second   Pittman, Don W. Soil—The Foundation of Agriculture. 1936. 202p.

Second   Plummer, Fred L., and Stanley M. Dore. Soil Mechanics and Foundations. New York and Chicago; Pitman, 1940. 473p.

First rank   Pochvennyli Institut Imeni Dokuchaeva. Studies in the Genesis and Geography of Soils. Moscow; Leningrad Academy of Sciences Press, 1935. 253p. (Text in German and English)

First rank   Polynov, Boris B. The Cycle of Weathering. London; T. Murby and Co., 1937–.

Second   Powers, Wilbur L., and T. A. H. Teeter. Land Drainage. New York; J. Wiley, 1922. 270p.

Second   Prange, Nettie M. G. Key to Success: General Principles of Soil Management. Jacksonville, Fla.; Wilson and Toomer Co., 1913. 86p.

First rank   Puri, Amar N. Soils, Their Physics and Chemistry. New York; Reinhold Publ. Corp., 1949. 550p.

Second   Quear, Charles L. Soils and Fertilizers, for Public Schools: A Discussion Upon the Nature and Treatment of Soils and the Value of Fertilizers, edited by O. L. Boor. Chicago, Ill.; E.F. Harmon and Co., 1915. 202p.

First rank   Ramann, Emil. The Evolution and Classification of Soils, translated by C. L. Whittles. Cambridge, U.K.; W. Hefer and Sons Ltd., 1928. 128p.

Second   Rastall, Robert H. Agricultural Geology. Cambridge, U.K.; University Press, 1916. 331p.

Second   Rather, Howard C. Field Crops. New York and London; McGraw-Hill, 1942. 454p.

First rank   Rayner, Mabel C. Mycorrhiza: An Account of Non-Pathogenic Infection by Fungi in Vascular Plants and Bryophytes. London; Wheldon and Wesley, Ltd., 1927. 246p.

First rank   Rayner, Mabel C., and W. Neilson-Jones. Problems in Tree Nutrition: An Account of Researches Concerned Primarily with the Mycorrhizal Habit . . . London; Faber and Faber Ltd., 1944. 184p. (1st published in Mcmxliv.)

Second   Remington, John S. The Manure Note Book. London; L. Hill, 1943. 58p. (4th ed. titled, The Manure and Fertilizer Note Book: A Handy Guide for Manure Manufacturers and Merchants, Farmers, Smallholders, and Market Gardeners, Agricultural and Horticultural, 1953. 123p.)

Second   Rennie, William. Successful Farming: How to Farm for Profit, the Latest Methods. Toronto; W. Rennie's Sons, 1900. 312p.

First rank   Rice, Clara M. Dictionary of Geological Terms (Exclusive of Stratigraphic Formations and Paleontologic Genera Snd species). Ann Arbor, Mich.; Edwards Brothers, Inc., 1943. (2d ed., 1951. 465p.)

Third   Richard, L. A. Diagnosis and Improvement of Saline and Alkali Soils. Riverside, Calif.; U.S. Regional Salinity Laboratory, 1947. 157p.

Second   Richards, Bertram D. Flood Estimation and Control. London; Chapman and Hall, 1944. 152p.

Second   Robbins, Wilfred W., Alden S. Crafts, and Richard N. Raynor. Weed Control: A Textbook and Manual. 1st ed. New York and London; McGraw-Hill, 1942. 543p. (3d ed., 1962. 660p.)

First rank   Roberts, Isaac P. The Fertility of the Land. New York; Macmillan Co., 1897. 415p. (11th ed. titled, The Fertility of the Soil: A Summary Sketch of the Relationship of Farm-Practice to the Maintaining and Increasing of the Productivity of the Soil. 1909. 421p.)

Second   Robertson, George S. Basic Slags and Rock Phosphates. Cambridge; The University Press, 1922. 120p.

Second   Robinson, Gilbert W. Mother Earth. Being letters on soil addressed to Professor R. G. Stapledon. London; T. Murby and Co., 1937. 202p.

First rank   Robinson, Gilbert W. Soils: Their Origin, Constitution, and Classification . . . London and New York; T. Murby and Co. and D. Van Nostrant Pub. Co., 1932. 390p. (3d ed. rev. and enl. 1949 titled: Soils, Their Origin, Constitution and Classification: An Introduction to Pedology. 573p.)

Second   Robinson, Solon. Guano: A Treatise of Practical Information for Farmers . . . New York, 1853. 96p.

Second   Rodale, Jerome I., ed. Compost, and How to Make It. Emmaus, Pa.; Organic Gardening, 1946. 63p.

Second   Rodale, Jerome I. The Organic Front. Emmaus, Pa.; Rodale Press, 1948. 199p.

First rank   Rodale, Jerome I. The Organic Method on the Farm. 1st ed. Emmaus, Pa.; The Organic Farmer, 1949. 128p. (5th ed., 1952. 128p.)

Second   Rodale, Jerome I. Pay Dirt: Farming and Gardening with Composts. New York; Devin-Adair Co., 1945. 242p.

Second   Roe, Jim, and William Raufer, eds. Successful Farming. A Better Living from Your Soil. Des Moines, Iowa, 1949. 106p.

First rank   Ruffin, Edmund. An Essay on Calcerous Manures. Petersburg. Va.; J. W. Campbell, 1832. 242p. (5th ed., Richmond, Va.; J. W. Randolph, 1852. 493p.)

First rank   Russell, Edward J. The Fertility of the Soil. Cambridge, U.K.; Cambridge University Press, 1913. 128p.

First rank   Russell, Edward J. Fertilizers in Modern Agriculture. London; H.M.S.O., 1939. 230p.

First rank   Russell, Edward J. The Micro-Organisms of the Soil. London and New York; Longmans, 1923. 188p. (Rothamsted Monograph)

First rank   Russell, Edward J. Soil Conditions and Plant Growth. London and New York; Longmans, Green, 1912. 168p. (11th ed. titled, Russell's Soil Conditions and Plant Growth, edited by Alan Wild. Burnt Mill, U.K. and New York; Longmans and J. Wiley, 1988. 991p.)

Second   Russell, Edward J. A Student's Book on Soils and Manures. Cambridge, U.K.; Cambridge University Press, 1915. 206p. (3d ed., 1940. 296p.)

Second   Sampson, Harry O. Effective Farming: A Text-Book for American Schools. New York; Macmillan, 1919. 490p. (Copyright 1918)

Second   Sanders, Harold G. An Outline of British Crop Husbandry. Cambridge, Eng.; Camabridge University Press, 1945, 1939. 348p. (3d ed., 1958. 345p.)

First rank   Sandon, H. The Composition and Distribution of the Protozoan Fauna of the Soil. Edinburgh; Oliver and Boyd, 1927. 237p.

Second   Saunders, William. Soil Culture, Cereals and Fruits. Ottawa, Canada; S. E. Dawson, 1900. 37p.

First rank   Schreiner, Oswald, and George H. Failyer. Colorimetric, Turbidity, and Titration Methods Used in Soil Investigations. Washington, D.C.; Govt. Print. Off., 1906. 60p.

Second   Scott Blair, George W. A Survey of General and Applied Rheology. London; Pitman and Sons, Ltd., 1944. 196p.

First rank   Sears, Paul B. Deserts on the March. Norman, Okla.; University of Oklahoma Press, 1935. 231p.

Second   Sempers, Frank W. Manures: How to Make and How to Use Them . . . Philadelphia, Pa.; W. A. Burpee, 1893. 218p. (4th ed., 1895. 218p.)

Second   Seymour, Edward L. D., ed. Farm Knowledge: A Complete Manual of Successful Farming. Garden City, N.Y.; Doubleday, Page, 1918. 4 vols. (Rev. ed., 1919.)

First rank   Shaler, Nathaniel S. Aspects of the Earth: A Popular Account of Some Familiar Geological Phenomena. New York; Scribner's Sons, 1889. 344p.

First rank   Shantz, Homer L., and C. F. Marbut. The Vegetation and Soils of Africa. New York; National Research Council and American Geographical Society, 1923. 263p.

Second   Shepard, Ward. Food or Famine: The Challenge of Erosion. New York; Macmillan, 1945. 225p.

Second   Shewell-Cooper, Wilfred E. Soil Humus and Health. London; J. Gifford, 1944. 84p.

First rank   Sigmond, Alexius A. J. de. The Principles of Soil Science = Altalanos Talajtan. London; T. Murby and Co., 1938. 362p. (Translated from the Hungarian by A.B. Yolland.)

Second   Smith, John W. Agricultural Meteorology: The Effect of Weather on Crops. New York; Macmillan, 1920. 304p.

Second   Smith, Raymond S., and D. C. Wimer. Laboratory Exercises in Soil Physics. University Park, Pa.; Pennsylvania State College, 1917.

Second   Smith, William C. How to Grow One Hundred Bushels of Corn per Acre on Worn Soils. Delphi, Ind.; Smith Pub. Co., 1910. 111p. (2d ed., rev. and enl., Cincinnati, Ohio; Stewart and Kidd Co., 1914. 188p.)

First rank   Snyder, Harry. The Chemistry of Soils and Fertilizers. Easton, Pa.; The Chemical Publishing Co., 1899. 277p. (2d ed. titled, Soils and Fertilizers, 1905. 294p.)

Second   Sornay, P. de. Green Manures and Manuring in the Tropics: Including an Account of the Economic Value of Leguminosae as Sources of Foodstuffs, Vegetable Oils, Drugs, etc. = Les Plantes Tropicales Alimentaires et Industrialles de la Famille de Laegumineuses. London; J. Bale, Sons and Danielsson, Ltd., 1916. 466p.

Second   Southwell, Richard V. Relaxation Methods in Engineering Science: A Treatise on Approximate Computation. London and New York; Oxford University Press, 1940. 252p.

Second   Southwell, Richard V. Relaxation Methods in Theoretical Physics: A Continuation of the Treatise, Relaxation Methods in Engineering Science. Oxford; Clarendon Press, 1946–56. 2 vols.

First rank   Spurway, Charles H. Soil Fertility Diagnosis and Control for Field, Garden, and Greenhouse Soils. East Lansing and Ann Arbor, Mich.; Charles Spurway and Edwards Brothers, Inc., 1948. 176p.

Second   Stephenson, John. Oligochaeta. London; Taylor and Francis, 1923. 518p. (New ed. titled, The Oligochaeta. Oxford, U.K.; Claredon Press, 1930. 978p.)

Second   Stephenson, Marjory. Bacterial Metabolism. London and New York; Longmans Green, 1930. 320p.

Second	Stevenson, William H., and I. O. Schaub. Soil Physics Laboratory Guide. New York; Orange Judd Co., 1905. 80p.
Second	Stewart, Alexander B. Report on Soil Fertility Investigations in India with Special Reference to Manuring: A Review of the Position to Date with Suggestions for the Planning and Conduct of Future Experiments. New Delhi, India; Army Press, 1947. 160p.
Second	Stewart, Henry. The Culture of Farm Crops. A Manual of the Science of Agriculture and a Handbook of Practice for American Farmers. Millington, N.J.; D. H. Nash, 1887. 334p.
First rank	Stiles, Walter. Trace Elements in Plants and Animals. Cambridge, U.K.; Cambridge University Press; and New York; Macmillan, 1946. 189p.
First rank	Stockbridge, Horace E. Rocks and Soils: Their Origin, Composition and Characteristics: Chemical, Geological and Agricultural. New York; J. Wiley, 1888. 239p. (2d ed, rev. and enl., 1906. 282p.)
Second	Stoddart, Laurence A., and Arthur D. Smith. Range Management. 1st ed. New York and London; McGraw-Hill, 1943. 547p. (3d ed., 1975. 532p.)
Second	Storer, Francis H. Agriculture in Some of its Relations with Chemistry. New York; C. Scribner's Sons, 1887. 2 vols.
Second	Stuhlman, Otto. An Introduction to Biophysics. New York; J. Wiley, and London; Chapman and Hall, 1943. 375p.
Second	Svedberg, Theodor. Colloid Chemistry. New York; Chemical Catalog, 1924. 265p.
Second	Sykes, Frank. Humus and the Farmer. London; Faber and Faber, 1946. 298p.
Second	Tannehill, Ivan R. Drought: It's Causes and Effects. Princeton, N.J.; Princeton University Press, 1947. 264p.
Second	Tauber, Henry. The Chemistry and Technology of Enzymes. New York; J. Wiley, 1949. 550p.
First rank	Taylor, Donald W. Fundamentals of Soil Mechanics. New York; J. Wiley, 1948. 700p.
Second	Taylor, Warren E. Soil Culture and Modern Farm Methods. Moiline, Ill.; Deere and Co., 1913. 278p.
First rank	Terzaghi, Karl. Theoretical Soil Mechanics. New York; J. Wiley, and London; Chapaman and Hall, 1943. 510p.
First rank	Terzaghi, Karl and Ralph B. Peck. Soil Mechanics in Engineering Practice. New York; J. Wiley, 1948. 566p.
First rank	Thaer, Albert D. The Principles of Agriculture, translated by William Shaw and Cuthbert W. Johnson. London; Ridgway, 1844. 2 vols.
Second	Thomas, Arthur W. S. Colloid Chemistry. 1st ed. New York and London; McGraw-Hill, 1934. 512p.
First rank	Thompson, Louis M. Soil Fertility. Dubuque, Iowa; W. C. Brown, 1949. 136p.
Second	Thompson, Louis M. Soils: Their Formation and Classification. Dubuque, Iowa; Wm. C. Brown, 1949. 75p.
Second	Thompson, Ralph S. Science in Farming: A Textbook on the Principles of Agriculture . . . Springfield, Ohio; The Farmers' Advance, 1882. 186p.
First rank	Thorne, Charles E. The Maintenance of Soil Fertility. New York; Orange Judd Pub. Co., Inc., and London; K. Paul, Trench, Trubner and Co., 1930. 332p.

First rank   Thorne, David W., and H. B. Peterson. Irrigated Soils, Their Fertility and Management. Philadelphia, Pa.; Blakiston Co., 1949. 288p. (2d ed., New York; Blakiston Co., 1954. 392p.)

First rank   Thorp, James. Geography of the Soils of China. Nanking, China; National Geological Survey of China, 1936. 552p.

First rank   Turrentine, John W. Potash in North America. New York; Reinhold Pub. Corp., 1943. 186p. (Previous ed. titled, Potash: A Review . . . , J. W. Turrentine, 1926.)

Second   Turrentine, John W. Potash: A Review, Estimate and Forecast. New York; Wiley, 1926. 188p.

First rank   United States. Dept. of Agriculture. Climate and Man. Washington, D.C.; U.S. Govt. Print. Off., 1941. 1248p.

First rank   United States. Dept. of Agriculture. Science in Farming. Washington, D.C.; U.S. Govt. Print. Off, 1947. 944p.

First rank   United States. Dept. of Agriculture. Soils and Men: Yearbook of Agriculture 1938. Washington, D.C.; Govt. Print. Off., 1938. 1232p.

First rank   Vageler, Paul W. E., and H. Greene. An Introduction to Tropical Soils. London; Macmillan, 1933. 240p.

Second   Van Dersal, William R., and Edward H. Graham. The Land Renewed: The Story of Soil Conservation. New York; Oxford University Press, 1946. 109p.

First rank   Van Hise, Charles R. The Conservation of Natural Resources in the United States. New York; Macmillan, 1910. 413p.

First rank   Van Slyke, Lucius L. Fertilizers and Crop Production. New York; Orange Judd Pub. Co., Inc., 1932. 493p. (This book is the successor to Fertilizers and Crops . . . , by L. L. Van Slyke, 1912.)

Second   Van Slyke, Lucius L. Fertilizers and Crops or The Science and Practice of Plant-Feeding . . . New York; Orange Judd Co., 1912. 734p.

Second   Van Vuren, Johannes P. J. Soil Fertility and Sewage: An Account of Pioneer Work in South Africa in the Disposal of Town Wastes. London; Faber and Faber, 1949. 236p.

Second   Vanstone, Ernest. Fertilisers and Manures: Their Manufacture Composition and Uses. London; Macmillan, 1947. 79p.

Second   Vanstone, Ernest. The Soil and the Plant. London; Macmillan, 1947. 71p.

Second   Vennard, John K. Elementary Fluid Mechanics. London; Chapman and Hall, ltd., 1940. 351p. (6th ed. by J. K. Vennard and Robert L. Street. New York; J. Wiley, 1982. 689p.)

First rank   Verwey, Evert J. W., and J. Th G. Overbeek. Theory of the Stability of Lyophobic Colloids: The Interaction of Soilparticles Having an Electric Double Layer. New York; Elsevier, 1948. 205p.

Second   Ville, Georges. On Artifical Manures: Their Chemical Selection and Scientific Application to Agriculture, translated and edited by William Crookes. London; Longmans, Green and Co., 1879. 450p. (New ed., rev. titled, Artificial Manures . . . , 1909. 347p.)

First rank   Vivian, Alfred. First Principles of Soil Fertility. New York; Judd Co., 1908. 265p. (2d ed., 1915. 265p.)

Second   Voorhees, Edward B. Fetilizers: The Source, Character and Composition of Natural, Home-Made and Manufactured Fertlizers . . . New York; Macmillan, 1898. 335p. (Rev. ed., 1916. 365p.)

First rank   Voorhees, Edward B. First Principles of Agriculture. New York; Silver, Burdett, 1896. 212p.

Second   Waggaman, William H. Phosphoric Acid, Phosphates, and Phosphatic Fertilizers. New York; Chemical Catalog Co., 1927. 370p.

First rank   Waksman, Selman A. Humus: Origin, Chemical Composition, and Importance in Nature. Baltimore, Md.; Williams and Wilkins, 1936. 494p. (2d ed. rev., 1938. 526p.)

First rank   Waksman, Selman A. Principles of Soil Microbiology. Baltimore, Md.; Williams and Wilkins Co., 1927. 897p. (2d ed. titled Soil Microbiology. New York; J. Wiley, 1952. 356p.)

First rank   Waksman, Selman A., and Robert L. Starkey. The Soil and the Microbe: An Introduction to the Study of the Microscopic Population of the Soil and Its Role in Soil Processes and Plant Growth. New York; J. Wiley, and London; Chapman and Hall, Ltd., 1931. 260p.

First rank   Walden, J. H. Soil Culture: Containing a Comprehensive View of Agriculture . . . New York; Saxton, Barker and Co., 1860. 444p.

Second   Wallace, Thomas. The Diagnosis of Mineral Deficiencies in Plants by Visual Symptoms; A Colour Atlas and Guide. London; H.M.S.O., 1944. 116p. (Exp. and updated ed. titled, Diagnosis of Mineral Disorders in Plants, edited by J. B. D. Robinson. 1st Amer. ed. New York; Chemical Pub. Co., 1984–87. 3 vols.)

Second   Ward, Robert De Courcy. Climate; Considered Especially in Relation to Man. New York; G. P. Putnam's Sons, 1908. 372p.

First rank   Ward, Robert De Courcy. The Climates of the United States. Boston, Mass. and New York; Ginn and Co., 1925. 518p.

Second   Waring, George. Elements of Agriculture: A Book for Young Farmers, with Questions Prepared for the Use of Schools. New York; D. Appleton and Co., 1854. 288p. (16th ed. titled, Fream's Agriculture: A Textbook, by William Fream. London; J. Murray, 1983. 816p.)

First rank   Warington, Robert. The Chemistry of the Farm. London; Bradbury and Sons, 1881. 128p.

First rank   Warington, Robert. Lectures on Some of the Physical Properties of Soil. Oxford, U.K.; Clarendon Press, 1900. 231p.

Second   Warren, George F. Elements of Agriculture. New York; Macmillan, 1909. 434p.

First rank   Weaver, John E. Root Development of Field Crops. 1st ed. New York; McGraw-Hill, 1926. 291p.

First rank   Weaver, John E., and William E. Bruner. Root Development of Vegetable Crops. 1st ed. New York; McGraw-Hill, 1927. 351p. (2d ed., Emmaus, Pa.; Organic Gardening, 1945. 64p.)

First rank   Weber, Gustavus A. The Bureau of Chemistry and Soils: Its History Activities and Organization. Baltimore, Md.; Johns Hopkins Press, 1928. 218p. (New York; AMS Press, 1974.)

First rank   Weir, Wilbert W. Productive Soils: The Fundamentals of Successful Soil Management and Profitable Crop Production. Philadelphia and London; J. B. Lippincott Co., 1920. 398p.

First rank   Weir, Wilbert W. Soil Science: Its Principles and Practice . . . Chicago and Philadelphia; J. B. Lippincott Co., 1936. 615p.

First rank   Weiser, Harry B. Inorganic Colloid Chemistry. New York; Wiley, and London; Chapman and Hall, 1933–38. 3 vols.

Second   Wheeler, Homer J. Manures and Fertilizers: A Text-Book for College Students . . . New York; Macmillan, 1913. 389p.

Second   Whiting, Albert L. Soil Biology: Laboratory Manual. 1st ed. New York; J. Wiley, 1917. 143p.

First rank   Whitney, Milton. Soil and Civilization: A Modern Concept of the Soil and the Historical Development of Agriculture. New York; Van Nostrand Reinhold, 1925. 278p.

First rank   Whitson, Andrew R. Soils and Soil Fertility. St. Paul, Minn.; Webb Publishing Co., 1912. 315p. (2d ed., 1915. Reprinted, 1917 and 1919.)

First rank   Widtsoe, John A. Dry-Farming: A System of Agriculture for Countries Under a Low Rainfall. New York; Macmillan, 1912. 445p.

First rank   Widtsoe, John A. The Principles of Irrigation Practice. New York; Macmillan, 1914. 496p.

First rank   Wilcox, Earley V. Tropical Agriculture: The Climate, Soils, Cultural Methods, Crops, Live Stock, Commercial Importance and Opportunities of the Tropics. New York and London; D. Appleton and Co., 1916. 373p.

First rank   Wilde, Sergius A. Forest Soils: Origin, Properties, Relation to Vegetation, and Silvicultural Management. 2d ed. Madison, Wis., 1942 319p. (Reproduced from typewritten copy.) (New ed. titled, Soil and Plant Analysis for Tree Culture. New Delhi; Oxford and IBH Pub. Co., 1981. 224p. (1st ed., 1941.)

First rank   Wilde, Sergius A. Forest Soils and Forest Growth. Waltham, Mass.; Chronica Botanica Co., 1946. 241p. (The first draft, originating from the author's lectures at the University of Wisconsin, was published in 1941 under title: Forest Soils: Origin, Properties, Relation to Vegetables, and Silvicultural Management.)

First rank   Wiley, Harvey W. Principles and Practice of Agricultural Analysis: A Manual for the Study of Soils, Fertilizers, and Agricultural Products . . . Eaton, Pa.; Chemical Pub. Co., 1894–1897. 3 vols. (3d ed., rev. and enl., 1931. 2 vols.)

Second   Willcox, Oswin W. ABC of Agrobiology: The Quantitative Science of Plant Life and Plant Nutrition for Gardeners, Farmers and General Readers. New York; W. W. Norton and Co., Inc., 1937. 323p.

Second   Wilson, Herbert M. Manual of Irrigation Engineering. 1st ed. London and New York; Longmans, Green, and Co., 1912. 504p. (7th ed., rev. and enl. by Arthur P. Davis and Herbert M. Wilson, titled, Irrigation Engineering. New York; J. Wiley, 1919. 640p.)

First rank   Wilson, Perry W. The Biochemistry of Symbiotic Nitrogen Fixation. Madison, Wis.; University of Wisconsin Press, 1940. 302p.

First rank   Winchell, N. H. and Alexander N. Winchell. Elements of Optical Mineralogy: An Introduction to Microscopic Petrography . . . New York; Van Nostrand, 1909, 1908. 502p. (4th and 5th ed. New York; J. Wiley, 1931–1951. 3 vols.)

Second   Wolfanger, L. A. The Major Soil Divisions of the United States. New York; Wiley, 1930. 139p.

Second   Woodward, Horace B. The Geology of Soils and Substrata, with Special Reference to Agriculture, Estates, and Sanitation. London; E. Arnold, 1912. 366p.

First rank   Worthen, Edmund L. Farm Soils. New York; J. Wiley, 1927. 410p. (4th
            ed. titled, Farm Soils: Their Management and Fertilization. 1948. 510p.)
Second   Wrench, Guy T. Reconstruction by Way of the Soil. London; Faber and
            Faber, 1946. 262p.
First rank   Wright, Charles H. Soil Analysis: A Handbook for Physical and Chemical
            Methods. New York; D. Van Nostrand, 1934. 236p. (2d ed., London; T.
            Murby and Co., 1939. 276p.)
Second   Wyatt, Francis. Modern High Farming. A Treatise on Soils, Plants, and
            Manures. New York; C. E. Bartholomew, 1886. 94p.

As stated earlier, the purpose of this historical analysis was to identify the core pre-1950 soils monographs and set priorities for their preservation. English language material only was included, in part because it was felt that non-English material should be preserved by the libraries of the countries which published the original works. Many titles came up in the analysis which were not in English and consequently were not ranked by our reviewers. The few titles in English, which were cited but not listed, usually were excluded because they did not fit into the discipline of soils. All of the foreign language titles are listed below in alphabetical order.

Historical Monograph Foreign Language Titles

Albareda, Josae M. Origen y Formación del Humus. Madrid, Spain; 1945. 92p.
Andre, Gustave. Chimie Agricole. Chimie du Sol. Paris; J.B. Baillière et Fils, 1913. 2 vols.
Beloch, Julius. Die Bevölkerung der Griechischrömischen Welt. Leipzig; Duncker and Humblot, 1886. 520p.
Bersch, Wilhelm. Handbuch der Moorkultur. Für Landwirte, Kulturtechniker und Studierende. Wien und Leipzig; W. Frick, 1909.
Berthelot, Marcellin. Chimie Végétale et Agricole. Paris; Masson, 1899. 4 vols.
Brohmer, P., P. Ehrmann and G. Ulmer. Die Tierwelt Mitteleuropas: Ein Handbuch zu ihrer Bestimmung als Grundlage für faunistisch-zoogeographische Arbeiten. Leipzig; Quelle and Meyer, 1927–. 5 vols.
Brooks, Charles F. and A. J. Connor, et al. Climatic Maps of North America. Cambridge, Mass.; Harvard University Press, 1936. 26p.
Dahl, Friedrich T. Vergleichende Physiologie und Morphologie der Spinnentiere unter Besonderer Berücksichtigung der Lebensweise. Jena, Germany; G. Fischer, 1913. 113p.
Davis, William M. Die Erklärende Beschreibung der Landformen. Leipzig and Berlin; B.G. Teubner, 1912. 565p.
Dehaerain, Pierre P. Traité de Chimie Agricole. 2nd ed. Paris; Masson et Cie, 1902. 969p.
Demolon, Albert. Dynamique du Sol. Représente la Matière d'une Partie des Leçons que Nous Avons Professées au Conservatoire des Arts et Métiers en 1930 Comme Suppléant du Professeur A.T. Scholoesing. 4th ed. Paris; Dunod, 1948. 487p.

Demolon, Albert. La Génétique des Sols et ses Applications. Paris; Presses Universitaires de France, 1949. 133p.

Demolon, Albert. Guide pur L'Etude Experimentale du Sol. Paris; Gauthier-Villars, 1933. 214p.

Ehrenberg, Paul. Die Bodenkolloide (der "Kolloide in Land und Forstwirtschaft" Erster Teil). Dritte auflage. Dresden and Leipzig; T. Steinkoppff, 1922. 717p.

Fallou, Friedrich A. Pedologie: Oder, allgemeiine und besondere bodenkunde. Dresden; G. Schonfeld (D.A. Werner), 1862. 487p.

Feher, Daniel. Untersuchungen über die Mikrobiologie des Waldbodens. Berlin; Springer, 1933. 272p.

Garrau, J. G. Les Engrais. Paris; Dunod, 1947. 220p.

Gedroits, Konstantin K. Chemische Bodenanalyse. Methoden und Anleitung zur Untersuchung von Boden im Laboratorium. Berlin; Gebrüder Borntraeger, 1926. 245p.

Gedroits, Konstantin K. Der Adsorbierende Bodenkomplex, und die Adsorbierten Bodenkationen als Grundlage der Genetischen Bodenklassifikation. Dresden and Leipzig; T. Steinkopff, 1929. 112p. (Translated from the Russian.)

Glinka, Konstantin D. Die typen der Bodenbildung, ihre Klassifikation und Geographische Verbreitung. Berlin; Gebruder Borntraeger, 1914.

Glinka, Konstantin D. Solontsy i Solonchaki Aziatskoi Chasti SSSR (Sibir' i Turkestan). Moscow; Novaia Derevnia, 1926.

Goedert, Paul. Les Sols de l'Afrique Centrale, Spécialement du Congo Belge, Caractéristiques Pédologiques. Fertilité. Introduction: Le Régime Pluvial au Congo Belge. Gembloux, Belgique; J. Duculot, 1938. 45p.

Grouven, Hubert. Ueber den Zusammenhang Zwischen Witterung, Boden und Düngung in Ihrem Einflüsse auf die Quantität und Qualität der Erndten. Glogau; C. Flemming, 1868. 364p.

Harroy, Jean Pual. Afrique, Terre qui Meurt: La degredation des sols Africains sous l'influence de la colonisation. 2nd ed. Brussels, Belgium; M. Hayez, 1949. 557p.

Haselhoff, Emil and E. Blanck. Lehrbuch der Agrikulturchemie. Berlin, Germany; Borntraeger, 1927. 4 vols.

Heinrich, Reinhold. Grundlagen zur Beurteilung der Ackerkrume in Beziehung auf Landwirtschaftliche Pflanzenproduktion. Wismar, Germany; Hinstorff, 1882. 244p.

Hellriegel, Hermann. Beitrage zu den Naturwissenschaftlichen Grundlagen des Ackerbaus mit Besonderer Berücksichtigung der Agrikultur-chemischen Methode der Sandkultur. Braunschweig; F. Vieweg, 1883. 796p.

Janke, Alexander. Arbeitsmethoden der Mikrobiologie, ein Praktikum der Allgemeinen, Landwirtschaflichen und Technischen Mikrobiologie. 2 umgearb und wesentlich erweiterte Aufl. Dresden; Steinkopff, 1946–. 2 vols. (1st ed. by A. Janke and Heinrich Zikes.)

Kappen, Hubert. Die Bodenaziditat nach Agrikulturchemischen Gesichtspunkten Dargestellt. Berlin; J. Springer, 1929. 363p.

Killian, Charles and D. Feher. Recherches sur la Microbiologie des Sols Desertiques: Resultats des missions sahariennes Killian-Feher. Paris; P. Lechevalier, 1939. 127p.

Konig, Joseph. Die Untersuchung Landwirtschaftlich und Gewerblich Wichtiger Stoffe. Praktisches Handbuch. 4th ed. Berlin; P. Parey, 1911. 1226p.

Kostyhev, Pavel A. Pochvy Chernozemnoi Oblasti Rossii ikh Proiskhozhdenie, Sostav i Svoistva. Moscow; Gos. Izdvo Selkhoz. Litry, 1949. 239p.

Kraus, Gregor. Boden und Klima auf Kleinstem Raum: Versuch einer exakten behandlung des standorts auf dem wellenkalk. Jena; G. Fischer, 1911. 184p.

Kubiena, Walter L. Entwicklungslehre des Bodens. Wien; Springer-Verlag, 1948. 215p. (Resume in English.)

Laatsch, Willy. Dynamik der Mitteleuropäischen Mineralboden. 3 Aufl. Dresden and Leipzig; T. Steinkopff, 1954. 277p. (Previous eds. published under title: Dynamik der Deutschen Acker- und Waldboden. 2nd ed., 1944. 289p.)

Lafaurie Acosta, Jose V. Clasificćion y Valoración de Tierras, Interpretación Ponderal del Suelo. Bogotá; Editorial Centro, Inst. Grafico Ltda., 1946. 320p.

Lohnis, Felix. Handbuch der Landwirtschaftlichen Bakteriologie. 2d ed. Berlin; Gerbrunder Borntraeger, 1935. 791p.

Lundegoardh, Henrik. Klima und Boden in ihrer Wirkung auf das Pflanzenleben. 3 verb. Aufl. Jena; G. Fischer, 1949. 484p.

Martonne, Emmanuel de. Traite de Géographie Physique. Paris; A. Colin, 1932–47. 3 vols.

Mayer, Adolf. Die Bodenkunde in Zehn Vorlesungen. Zum Gebrauch an Universitäten und Höheren Landwirtschaftlichen Lehranstalten Sowie Zmselbststudium. 5th ed. Heidelberg; Carl Winter's Universitätsbuchhandlung, 1901. 174p.

Mitscherlich, Eilhard A. Bodenkunde für Land und Forstwirte. Berlin; P. Parey, 1905. 364p. (Rev. ed., 1950. 371p. 5th and following eds. titled: Bodenkunde für Landwirte, Forstwirte und Gartner in Pflanzenphysiologischer Ausrichtung und Auswertung.)

Nowacki, Anton J. Praktische Bodenkunde. Anleitung zur untersuchung, klassifikation und kartierung des bodens. Berlin; Verlagsbuchhandlung Paul Parey, 1910. 216p.

Penard, Eugene. Faune Rhizopodique du Bassin du Léman. Geneve; H. Kundig, 1902. 714p.

Pochon, Jacques and Yao-tseng Tchan. Précis de Microbiologie du Sol: Principes, techniques, place dans les cycles géobiologiques. Paris; Masson, 1948. 222p.

Puchner, Heinrich. Bodenkunde für Landwirte unter Berücksichtigung der Benützung des Bodens auf Pflanzenstandort, Baugrund und Technisches Material . . . Stuttgart; F. Enke, 1923. 710p.

Puchner, Heinrich. Der Torf. Stuttgart; F. Enke, 1920. 355p.

Ramann, Emil. Bodenkunde. 2d ed. Berlin; J. Springer, 1905. 431p.

Ramann, Emil. Forstliche Bodenkunde und Standortslehre. Berlin; J. Springer, 1893. 479p.

Risler, Eugene. Géologie Agricole: Premiére Partie du cours d'Agriculture Comparée fait a l'Institut National Agronomique. Paris; Berger-Levrault, 1889–98. 4 vols. (Vol. 1, 2d ed., 1889.)

Schmalfuss, Karl. Pflanzenernahrung und Bodenkunde. Mit 27 Abbildungen. Stuttgart; S. Hirzel, 1948. 274p.

Schumacher, Wilhelm. Die Physik des Bodens in ihren Theoretischen und Practischen . . . Berlin; Wiegandt and Hempel, 1864. 505p.

Steboutte, Alexandre. Lehrbuch der Allgemeinen Bodenkunde: Der boden als dynamisches system. Berlin; Gebrüder Borntraeger, 1930. 518p.

Stremme, Hermann. Grundzüge der Praktischen Bodenkunde . . . Berlin; Gerbrüder Borntraeger, 1926. 332p.

Tondeur, G. Erosion du Sol Specialement au Congo Belge. 1st ed. Bruxelles; Services de l'Agriculture du Ministère des Colonies, 1947. 104p. (3d ed., 1954. 240p.)

Trewartha, Glenn T. and Lyle H. Horn. An Introduction to Weather and Climate. New York and London; McGraw-Hill, 1937. 373p. (5th ed. titled: An Introduction to Climate, 1980. 416p.)

Truffaut, Georges. Sols, Terres et Composts Utilises por l'Horticulture. Paris; O. Doin, 1896. 308p.

Tschapek, Marcos W. Quimica Coloidal del Suelo. Buenos Aires; Editora Coni, 1949–.

Vageler, Paul W. E. Grundriss der Tropischen und Subtropischen Bodenkunde, für Pflanzer und Studicrende. 2d ed. Berlin; Verlagsgesellschaft für Ackerbaum, 1938. 252p.

Villar, Emilio H. del. Types de Sol de l'Afrique du Nord. Rabat, 1947. 2 vols.

Wiegner, Georg. Boden und Bodenbildung in Kolloidchemischer Betrachtung. Dresden and Leipzig; T. Steinkopff, 1918. 98p.

Winogradsky, Serge. Microbiologie du Sol: Problemes et Méthodes; cinquante ans de recherches. Paris; Masson, 1949. 861p.

Wollny, Ewald. Der Einfluss, der Pflanzendecke und Beschattung auf die Physikalischen Eigenschaften und die Fruchtbarkeit des Bodens. Berling; Wiegandt, Hempel and Parey, 1877. 197p.

Wollny, Ewald. Die Zersetzung der Organischen Stoffe und die Humusbildungen, mit Rucksicht auf die Bodenkultur. Heidelberg; Carl Winter's Universitatsbuchhandlung, 1897. 479p.

Index

Authors and titles in the core list of monographs (pp. 181–259), the core list of journals (pp. 271–274), the soil maps catalog (pp. 318–369), the reference update (pp. 371–380), and the historically significant monographs (pp. 411–433) are *not* included in this index.

Library of Congress Cataloging-in-Publication Data

The Literature of soil science / edited by Peter McDonald.
    p.   c.m — (The Literature of the agricultural sciences)
    Includes bibliographical references and index.
    ISBN 0-8014-2921-8 (alk. paper)
    1. Soil science literature.   I. McDonald, Peter, 1952–   .   II. Series.
S590.45.L58   1994
631.4—dc20   93-27394